UNITEXT for Physics

UNITEXT for Physics series publishes textbooks in physics and astronomy, characterized by a didactic style and comprehensiveness. The books are addressed to upper-undergraduate and graduate students, but also to scientists and researchers as important resources for their education, knowledge, and teaching.

More information about this series at http://www.springer.com/series/13351

Salvatore Capozziello · Wladimir-Georges Boskoff

A Mathematical Journey
to Quantum Mechanics

 Springer

Salvatore Capozziello
Dipartimento di Fisica
INFN Sezione di Napoli - Università degli
Studi "Federico II"
Napoli, Italy

Wladimir-Georges Boskoff
Department of Mathematics
and Informatics
Universitatea Ovidius Constanţa
Constanta, Romania

ISSN 2198-7882
UNITEXT for Physics
ISBN 978-3-030-86100-1
https://doi.org/10.1007/978-3-030-86098-1

ISSN 2198-7890   (electronic)

ISBN 978-3-030-86098-1   (eBook)

This Springer imprint is published by the registered company Springer Nature Switzerland AG
The registered company address is: Gewerbestrasse 11, 6330 Cham, Switzerland

I cannot believe God plays dice with the Cosmos!

—Albert Einstein

Einstein, stop telling God what to do!

—Niels Bohr

Dedicated to all people who honestly try to interpret the Reality and who humbly understand that it is greater than them.

Preface

Someone could believe that Quantum Mechanics has to be presented only in a modern approach, highlighting from the beginning Principles written in the language of Functional Analysis. Hilbert spaces, Hermitian and unitary operators, eigenvalues and eigenvectors are part of this modern language from which emerges a frame to understand what we call reality. Ladder operators help us to create and destroy matter. We jump between energy levels in this way. There are quantum numbers which describe at a glance the properties of a quantum mechanical system. Some other operators, which are the generalization of unitary operators in four dimensions, describe both the polarization of photons and the electron spin. In this modern language, even the impossibility to predict at the same time the position and the momentum of a particle is, in fact, an operatorial inequality whose skeleton is the basic Cauchy-Buniakowski-Schwarz inequality. The atom can be described by the Schrödinger equation, written in spherical coordinates and solved by some advanced mathematical technique. This modern language allows analogies with Classical Mechanics considered in its different formulations. For example, starting from the Hamilton-Jacobi formula in Classical Mechanics, it exists a counterpart in Quantum Mechanics. Only terms containing $\hbar$ make the difference between classic and quantum formulations.

This advanced language started from Dirac, Heisenberg, Pauli, Ehrenfest, Schrödinger and all the others who contributed to its development. We believe that is important to be known by students and it is presented in this book.

However we also believe that it is important, for people coming into the territory of Quantum Mechanics, to know how this language has been formulated to approach this unintuitive part of Physics. In this book, without claiming for completeness, we offer a road mainly devoted to undergraduate students. The task is very difficult because the Mathematics of Quantum Mechanics is not simple even before the abstract picture we mentioned before. The creative fathers invented notions and descriptions which, time by time, have to be confronted with self-consistent concepts and experiments.

We want to present the evolution of quantum mechanical concepts based on experimental facts and to develop the related Mathematics accordingly. In other words, we want to improve, step by step, the mathematical contents with the aim to guide students along the conceptual path of Quantum Mechanics because we realized, as teachers, that often some concepts are not fully understood, even if technically used.

For any statement, we give the complete proof. All material is explained in a friendly manner for students in Physics, Mathematics, Chemistry and Engineering. If someone reads from the beginning to the end, she/he has the possibility to understand step by step. For example, we develop the harmonic oscillator in Newton mechanics, and then pass to the Lagrange formalism, and finally in Hamilton formalism to step in the basic ideas of Quantum Mechanics in view to obtain a first solution and then quantize the energy levels depending on the Hermite polynomials. The idea is to look at the Mathematics behind the Hermite polynomials and to begin to see structures which lead to the Hilbert spaces. After, we develop other lectures to move toward the axiomatic formalism of Quantum Mechanics. In Dirac notations with ladder operators, we have a sort of *surrealistic view* on the harmonic oscillator, if we compare it to the solution found with respect to Hermite polynomials. In this last case, the quantized energy levels were necessary to obtain the solution depending on Hermite polynomials. The "surrealistic view" in modern mathematical language makes the ladder operators responsible for the quantized energy levels.

The same approach can be realized when we look at hydrogen atom: there is a huge gap between the Bohr model and the Schrödinger model of atom derived in spherical coordinates. Here we obtain the spherical harmonics and the quantum numbers n, l, m; the solution depends on Legendre's polynomials and we present all these parts giving the complete solutions. The philosophy is that passing from the Bohr model to the Schrödinger solution, the accuracy of the latter is achieved thanks to the more advanced Mathematics.

Some words on Heisenberg's uncertainty principle are necessary because this topic can also be seen as an "evolutionary" concept along this book. We meet firstly this subject when we study the Gauss wave packets. Again, the Mathematics behind oblige us to come step into the world of Fourier transforms. Here we find that Heisenberg's principle can be formulated under the standard of Fourier transforms: its equivalent is the well-known Pinsky's theorem. When we develop the Mathematics for Quantum Mechanics, it is possible to identify the Heisenberg uncertainty principle with the Cauchy-Buniakowsky-Schwarz formula for some special operators. Also in this case, the mathematical evolution points out the Physics of the quantum mechanical objects. Several other examples can be developed in this perspective.

We can use the Dirac words: *God used beautiful Mathematics in creating the World*. They reflect exactly what we are humbly trying to present here: the beautiful Mathematics of Quantum Mechanics. This is the *Mathematical Journey to Quantum Mechanics* we propose.

The main point of this book is that we try to make it readable. Having complete proofs with detailed calculations, it gives chances to all readers to understand some difficult mathematical parts necessary to have a comprehensive picture of the amazing subject of Quantum Mechanics. The philosophy we followed in writing this book is the same as that used for the companion book *A Mathematical Journey to Relativity* [1] where we arrived to General Relativity starting from the concepts of basic geometry. The ultimate aim is to show that all the deep concepts of modern physics can be better understood if one considers in detail the mathematical language in which they are formulated.

Napoli, Italy   Salvatore Capozziello
Costanta, Romania   Wladimir-Georges Boskoff
July 2021

Contents

Chapter 1
Introduction: How to Read This Book

All Science is either Physics
or stamp collecting.

Ernest Rutherford

A mathematical journey towards Quantum Mechanics would like to be an itinerary moving from Classical Mechanics to Special Relativity, then continuing to Quantum Mechanics. This book consists of self-contained short lectures, the first thirteen being a necessary revision of some basic concepts of Classical Mechanics and Special Relativity.

However, if someone is already familiar with both Lagrangian and Hamiltonian formulations of Classical Mechanics or with Lorentz transformations in Special Relativity and their consequences, it is possible to start directly from the chapter "Why Quantum Mechanics?".

But, if you read the following lines and they seem vague statements, it is better to start from the beginning:
"The Hamiltonian written in the form

$$H = \frac{1}{2m}(p_x^2 + p_y^2 + p_z^2) + V(t, x, y, z),$$

is transformed into the Hamilton–Jacobi equation

$$-\frac{\partial S}{\partial t} = \frac{1}{2m}\left[\left(\frac{\partial S}{\partial x}\right)^2 + \left(\frac{\partial S}{\partial y}\right)^2 + \left(\frac{\partial S}{\partial z}\right)^2\right] + Vt, x, y, z).$$

© The Author(s), under exclusive license to Springer Nature Switzerland AG 2021
S. Capozziello and W.-G. Boskoff, *A Mathematical Journey to Quantum Mechanics*,
UNITEXT for Physics, https://doi.org/10.1007/978-3-030-86098-1_1

Its equivalent, in Quantum Mechanics, is

$$-\frac{\partial S}{\partial t} = \frac{1}{2m}\left[\left(\frac{\partial S}{\partial x}\right)^2 + \left(\frac{\partial S}{\partial y}\right)^2 + \left(\frac{\partial S}{\partial z}\right)^2\right] - \frac{i\hbar}{2m}\,\nabla^2 S + V(x,y,z,t).$$

The classical Hamilton–Jacobi equation is recovered when $\hbar$ is negligible."

"The Hamiltonian formalism of Classical Mechanics gives formulas like

$$[F,H] := \sum_{i=1}^{n}\left[\frac{\partial F}{\partial q^i}\frac{\partial H}{\partial p_i} - \frac{\partial F}{\partial p_i}\frac{\partial H}{\partial q^i}\right],$$

or

$$\frac{dF}{dt} = [F,H]$$

where $F = F(p_i, q_i)$, $i \in \{1, 2, \ldots, n\}$ is a real function, $H := H(q^i, p_i)$ is the Hamiltonian function depending on the generalized coordinates and the first formula is the definition of the Poisson brackets of the functions F and H. The consequences

$$[q^i, q^j] = [p_i, p_j] = 0, \ [q^i, p_j] = \delta^i_j$$

are similar to the formulas

$$[\hat{p}_j, \hat{x}_k] = -i\hbar\delta_{jk}, \ j,k = 1,2,3$$

which involve the quantum mechanical operators momenta and positions. At a first glance

$$\frac{dF}{dt} = [F,H]$$

has its equivalent in Quantum Mechanics thanks to the Ehrenfest Theorem, that is

$$\frac{d}{dt}\left\langle \hat{A} \right\rangle = \frac{1}{i\hbar}\left\langle [\hat{A}, \hat{H}] \right\rangle ."$$

So, if the mathematical language of the two above statements is unknown for you, it is better to start from the Lagrangian and Hamiltonian formalisms of Classical Mechanics and after to step into the language of Quantum Mechanics.

The same, if the following paragraph where it is shows that light, considered as an electromagnetic wave, is invariant under Lorentz transformations and it is not invariant under Galilei transformations:

"Denote by

$$E = E(t,x,y,z) := (E_x(t,x,y,z), E_y(t,x,y,z), E_z(t,x,y,z))$$

the electric field vector and by

$$B = B(t, x, y, z) := (B_x(t, x, y, z), B_y(t, x, y, z), B_z(t, x, y, z))$$

the magnetic field vector. In geometric units, i.e. $c = 1$, the *Maxwell equations in vacuum* are

$$\begin{cases} \nabla \cdot E = 0 \\ \nabla \times E = -\dfrac{\partial B}{\partial t} \\ \nabla \cdot B = 0 \\ \nabla \times B = \dfrac{\partial E}{\partial t} \end{cases}$$

If

$$\Box := \frac{\partial^2}{\partial t^2} - \nabla^2$$

is d'Alembert operator, after some calculations, the previous equations can be written in the form

$$\begin{cases} \Box E = 0 \\ \Box B = 0. \end{cases}$$

Here, the d'Alembert operator is used to describe the wave equations. The obtained form is used to show the invariance of Maxwell equations under the Lorentz transformations.

For the system at rest, here denoted $R(t, z)$, each component of the first equation is

$$\frac{\partial^2 \mathbb{E}}{\partial t^2} - \frac{\partial^2 \mathbb{E}}{\partial z^2} = 0,$$

that is we prefer to denote by $\mathbb{E}$ each of the components E_k, $k = x, y, z$ of the electric field E. How this simple equation looks like in $S(\tau, \bar{z})$, if S is a moving system supposed to move at constant speed v along the z axis in R?

To answer, we have to use the Lorentz inverse transformation L_{-v}, that is

$$\begin{cases} \tau = \dfrac{t - z\, v}{\sqrt{1 - v^2}} \\ \bar{z} = \dfrac{-t\, v + z}{\sqrt{1 - v^2}}. \end{cases}$$

Let's denote by $\bar{\mathbb{E}}(\tau, \bar{z}) = \bar{\mathbb{E}}\left(\dfrac{t - z\, v}{\sqrt{1 - v^2}}, \dfrac{-t\, v + z}{\sqrt{1 - v^2}} \right) := \mathbb{E}(t, z)$ the corresponding component of the electric field in S, which, obviously has to be the same as in R. It remains to prove the equality

$$\frac{\partial^2 \mathbb{E}}{\partial t^2} - \frac{\partial^2 \mathbb{E}}{\partial z^2} = \frac{\partial^2 \bar{\mathbb{E}}}{\partial \tau^2} - \frac{\partial^2 \bar{\mathbb{E}}}{\partial \bar{z}^2}.$$

After simple algebra, the equality is proved. It is important to say that if we try with Galilean transformation, the equality does not hold, therefore light, considered as an electromagnetic phenomenon, cannot be framed in Classical Mechanics".

Again, if the language and the computations seem to be difficult or if the passage from Maxwell's equations to d'Alembert's equations is not familiar, it is better to start from the below lectures in Special Relativity.

Furthermore, in the evolution of mathematical language for Quantum Mechanics, we are looking also at the first experiments which established the atomic structure. It is worth noticing that several experiments proved the wave nature of light according to Young, Maxwell, and others. We are going to present here the mathematical structure of these experiments.

Besides, starting from the statement that light is made by particles, as said by Descartes, Newton, Einstein, and others, the description of photoelectric effect, obtained by Einstein, proved the existence of photons. Specifically Einstein used Planck's formula $E = h\nu$ to explain how electrons are released from a metal when ultraviolet high frequency light hits the surface of the metal. In fact, Planck's formula $E = h\nu$ expresses quanta of energy which represent photons. Photons hit the metal and extract electrons. Planck's formula can be considered the fundamental step in the formulation of Quantum Mechanics. We have dedicated a lecture to Planck's idea of energy quantum packets and another lecture to photoelectric effect.

Therefore we have to accept the idea that light is constituted by waves and particles at the same time. This means that LIght has a dual behavior.

Louis de Broglie was the first who considered possible to think that electrons have a dual behavior, too. This statement helped Bohr's model of hydrogen atom (now obsolete) to survive at least at the level of its first axiom. The electron is seen as a standing wave but also as a deterministic particle involved in a circular orbit around the nucleus, as Rutherford claimed. And this is the major problem that made physicist to cancel this model, even if it correctly predicts the Rydberg series. We have a series of lectures dedicated to these subjects.

How Mathematics adapted to the evolution of atomic conception?

Waves in the one dimensional case can be written in the form

$$\Psi(t, x) = Ae^{i(kx-wt)} = A[\cos(kx - wt) + i\sin(kx - wt)].$$

This is a traveling monochromatic wave which verifies the d'Alembert equation

$$\Box\Psi = 0.$$

Schrödinger conceived another equation to describe the subatomic world. We can only speculate that, in his picture, d'Alembert equation is a deterministic classical equation while Schrödinger time dependent equation

$$i\hbar\frac{\partial\Psi}{\partial t}(t, x) = -\frac{\hbar^2}{2m}\frac{\partial^2\Psi}{\partial x^2}(t, x) + V(t, x)\Psi(t, x).$$

seems more related to a theory formulated for subatomic particles. If $V = 0$, we obtain the equation of a particle which travels freely in the empty space. The standing waves can be studied easier using the Schrödinger time independent equation

$$H\Psi = -\frac{\hbar^2}{2m}\frac{d^2\Psi}{dx^2} + V(x)\Psi$$

where H is the total energy of the described system. The examples we study below, that is the free particle, the particle in a box and the harmonic oscillator give rise to a new mathematical world to be involved in the description of Quantum Mechanics. If we look only at the harmonic oscillator, results can be summarized in the following way. The equation is

$$-\frac{\hbar^2}{2m}\frac{d^2\Psi}{dx^2}(x) + \frac{1}{2}m\omega^2 x^2 \Psi(x) = H\Psi(x).$$

If we denote $u = \sqrt{\frac{m\omega}{\hbar}}x$ and $\varepsilon = \frac{2H}{\hbar\omega}$, we obtain the equation in the variable u

$$\frac{d^2\Psi}{du^2}(u) + (\varepsilon - u^2)\Psi(u) = 0.$$

The solutions are obtained if $\varepsilon = 2n + 1$, $n \in \mathbb{N}$ and depends on the Hermite polynomials. The form of the solutions is

$$\Psi_n(u) = A_n H_n(u)e^{-u^2/2},$$

where $H_n(u)$ are the Hermite polynomials and A_n is a constant we find with the constraint

$$\int_{-\infty}^{\infty} |\Psi_n(u)|^2 du = 1.$$

Since

$$\varepsilon = \varepsilon_n = 2n + 1 = \frac{2H}{\hbar\omega}$$

we find that the total energy H is quantized, depending on n, being H_n. There is a notation risk with respect to the Hermite polynomials denoted by H_n, too. Therefore, we denote in this example the total energy by the letter E and we obtain the n-level quantized energy as

$$E_n = \left(n + \frac{1}{2}\right)\hbar\omega.$$

If we return to the variable x, after all computations, we find

$$\Psi_n(x) = \frac{1}{\sqrt{2^n n!}} \left(\frac{m\omega}{\pi\hbar}\right)^{1/4} H_n\left(\sqrt{\frac{m\omega}{\hbar}}x\right) e^{-m\omega x^2/2\hbar}$$

and

$$\Psi_n(t, x) = e^{-i E_n t/\hbar} \frac{1}{\sqrt{2^n n!}} \left(\frac{m\omega}{\pi\hbar}\right)^{1/4} H_n\left(\sqrt{\frac{m\omega}{\hbar}}x\right) e^{-m\omega x^2/2\hbar}.$$

Now, the last step: since Schrödinger equation is a linear one, the general solution is described by the sum of all modes of n-oscillations, that is

$$\Psi(t, x) = \sum_n c_n e^{-i E_n t/\hbar} \Psi_n(x).$$

The meaning of this sum is related to the Mathematics behind the Hermite polynomials, needed to describe the harmonic oscillator. The mathematical language and even the meaning of the mathematical objects involved in Quantum Mechanics have to evolve in an appropriate way. Therefore the next lectures are devoted both to the Hermite polynomials and to the theory of Hilbert spaces. Of course, this book cannot substitute a book of Functional Analysis. We intended to offer just basic aspects necessary to understand Quantum Mechanics. Technical aspects of some results have to be studied by the readers in specialized textbooks. We do not step into the theory of L^p spaces, which are Hilbert spaces if and only if $p = 2$ because we need only L^2 for our purposes. All the ingredients of this theory is already part of the harmonic oscillator mathematics.

Considering the ladder operators and Dirac vision of harmonic oscillator, we step into "another quantum mechanics description" which gives the possibility to "create" or "annihilate" particles.

The reader will find in this book carefully proofs of basic facts of Quantum Mechanics. To prevent confusions, we wish to present in this Introduction some theoretical facts we will need later.

For example, let us consider a function $\Psi(x)$ constructed in the following way: for each integer k such that $|k| \geq 3$ we consider the interval $\left[k - \frac{1}{k^4}, k + \frac{1}{k^4}\right]$. The function is 0 outside these intervals. We have $\Psi(k) = 1$ for all k considered in the description. On each interval $\left[k - \frac{1}{k^4}, k + \frac{1}{k^4}\right]$ the function has the graph described by two segments, the first one lying from $\left(k - \frac{1}{k^4}, 0\right)$ to $(k, 1)$, the second one lying from $(k, 1)$ to $\left(k + \frac{1}{k^4}, 0\right)$.

Three mathematical facts are obvious.

The first one is related to the smoothness of this function. This function does not belong to $\mathcal{C}^\infty(\mathbb{R})$.

The second one is related to the fact that

$$\int_{-\infty}^{\infty} |\Psi(x)|^2 dx < 2 \sum_{k \in \mathbb{N},\ k \neq 0} \frac{1}{k^4} < \infty,$$

i.e. $\Psi \in L^2(\mathbb{R})$. The third one is related to the fact that functions in $L^2(\mathbb{R})$ not necessarily approach to 0 at plus and minus ∞. Let us consider the sets $\{k\}_{k \in \mathbb{N}}$ and $\left\{ k + \frac{1}{k^4} \right\}_{k \in \mathbb{N}}$ for $k > 3$. Since $\Psi(k) = 1$ and $\Psi\left(k + \frac{1}{k^4}\right) = 0$, it results that the function Ψ has not a limit at infinity.

This example shows that it is not true the fact that if a function belongs to $L^2(\mathbb{R})$, it approaches 0 at infinity.

Now, we can construct in the same way another function ϕ with the only difference that, at each k as above, $\phi(k) = 2$. This function also belongs to $L^2(\mathbb{R})$ without being smooth.

For such functions we cannot write the equality

$$\left\langle \Psi \left| \frac{d^2}{dx^2} \phi \right.\right\rangle_{L^2(\mathbb{R})} = \int_{-\infty}^{\infty} \Psi^*(x) \frac{d^2\phi}{dx^2}(x)dx = \int_{-\infty}^{\infty} \frac{d^2\Psi^*}{dx^2}(x)\phi(x)dx = \left\langle \frac{d^2}{dx^2}\Psi \left| \phi \right.\right\rangle_{L^2(\mathbb{R})}$$

which leads to the fact that the one dimensional Hamiltonian operator is a Hermitian one.

We need something else to prove the above equality: $L^2(\mathbb{R})$ contains a very important dense subset, $C_0^\infty(\mathbb{R})$.

$C_0^\infty(\mathbb{R})$ is the set of all infinitely differentiable functions real or complex valued having a compact support. The support of a function is the topological closure of the set of $x \in \mathbb{R}$ such that $f(x) \neq 0$. The previous Ψ and ϕ are approximated by Ψ_n and ϕ_n, sets from this important subset. Now, we have in mind that on a compact subset of $\mathbb{R}$, which include the union of the supports of Ψ_n and ϕ_n, it makes sense using twice the integration by parts, that is

$$\left\langle \Psi_n \left| \frac{d^2}{dx^2} \phi_n \right.\right\rangle_{L^2(\mathbb{R})} = \int_{-\infty}^{\infty} \Psi_n^*(x) \frac{d^2\phi_n}{dx^2}(x)dx = \int_{-\infty}^{\infty} \frac{d^2\Psi_n^*}{dx^2}(x)\phi_n(x)dx = \left\langle \frac{d^2}{dx^2}\Psi_n \left| \phi_n \right.\right\rangle_{L^2(\mathbb{R})}$$

and this equality is preserved at the limit. In this way, we avoid the false claim about functions in $L^2(\mathbb{R})$ presented above. That is, we obtain the equality

$$\left\langle \Psi \left| \frac{d^2}{dx^2} \phi \right.\right\rangle_{L^2(\mathbb{R})} = \left\langle \frac{d^2}{dx^2}\Psi \left| \phi \right.\right\rangle_{L^2(\mathbb{R})}$$

which shows that the Hamilton operator is a Hermitian one on $L^2(\mathbb{R})$.

Of course, this particular proof holds for $V(x) = 0$. If $V(x) \neq 0$, we have to take into account that it is real valued. Therefore

$$\langle \Psi \mid V(x)\phi \rangle_{L^2(\mathbb{R})} = \langle V(x)\Psi \mid \phi \rangle_{L^2(\mathbb{R})}$$

holds. Together, the two results show that the Hamiltonian is a Hermitian operator, i.e.

$$\langle \Psi \mid H\phi \rangle_{L^2(\mathbb{R})} = \left\langle \Psi \mid \left(\frac{d^2}{dx^2} + V(x) \right) \phi \right\rangle_{L^2(\mathbb{R})} = \left\langle \left(\frac{d^2}{dx^2} + V(x) \right) \Psi \mid \phi \right\rangle_{L^2(\mathbb{R})} = \langle H\Psi \mid \phi \rangle_{L^2(\mathbb{R})}$$

These are the details one needs when we discuss about the fact that the Hamiltonian is a Hermitian operator.

If we are working in $L^2([0, L])$, as it happens in Lecture 31, the dense set is $\mathcal{C}^\infty([0, L])$ and more, if $\Psi \in L^2(\mathbb{R})$ and $\phi \in L^2(\mathbb{R})$ have the properties $\Psi(0) = \phi(0) = \Psi(L) = \phi(L) = 0$, we can choose sets Ψ_n, $\phi_n \in \mathcal{C}^\infty([0, L])$ such that $\Psi_n(0) = \phi_n(0) = \Psi_n(L) = \phi_n(L) = 0$ for each $n \in \mathbb{N}$. The equality which is preserved at the limit is

$$\left\langle \Psi_n \mid \frac{d^2}{dx^2} \phi_n \right\rangle_{L^2([0,L])} = \int_0^L \Psi_n^*(x) \frac{d^2\phi_n}{dx^2}(x)dx = \int_0^L \frac{d^2\Psi_n^*}{dx^2}(x)\phi_n(x)dx = \left\langle \frac{d^2}{dx^2} \Psi_n \mid \phi_n \right\rangle_{L^2([0,L])}.$$

Of course, these equalities are obtained applying twice the integration by parts. It is important to underline that in $L^2(\mathbb{R}^3)$ the dense set is $\mathcal{C}_0^\infty(\mathbb{R}^3)$. This will be used in Lecture 39.

Why all these discussions? When we are talking about observables, we are dealing with Hermitian operators. Among the observables of a quantum systems, the Hamiltonian plays a major role. So, at least in particular cases, the Hamiltonian has to be a Hermitian operator and we have to be able to prove this property.

We left as a final consideration the definition of complex numbers. We do not totally like to say that

$$\mathbb{C} := \{z \mid z = a + ib, \ a, \ b \in \mathbb{R}, \ i^2 = -1\}.$$

The operations seen in the previous definition are an addition and a multiplication. Are these operations those we know from real numbers? The answer is no, of course, but it is better to say why to prevent undesirable confusions.

Consider

$$\mathbb{R}^2 := \{(a, b) \mid a \in \mathbb{R}, \ b \in \mathbb{R}\}$$

and two operations $\oplus$ and $\odot$ constructed with respect the addition and multiplication on $\mathbb{R}$:

$$(a, b) \oplus (c, d) := (a + c, b + d) \text{ and}$$

$$(a, b) \odot (c, d) := (ac - bd, ad + bc).$$

The reader can prove that a field structure is highlighted by $(\mathbb{R}^2, \oplus, \odot)$.

A simple exercise shows that $(0, 1) \odot (0, 1) = (-1, 0)$.

It is only a notation to write $(0, 1)^2 = (-1, 0)$. Therefore we can write

$$(a, b) = (a, 0) \oplus (0, b) = (a, 0) \oplus (0, 1) \odot (b, 0).$$

If we identify a complex number as a pair of the previous field, i.e. $z := (a, b)$ and each pair $(a, 0)$ by the real number a (this happens because the function $f :$ $\{(x, 0)| \; x \in \mathbb{R}\} \to \mathbb{R}$ is a bijective one), the pair $(0, 1)$ remains as a mathematical object which does not correspond to a real number because its form $(0, 1)$ differs by $(x, 0)$. If we denote it by i, the previous exercise tells us that $i^2 = -1$ (at the level of our previous identifications). Finally we can "write" $z = a + ib$, but we have to look at this new entity, the complex number z, with respect to our explanations given before. In this book, we use the notation $z^* = a - ib$ for the conjugate of the complex number $z = a + ib$.

The above considerations clearly point out how it is important the mathematical language in the formulation of any theory.

A final remark is necessary for the reported Bibliography. This book is meant to be as self-consistent as possible considering the evolution of mathematical language of Quantum Mechanics. However, we point out some important key books which can be used for further information and discussions. Refs. [1–3] can be used for the thirteen lectures of the first two Chapters, while Refs. [4–9] can be considered in view of mathematical description of Quantum Mechanics, in particular for Chaps. 6, 8, 9 and 10. References from [10] to [25] are related to each Chapter starting from the third one. In particular, Refs. [13–15, 22] and [23] are for Chaps. 8, 9 and 10 being related to the modern approach to Quantum Mechanics. Some basic concepts of Relativistic Quantum Mechanics are reported in Chap. 11 but a full development of this subject can be found in [28, 29]. The ideas from which this book has been developed are suggested in [26, 27].

Chapter 2
Newtonian, Lagrangian and Hamiltonian Mechanics

Language is only the instrument of Science,
and words are but the signs of ideas.

Samuel Johnson

2.1 Lecture 1: A Summary of the Principles of Newtonian Mechanics

Newtonian Mechanics [2, 3] is the branch of Physics which studies the way in which bodies are changing their position in space and time. Space where the objects are at rest or in motion is the Euclidean 3-dimensional space E^3. Time is represented by the real axis $\mathbb{R}$. It is absolute and the same for any observer. All objects, regardless of size, can be identified as points with a given mass in the previous E^3 space. So, the Euclidean frame of coordinates $Oxyz$ becomes the absolute place where all is happening. Forces are seen as vectors. For a given point M in space, the vector $\vec{X} = O\vec{M}$ is called a position vector. If the point evolves in time, we write this as

$$\vec{X}(t) = (x(t), y(t), z(t)).$$

The velocity vector is

$$\dot{\vec{X}} = (\dot{x}(t), \dot{y}(t), \dot{z}(t))$$

and the acceleration vector is

$$\ddot{\vec{X}} = (\ddot{x}(t), \ddot{y}(t), \ddot{z}(t)).$$

© The Author(s), under exclusive license to Springer Nature Switzerland AG 2021
S. Capozziello and W.-G. Boskoff, *A Mathematical Journey to Quantum Mechanics*,
UNITEXT for Physics, https://doi.org/10.1007/978-3-030-86098-1_2

Of course, we make the assumption that the coordinates functions are infinitely differentiable on their domain of definition which differs from a model to another. The foundations of Newtonian Mechanics are based on three fundamental principles, the so called Newton's Laws of motion. They were introduced by Isaac Newton in "*Philosophiae Naturalis Principia Mathematica*", book published in 1687. The Principle of Inertia, or the first law, asserts: "*A physical body preserves its rest state or will continue moving at its current velocity conserving its direction, until a force causes a change in its state of moving or rest. The physical body will change the velocity and the direction according to this force.*" A particular case is related to the rectilinear uniform motion, when the body is moving on a straight line at constant speed. The frames where this principle is available are called inertial frames. These frames are at rest or they move rectilinear at constant speed. This fundamental principle was first enunciated by Galileo Galilei. We can say that this principle tells us where, according to Newton, the two others fundamental principles make sense: in inertial frames. At the same time, it tells us that it is impossible to make a distinction between the state "*at rest*" and the state of "*rectilinear motion at constant speed.*" Imagine you are in the bowl of a ship and you have no possibility to observe outside. You slept and you waked up. You cannot distinguish between the two states without an observation, a possible comparison. You will play table tennis alike in both states, the object fall down in same way in both states, etc. The two states are equivalent for you in the given conditions.

Newton introduces a concept, the quantity of motion of a body as the product between the mass m and its velocity $\vec{v}$. This quantity of motion is known today as momentum and it is denoted by $\vec{p}$, therefore $\vec{p} := m\,\vec{v}$. The second law asserts: "*The force who acts on a body is the variation in time of the quantity of motion.*" Its differential form is $\vec{F} = \dfrac{d\,\vec{p}}{dt}$. If m does not depend on time, then

$$\vec{F} = \frac{d\,\vec{p}}{dt} = m\frac{d\,\vec{v}}{dt} = m\,\vec{a},$$

that is the force who acts on a body is proportional to the body acceleration through its mass.

Newton's third law states: "*When a body acts on a second body by the force $\vec{F}$, the second body simultaneously reacts on the first body by the force $-\vec{F}$.*"

Example 2.1.1 *Let's consider an example which will be related to Quantum Mechanics: the harmonic oscillator. A mass attached to a spring, a small period pendulum, an acoustic wave are examples of harmonic oscillators in Classical Mechanics. The restoring force is proportional to the displacement x,*

$$\vec{F} = -k\,\vec{x},$$

where k is a positive constant. Therefore the differential equation to solve is stated by Newton's second law,

$$m\ddot{x}(t) = -kx(t).$$

The solution is

$$x(t) = A\cos(\omega t + \phi),$$

where A is a positive number which expresses the amplitude and $\omega = \sqrt{\dfrac{k}{m}}$. *Here ϕ is the phase. The solution is a periodic function, therefore the oscillatory movement has a period, $T = \dfrac{2\pi}{\omega}$. This problem allows a view in respect of the conservative force derived from a potential energy. The potential energy of the harmonic oscillator is*

$$V(x) := \frac{1}{2}kx^2.$$

Summary of Lecture 1. The Classical Mechanics was developed in the 17th century by Isaac Newton. It is a deterministic theory for macroscopic objects which allows, from a given state of a system, to predict its evolution in the future but also to understand how it moved in the past. There is an absolute space and an absolute time given by an "universal clock". The absolute space is the Euclidean one in which all objects evolves under some dynamical rules, the principles stated by Newton.

The first principle is called Principle of Inertia and states how the objects are moving in absence of forces or if a force acts on the objects.

The second principle, known as Newton's second law, describes the motion in the absolute time of an object and it can be simply written as $\vec{F} = \dfrac{d\vec{p}}{dt}$. If m does not depend on time, then

$$\vec{F} = \frac{d\vec{p}}{dt} = m\frac{d\vec{v}}{dt} = m\vec{a},$$

while the third principle describes the response of a physical system against the object which exerted the force responsible of the change of the state of the system. It can be described as $\vec{F}_{AB} = -\vec{F}_{BA}$. The example of the harmonic oscillator is presented starting from the restoring force which is proportional to the displacement x, i.e.

$$\vec{F} = -k\vec{x},$$

where k is a positive constant. It results the equation of the harmonic oscillator

$$m\ddot{x}(t) = -kx(t).$$

The solution is

$$x(t) = A\cos(\omega t + \phi),$$

where A is a positive number which expresses the amplitude and $w = \sqrt{\dfrac{k}{m}}$; ϕ is the phase. The solution is a periodic function with period $T = \dfrac{2\pi}{\omega}$. This problem represents a straightforward picture for a conservative force derived from a potential energy. The potential energy of the harmonic oscillator is

$$V(x) := \frac{1}{2}kx^2.$$

The way we see the harmonic oscillator will change with respect to the formalism for the quantum harmonic oscillator. In this last case, we shall adopt a different perspective by which Quantum Mechanics can be formulated.

2.2   Lecture 2: The Mechanical Lagrangian

In a system of coordinates (t, x), let $(t, x(t))$ be the trajectory of a particle of mass m moving under the influence of a force derived from a time independent *potential* V. Since V depends only on the position, we denote this by $V := V(x)$.

Newton's equation of motion is

$$m\ddot{x}(t) = F(x),$$

where the force acting on the particle is $F(x) = -\dfrac{dV}{dx}$.

Given initial conditions, the trajectory $(t, x(t))$ is comprised between the initial point $(t_1, x(t_1))$ and the final point $(t_2, x(t_2))$.

Let us stress that this trajectory is the expression of the force acting on the particle under some initial conditions. Therefore, there is a single trajectory determined by the force and the initial conditions.

Now let us consider all the paths connecting $(t_1, x(t_1))$ and $(t_2, x(t_2))$. They can be thought as $y(t) + \eta(t)$, with $y(t_1) = x(t_1)$, $y(t_2) = x(t_2)$, $\eta(t_1) = \eta(t_2) = 0$.

Having all these paths, what we need to discover the original path described by the Newton equation of motion?

To answer this question, we need some technical details.

We have used V such that $F = -\dfrac{dV}{dx}$. We defined V as a potential (function of space coordinates) connected with the force F, by $dV = -Fdx$.

Consider a body of mass M at the origin O of a line whose spatial coordinate is denoted by x. Suppose tha,t at point $N(x)$, a body of mass m exists. The gravitational force in this case has the intensity $F = \dfrac{GMm}{x^2}$ where G is the Newton gravitational constant. The work done by the body of mass M to move the body of mass m from x to $x - dx$ is $-Fdx$. There is an energy E transferred to make this work. Its variation ΔE is $-Fdx$. By definition, the *gravitational potential energy* $V(r)$ is related to the work made to move the body of mass m from the infinity to the point having the coordinate r, that is

$$V(r) = \int_\infty^r Fdx = \int_\infty^r \frac{GMm}{x^2}dx = -\frac{GMm}{r}.$$

The potential energy can be denoted by V. If one looks at the formula obtained and takes into account the formula of the gravitational potential $-\dfrac{GM}{r}$, we obtain the relation

$$V = m\Phi.$$

Therefore, the gravitational potential represents the work (energy transferred) per unit mass necessary to move a body from infinity to the point having the coordinate r; that is

$$\Phi(r) = \frac{1}{m}\int_\infty^r Fdx = \frac{1}{m}\int_\infty^r \frac{GMm}{x^2}dx = -\frac{GM}{r}.$$

If we consider the constant gravitational field determined by the constant acceleration g between the origin O and a point H at the coordinate h, the potential energy is expressed by the formula $V = mgh$. The explanation is related to the difference of formal integrals

$$V := V(h) - V(0) = \int_\infty^h gmdx - \int_\infty^0 gmdx = \int_0^h gmdx = gmh$$

which describes the amount of energy necessary to move the body at h to 0.

In the same way, we can define the kinetic energy. Let us start from $F = ma = m\dfrac{dv}{dt}$ written in its discrete form, $F = m\dfrac{\Delta v}{\Delta t}$. If we multiply by Δr, we obtain $F\Delta r = m\dfrac{\Delta v}{\Delta t}\Delta r = m\dfrac{\Delta r}{\Delta t}\Delta v = mv\Delta v$, which can be written in the differential way as

$$Fdr = mvdv.$$

Now, the amount of energy necessary to bring a body initially at rest to the speed v is

$$T(v) = \int_0^v F dx = \int_o^v m x dx = m\frac{v^2}{2}.$$

Since v can be seen as $\dot{x}(t)$, we may consider the *kinetic energy of the mechanical system* defined by the formula $T = T(\dot{x}) := \frac{1}{2}m(\dot{x}(t))^2$. Another possible notation is K_E. Here, by *mechanical system*, we intend a system of material elements that interact by mechanical principles. A material point and a force which acts on it is a possible example. Two materials points which interact through the gravitational force offer another example. In this perspective, the next exercise has important consequences in Newtonian Mechanics.

Exercise 2.2.1 *Consider a mechanical system whose kinetic energy is $T(\dot{x}) := \frac{1}{2}m(\dot{x}(t))^2$ and its potential energy is V (such that the force which acts is $F(x) = -\frac{dV}{dx}$). Show that the total energy of the system, $T + V$, is a constant.*

Hint. If we derive with respect to t the total energy, we obtain

$$\frac{d}{dt}(T + V) = \left(m\dot{x}(t)\ddot{x}(t) + \frac{dV}{dx}\frac{dx}{dt}\right) = (m\ddot{x}(t) - F)\,\dot{x}(t) = 0,$$

that is $T + V$ is a constant.

This total energy can be called the Hamiltonian of the mechanical system. In the case when the system is reduced at a single particle of mass m, $H = T + V$, $T = \frac{p^2}{2m}$, $V = V(q)$. Here $q := x$, a notation that we will adopt below.

We define the *mechanical Lagrangian* of the system by

$$L = L(x, \dot{x}) := T - V = \frac{1}{2}m(\dot{x}(t))^2 - V(x).$$

Let us observe that x and $\dot{x}$ depends on t, that is the Lagrangian is implicitly a function of time.

In this formalism, it makes sense to consider a functional called *action*,

$$S[y] = \int_{t_1}^{t_2}\left[\frac{1}{2}m(\dot{y}(t))^2 - V(y)\right]dt$$

which exists for any path $y(t)$, not only for the "physical one" where $x(t)$ is.

Now consider the action corresponding to $y(t) + \eta(t)$,

$$S[y + \eta] = \int_{t_1}^{t_2}\left[\frac{1}{2}m(\dot{y}(t) + \dot{\eta}(t))^2 - V(y(t) + \eta(t))\right]dt.$$

We have, after expanding V in Taylor series with respect to $y(t)$,

$$S[y + \eta] = S[y] + \int_{t_1}^{t_2} \left[m\dot{y}(t)\dot{\eta}(t) - \frac{dV}{dy}(y(t))\eta(t) \right] dt + \mathbb{O}(\eta^2),$$

where $\mathbb{O}(\eta^2)$ are terms depending on η or $\dot{\eta}$ for powers greater than 2. We can write

$$S[y + \eta] = S[y] + \delta S + \mathbb{O}(\eta^2),$$

where

$$\delta S = \int_{t_1}^{t_2} \left[m\dot{y}(t)\dot{\eta}(t) - \frac{dV}{dy}(y(t))\eta(t) \right] dt$$

is called the *first order variation of the action S*. Since $\eta(t_1) = \eta(t_2) = 0$, we obtain

$$\delta S = \int_{t_1}^{t_2} \left[m\dot{y}(t)\dot{\eta}(t) - \frac{dV}{dy}(y(t)\eta(t)) \right] dt =$$

$$= \int_{t_1}^{t_2} \left[m\frac{d(\dot{y}(t)\eta(t))}{dt} - m\ddot{y}(t)\eta(t) - \frac{dV}{dy}(y(t)\eta(t)) \right] dt =$$

$$= \cancel{m\dot{y}(t_2)\eta(t_2)} - \cancel{m\dot{y}(t_1)\eta(t_1)} - \int_{t_1}^{t_2} \left[m\ddot{y}(t) + \frac{dV}{dy}(y(t)) \right] \eta(t)dt =$$

$$= -\int_{t_1}^{t_2} \left[m\ddot{y}(t) + \frac{dV}{dy}(y(t)) \right] \eta(t)dt.$$

Therefore, $\delta S \equiv 0$ means

$$\int_{t_1}^{t_2} \left[m\ddot{y}(t) + \frac{dV}{dy}(y(t)) \right] \eta(t)dt = 0$$

for every η, and it happens if and only if $m\ddot{y}(t) + \dfrac{dV}{dy}(y(t)) = 0$, i.e. exactly for the path $y(t) = x(t)$. We have proved:

Theorem 2.2.2 *The first order variation of the action S vanishes, i.e.*

$$\delta S = \int_{t_1}^{t_2} \left[m\dot{y}(t)\dot{\eta}(t) - \frac{dV}{dy}(y(t))\eta(t) \right] dt \equiv 0$$

if and only if $y(t)$ satisfies Newton's equation of motion

$$m\ddot{x}(t) - F(x) = 0.$$

So, the answer is: *The "physical right path" happens when the first order variation* δS *vanishes*. Therefore the right path is described by the condition $\delta S \equiv 0$. This is known as the *Hamilton's stationary action principle*.

Summary of Lecture 2. The *mechanical Lagrangian* of a mechanical system,

$$L = L(x, \dot{x}) := T - V = \frac{1}{2}m(\dot{x}(t))^2 - V(x),$$

offers another view of Newton's second law.

Consider a solution under initial conditions of Newton's law of motion $m\ddot{x}(t) = -\dfrac{dV}{dx}$. Let $(t, y(t))$ be the generic form of all paths connecting the initial point $(t_0, x(t_0))$ to a given point of the solution-path $(t, x(t))$, say $(t_1, x(t_1))$, and the functional called action

$$S[y] := \int_{t_1}^{t_2} \left[\frac{1}{2}m(\dot{y}(t))^2 - V(y) \right] dt.$$

What condition, written in terms of S, describes the initial path $(t, x(t))$? This happens through the vanishing of the first order action δS written as

$$\delta S = \int_{t_1}^{t_2} \left[m\dot{y}(t)\dot{\eta}(t) - \frac{dV}{dy}(y(t))\eta(t) \right] dt \equiv 0.$$

It is proved that the previous condition is satisfied only for $y(t)$ which satisfies the equation of motion

$$m\ddot{x}(t) - F(x) = 0.$$

This point of view will be extended to more general Lagragians giving rise to the *Lagrangian Mechanics*.

2.3 Lecture 3: The Euler–Lagrange Equations

Let us now consider another problem.

Can we find an equation, described with respect a general function $L(x, \dot{x})$ *such that the function* $x = x(t)$, *which connects the given points* $(t_1, x(t_1))$; $(t_2, x(t_2))$ *extremizes the functional*

$$S[x] = \int_{t_1}^{t_2} L(x(t), \dot{x}(t))dt \ ?$$

Let us discuss first what is the mathematical meaning of the sentence "extremizes the functional S." Consider the perturbation of $x(t)$,

$$y_\lambda(t) = x(t) + \lambda \eta(t), \ \lambda \in \mathbb{R}$$

which preserves the endpoints $(t_1, x(t_1))$; $(t_2, x(t_2))$, that is $\eta(t_1) = \eta(t_2) = 0$ and let us construct the action

$$S[y_\lambda] = \int_{t_1}^{t_2} L(y_\lambda(t), \dot{y}_\lambda(t))dt = \int_{t_1}^{t_2} L(x(t) + \lambda\eta(t), \dot{x}(t) + \lambda\dot{\eta}(t))dt.$$

Extremizing the functional $S[x]$ means whether $S[y_\lambda] \geq S[x]$ for all $\lambda \in \mathbb{R}$, or $S[y_\lambda] \leq S[x]$ for all $\lambda \in \mathbb{R}$, where the equality works if and only if $\lambda = 0$.

Therefore, *extremizing the functional $S[x]$ implies the condition* $\left. \dfrac{dS}{d\lambda} \right|_{\lambda=0} \equiv 0$.

Since

$$\frac{dL}{d\lambda} = \frac{\partial L}{\partial y_\lambda}\frac{\partial y_\lambda}{\partial \lambda} + \frac{\partial L}{\partial \dot{y}_\lambda}\frac{\partial \dot{y}_\lambda}{\partial \lambda} = \frac{\partial L}{\partial y_\lambda}\eta(t) + \frac{\partial L}{\partial \dot{y}_\lambda}\dot{\eta}(t),$$

it results

$$\left. \frac{dL}{d\lambda} \right|_{\lambda=0} = \frac{\partial L}{\partial x}\eta(t) + \frac{\partial L}{\partial \dot{x}}\dot{\eta}(t),$$

therefore the condition $\left. \dfrac{dS_\lambda}{d\lambda} \right|_{\lambda=0} \equiv 0$ is written as

$$\left. \frac{dS_\lambda}{d\lambda} \right|_{\lambda=0} = \int_{t_1}^{t_2} \left[\frac{\partial L}{\partial x}\eta(t) + \frac{\partial L}{\partial \dot{x}}\dot{\eta}(t) \right] dt \equiv 0.$$

Definition 2.3.1 *The curve $x = x(t)$ which extremizes the functional*

$$S[x] = \int_{t_1}^{t_2} L(x(t), \dot{x}(t))dt$$

is called a stationary point of the functional $S[x]$. '

Theorem 2.3.2 (Euler–Lagrange equation) *The curve $x = x(t)$ which connects the given points $(t_1, x(t_1))$, $(t_2, x(t_2))$ satisfies the Euler–Lagrange equation*

$$\frac{d}{dt}\left(\frac{\partial L}{\partial \dot{x}} \right) - \frac{\partial L}{\partial x} = 0$$

if and only if it is a stationary point of the functional

$$S[x] = \int_{t_1}^{t_2} L(x(t), \dot{x}(t))dt.$$

Proof Let us first observe that the integration by parts leads to

$$\int_{t_1}^{t_2} \frac{\partial L}{\partial x}\eta(t)dt + \int_{t_1}^{t_2} \frac{\partial L}{\partial \dot{x}}\dot{\eta}(t)dt =$$

$$= \int_{t_1}^{t_2} \frac{\partial L}{\partial x}\eta(t)dt + \frac{\partial L}{\partial \dot{x}}\eta(t_2) - \frac{\partial L}{\partial \dot{x}}\eta(t_1) - \int_{t_1}^{t_2} \frac{d}{dt}\left(\frac{\partial L}{\partial \dot{x}}\right)\eta(t)dt =$$

$$= \int_{t_1}^{t_2} \left[\frac{\partial L}{\partial x} - \frac{d}{dt}\left(\frac{\partial L}{\partial \dot{x}}\right)\right]\eta(t)dt.$$

The condition $\left.\dfrac{dS}{d\lambda}\right|_{\lambda=0} \equiv 0$ means

$$\int_{t_1}^{t_2} \left[\frac{d}{dt}\left(\frac{\partial L}{\partial \dot{x}}\right) - \frac{\partial L}{\partial x}\right]\eta(t)dt = 0$$

for every function η. We obtain

$$\frac{d}{dt}\left(\frac{\partial L}{\partial \dot{x}}\right) - \frac{\partial L}{\partial x} = 0.$$

$\square$

Example 2.3.3 *Let us consider again the harmonic oscillator adopting the Euler–Lagrange equation. In the case of harmonic oscillator, the potential energy is $V(x) :=$* $\dfrac{1}{2}kx^2$, *therefore we have the Lagrangian*

$$L(x, \dot{x}) := m\frac{\dot{x}^2}{2} - \frac{1}{2}kx^2.$$

If we calculate

$$\frac{d}{dt}\left(\frac{\partial L}{\partial \dot{x}}\right) - \frac{\partial L}{\partial x} = 0$$

for the considered Lagrangian, we obtain the original equation of motion

$$m\ddot{x}(t) = -kx(t)$$

as expected. In this case, the Euler–Lagrange equation offers only another view for Newton's second law.

The previous proof of Euler–Lagrange equation can be generalized.

For the Lagrangian $L = L(x^1, \ldots, x^n, \dot{x}^1, \ldots, \dot{x}^n)$, we proceed as before on each pair of variables $x^k, \dot{x}^k$, $k = 1, \ldots, n$. When we extremize the functional

$$S[x] = \int_{t_1}^{t_2} L(x^1(t), \dot{x}^1(t), \ldots, x^n(t), \dot{x}^n(t))dt$$

we have, as we explained, to construct

$$S[y_\lambda] = \int_{t_1}^{t_2} L(y_\lambda(t), \dot{y}_\lambda(t))dt =$$

$$= \int_{t_1}^{t_2} L(x^1(t) + \lambda\eta_1(t), \dot{x}^1(t) + \lambda\dot{\eta}_1(t), \ldots, x^n(t) + \lambda\eta_n(t), \dot{x}^n(t) + \lambda\dot{\eta}_n(t))dt$$

The condition

$$\left.\frac{dS}{d\lambda}\right|_{\lambda=0} = \int_{t_1}^{t_2} \sum_{k=1}^{n} \left[\frac{d}{dt}\left(\frac{\partial L}{\partial \dot{x}^k}\right) - \frac{\partial L}{\partial x^k}\right] \eta_k(t)dt \equiv 0$$

for every function η_k, $k = 1, \ldots, n$ leads to the general Euler–Lagrange equations

$$\begin{cases} \dfrac{d}{dt}\left(\dfrac{\partial L}{\partial \dot{x}^1}\right) - \dfrac{\partial L}{\partial x^1} = 0 \\ \dfrac{d}{dt}\left(\dfrac{\partial L}{\partial \dot{x}^2}\right) - \dfrac{\partial L}{\partial x^2} = 0 \\ \cdots\cdots\cdots\cdots\cdots\cdots\cdots \\ \dfrac{d}{dt}\left(\dfrac{\partial L}{\partial \dot{x}^n}\right) - \dfrac{\partial L}{\partial x^n} = 0. \end{cases}$$

First at all we can write in a simpler form the previous system, i.e.

$$\frac{d}{dt}\left(\frac{\partial L}{\partial \dot{x}^k}\right) - \frac{\partial L}{\partial x^k} = 0, \ k = 1, 2, \ldots, n.$$

If you look carefully at the way we derived the Euler–Lagrange equations, you may understand that all remain the same if the Lagrangian is depending only implicitly on t in the variables $x^i(t)$ and $\dot{x}^i$. Another proof can be done if the Lagrangian depends explicitly on t, that is if $L = L(x^1, \ldots, x^n, \dot{x}^1, \ldots, \dot{x}^n, t)$.

Summary of Lecture 3. In this lecture, we studied the Euler–Lagrange equation

$$\frac{d}{dt}\left(\frac{\partial L}{\partial \dot{x}}\right) - \frac{\partial L}{\partial x} = 0$$

whose solution is a curve $x = x(t)$ which connects two given points $(t_1, x(t_1))$, $(t_2, x(t_2))$. This one becomes the general law of motion and it is obtained looking at a stationary point of the functional

$$S[x] = \int_{t_1}^{t_2} L(x(t), \dot{x}(t)) dt.$$

The result can be extended to Lagrangians depending on n variables as described as described above. The example of harmonic oscillator is presented in this framework with the potential energy is $V(x) := \frac{1}{2}kx^2$. Therefore we have the Lagrangian

$$L(x, \dot{x}) := m\frac{\dot{x}^2}{2} - \frac{1}{2}kx^2.$$

If we compute

$$\frac{d}{dt}\left(\frac{\partial L}{\partial \dot{x}}\right) - \frac{\partial L}{\partial x} = 0$$

for the considered Lagrangian, we obtain the original equation of motion

$$m\ddot{x}(t) = -kx(t)$$

as expected. As final remark, we can say that Lagrangian Mechanics is the generalization of Newtonian Mechanics.

2.4 Lecture 4: The Mechanical Hamiltonian

In the 1-dimensional case, let us denote with $q := x$ the space coordinate and $p := m\dot{x} = m\dot{q}$ the momentum coordinate. The *mechanical Hamiltonian* is

$$H(x, \dot{x}) = T(\dot{x}) + V(x)$$

where

$$T := \frac{1}{2}m\dot{x}^2 \; ; \; V := V(x).$$

According to the previous notations, we obtain $T = T(p) = \frac{p^2}{2m}$ and $V = V(q)$. It results

$$\frac{\partial H}{\partial p} = \frac{\partial(T+V)}{\partial p} = \frac{\partial T}{\partial p} = \frac{p}{m} = \dot{q}.$$

Then, using Newton's second law we have

$$\frac{\partial H}{\partial q} = \frac{\partial (T+V)}{\partial q} = \frac{\partial V}{\partial q} = -m\ddot{q} = -\dot{p}.$$

Therefore, in 1-dimensional case, the Hamilton equations are

$$\begin{cases} \dot{q} = \dfrac{\partial H}{\partial p} \\ \dot{p} = -\dfrac{\partial H}{\partial q}. \end{cases}$$

Let us stress the following. The Hamiltonian is connected to the Lagrangian through the formula

$$H = \dot{q}\frac{\partial L}{\partial \dot{q}} - L.$$

To prove this simple assertion, we use $L = T - V$ and $H = T + V$. It results

$$L + H = 2T = \frac{p^2}{m}.$$

At the same time, $\dot{q}\dfrac{\partial L}{\partial \dot{q}} = \dot{q}\dfrac{\partial T}{\partial \dot{q}} = m\dot{q}^2 = \dfrac{p^2}{m}$, that is

$$L + H = \frac{p^2}{m} = \dot{q}\frac{\partial L}{\partial \dot{q}}.$$

This equality, called *Legendre's transform*, is very important because it allows to describe a Hamiltonian starting from a given Lagrangian as functions of any number of variables.

Example 2.4.1 *Let us consider again the harmonic oscillator, now using Hamilton's equations and the new notations. In this case, the potential energy is $V(q) := \frac{1}{2}kq^2$ and the kinetic energy is $T(p) = \dfrac{p^2}{2m}$, where $p = m\dot{x} = m\dot{q}$. The Hamiltonian can be written in the form*

$$H(q, p) := \frac{p^2}{2m} + \frac{1}{2}kq^2.$$

If we compute

$$\begin{cases} \dot{q} = \dfrac{\partial H}{\partial p} \\ \dot{p} = -\dfrac{\partial H}{\partial q}. \end{cases}$$

for the considered Hamiltonian, we obtain the original equation of motion from the second equation,

$$m\ddot{q}(t) = -kq(t)$$

as expected, while the first equation represents the definition of the momentum.

The Hamiltonian equations offer another view for Newton's second law. As we will see, Hamiltonians are extremely useful in Quantum Mechanics.

The n-dimensional mechanical Hamiltonian can be defined with respect to both the generalized coordinates $q := (q^1, q^2, \ldots, q^n)$, representing the position in the n-dimensional Euclidean space, and the generalized momenta $p := m\dot{q}$, that is $(p_1, p_2, \ldots, p_n) := (m\dot{q}^1, m\dot{q}^2, \ldots, m\dot{q}^n)$.

Therefore the kinetic energy formula is

$$T(p) = T(p_1, p_2, \ldots, p_n) = \frac{p^2}{2m} = \frac{p_1^2 + p_2^2 + \cdots + p_n^2}{2m}.$$

Exactly as in the one-dimensional case, the potential energy is a function V depending only on the position q.

The mechanical Hamiltonian is, in this case,

$$H(q, p) := T(p) + V(q) = \frac{p^2}{2m} + V(q).$$

Since

$$\frac{\partial H(q, p)}{\partial p_k} = \frac{\partial T(p)}{\partial p_k} = \frac{p_k}{m} = \dot{q}^k$$

and Newton's second law leads to

$$\frac{\partial H(q, p)}{\partial q^k} = \frac{\partial V(q)}{\partial q^k} = -m\ddot{q}^k = -\dot{p}_k.$$

Adopting convenient notations

$$\frac{\partial H}{\partial q} = \left(\frac{\partial H}{\partial q^1}, \ldots, \frac{\partial H}{\partial q^n}\right);$$

$$\frac{\partial H}{\partial p} = \left(\frac{\partial H}{\partial p_1}, \ldots, \frac{\partial H}{\partial p_n}\right),$$

the *Hamilton equations* are

$$\begin{cases} \dot{q} = \dfrac{\partial H}{\partial p} \\ \dot{p} = -\dfrac{\partial H}{\partial q}. \end{cases}$$

This is a system of $2n$ equations which makes the same predictions as the Euler–Lagrange one in the case of mechanical Lagrangian. This represents another way to express the evolution of a material point or of a system of material points in Mechanics as in the case of second Newton's law. It is worth observing that there are two further formalisms for Newtonian mechanics, the Euler–Lagrange's one and the Hamiltonian's one. The second one has the advantage of working with respect the total energy of the mechanical system.

Summary of Lecture 4. Another possible formalism for Mechanics is the Hamiltonian one. In 1-dimensional case, if $q := x$ is the space coordinate and $p := m\dot{x} = m\dot{q}$ is the momentum coordinate and

$$H(x, \dot{x}) := T(\dot{x}) + V(x)$$

is the Hamiltonian, the equation of motion is described by the system

$$\begin{cases} \dot{q} = \dfrac{\partial H}{\partial p} \\ \dot{p} = -\dfrac{\partial H}{\partial q}. \end{cases}$$

The connection with the Lagrangian formalism is achieved through the Legendre transformation

$$H = \dot{q}\frac{\partial L}{\partial \dot{q}} - L.$$

Again, we insist on the harmonic oscillator seen through the Hamilton equations: The potential energy is $V(q) := \dfrac{1}{2}kq^2$ and the kinetic energy is $T(p) := \dfrac{p^2}{2m}$, where $p = m\dot{x} = m\dot{q}$. We have the Hamiltonian

$$H(q, p) := \frac{p^2}{2m} + \frac{1}{2}kq^2.$$

If we compute

$$\begin{cases} \dot{q} = \dfrac{\partial H}{\partial p} \\ \dot{p} = -\dfrac{\partial H}{\partial q}. \end{cases}$$

for the considered Hamiltonian, we obtain the original equation of motion from the second equation,

$$m\ddot{q}(t) = -kq(t)$$

as expected, while the first equation represents the definition of the momentum. The case of multiple variables Hamiltonians is also presented.

2.5 Lecture 5: The Hamilton Equations

Consider the generalized coordinates $(q^1, q^2, \ldots, q^n)$, each one depending on t, and the associated generalized velocities $(\dot{q}^1, \dot{q}^2, \ldots, \dot{q}^n)$. A general Lagrangian is seen as a function depending on position and velocities, that is $L = L(q^1, q^2, \ldots, q^n, \dot{q}^1, \dot{q}^2, \ldots, \dot{q}^n)$. Let us introduce the notation $L = L(q^i, \dot{q}^i)$. However, a Lagrangian can depend explicitly on t, so the most general form is $L = L(q^i, \dot{q}^i, t)$.

We define a general Hamiltonian $H = H(q^i, \dot{q}^i, t)$ by the Legendre transformation formula

$$H(q^i, \dot{q}^i, t) := \sum_{i=1}^{n} \dot{q}^i \frac{\partial L}{\partial \dot{q}^i} - L(q^i, \dot{q}^i, t),$$

or in a simpler notation,

$$H := \sum_{i=1}^{n} \dot{q}^i \frac{\partial L}{\partial \dot{q}^i} - L.$$

By definition, the generalized momenta are

$$p_i(q^i, \dot{q}^i, t) := \frac{\partial L}{\partial \dot{q}^i}.$$

The previous formula becomes

$$H = \sum_{i=1}^{n} p^i \dot{q}^i - L.$$

The simplest derivation of Hamilton's equations starts from the total derivative of the Lagrangian L. We have

$$dL = \sum_{i=1}^{n} \left[\frac{\partial L}{\partial q^i} dq^i + \frac{\partial L}{\partial \dot{q}^i} d\dot{q}^i \right] + \frac{\partial L}{\partial t} dt = \sum_{i=1}^{n} \left[\frac{\partial L}{\partial q^i} dq^i + p_i d\dot{q}^i \right] + \frac{\partial L}{\partial t} dt,$$

that is

$$dL = \sum_{i=1}^{n} \left[\frac{\partial L}{\partial q^i} dq^i + d(p_i \dot{q}^i) - \dot{q}^i dp_i \right] + \frac{\partial L}{\partial t} dt.$$

This last equality can be written in the form

$$d\left(\sum_{i=1}^{n} p_i \dot{q}^i - L\right) = \sum_{i=1}^{n}\left[-\frac{\partial L}{\partial q^i}dq^i + \dot{q}^i dp_i\right] - \frac{\partial L}{\partial t}dt$$

and taking into account the definition of Hamiltonian, we have

$$dH = \sum_{i=1}^{n}\left[-\frac{\partial L}{\partial q^i}dq^i + \dot{q}^i dp_i\right] - \frac{\partial L}{\partial t}dt.$$

Directly, it is

$$dH = \sum_{i=1}^{n}\left[\frac{\partial H}{\partial q^i}dq^i + \frac{\partial H}{\partial \dot{q}^i}d\dot{q}^i\right] + \frac{\partial H}{\partial t}dt.$$

We obtain, by associating the corresponding terms from the above right side definitions of dH, that

$$\frac{\partial H}{\partial q^i} = -\frac{\partial L}{\partial q^i}; \quad \frac{\partial H}{\partial p_i} = \dot{q}^i; \quad \frac{\partial H}{\partial t} = -\frac{\partial L}{\partial t}.$$

Now, using the Euler–Lagrange equations

$$\frac{d}{dt}\left(\frac{\partial L}{\partial \dot{q}^i}\right) - \frac{\partial L}{\partial q^i} = 0$$

the first equality becomes

$$\dot{p}_i = -\frac{\partial H}{\partial q^i},$$

so the *general Hamilton's equations* are

$$\dot{p}_i = -\frac{\partial H}{\partial q^i}; \quad \frac{\partial H}{\partial p_i} = \dot{q}^i; \quad \frac{\partial H}{\partial t} = -\frac{\partial L}{\partial t}.$$

If the Lagrangian does not depend on t, the Hamiltonian will not depend on t, therefore Hamilton's equations are only

$$\dot{p}_i = -\frac{\partial H}{\partial q^i}; \quad \frac{\partial H}{\partial p_i} = \dot{q}^i.$$

In the next section, we present Poisson's brackets and formulas involving Hamiltonians. The connection and the consequences with the formalism of Quantum Mechanics is presented below, in particular in Lectures 42 and 43.

Summary of Lecture 5. Consider generalized coordinates denoted by q^i, $i \in \{1, 2, \ldots, n\}$, each one depending on t, and the associated generalized velocities denoted by $\dot{q}^i$. If the general Lagrangian is a function depending on position and velocities, $L = L(q^i, \dot{q}^i)$, we define the general Hamiltonian

$$H := \sum_{i=1}^{n} \dot{q}^i \frac{\partial L}{\partial \dot{q}^i} - L$$

and the generalized momenta

$$p_i(q^i, \dot{q}^i, t) := \frac{\partial L}{\partial \dot{q}^i}.$$

The corresponding Hamilton equations are

$$\dot{p}_i = -\frac{\partial H}{\partial q^i}; \quad \frac{\partial H}{\partial p_i} = \dot{q}^i.$$

2.6 Lecture 6: Poisson's Brackets in Hamiltonian Mechanics

Consider a real function F whose variables are p_i, q_i, $i \in \{1, 2, \ldots, n\}$. Let us take into account a time independent Hamiltonian $H := H(q^i, \dot{q}^i)$ which can also be written as $H := H(q^i, p_i)$, being $p_i = m\dot{q}^i$. Here q^i and p_i depend on t. Therefore we can write

$$\frac{dF}{dt} = \sum_{i=1}^{n} \left[\frac{\partial F}{\partial q^i} \dot{q}^i + \frac{\partial F}{\partial p_i} \dot{p}_i \right] = \sum_{i=1}^{n} \left[\frac{\partial F}{\partial q^i} \frac{\partial H}{\partial p_i} - \frac{\partial F}{\partial p_i} \frac{\partial H}{\partial q^i} \right] := [F, H].$$

Here $[F, H]$ denotes *Poisson's brackets* of the functions F and H.

If $F = H$, we obtain $\dfrac{dH}{dt} = [H, H] = 0$ which highlights that we are working with a time independent Hamiltonian.

If $F = q^j$, we obtain

$$\dot{q}^j = \frac{dq^j}{dt} = [q^j, H] = \sum_{i=1}^{n} \left[\frac{\partial q^j}{\partial q^i} \frac{\partial H}{\partial p_i} - \frac{\partial q^j}{\partial p_i} \frac{\partial H}{\partial q^i} \right] = \frac{\partial H}{\partial p_j}.$$

If $F = p_j$, we obtain

$$\dot{p}_j = \frac{dp_j}{dt} = [p_j, H] = \sum_{i=1}^{n} \left[\frac{\partial p_j}{\partial q^i} \frac{\partial H}{\partial p_i} - \frac{\partial p_j}{\partial p_i} \frac{\partial H}{\partial q^i} \right] = -\frac{\partial H}{\partial q^j}.$$

Therefore the relation

$$\frac{dF}{dt} = [F, H],$$

involving Poisson's brackets, gives Hamilton's equations for particular F.

Let us now concentrate on variables only. We are interested to compute $[q^i, q^j]$, $[p_i, p_j]$, $[q^i, p_j]$. Firstly we prove that $[q^i, q^j] = 0$. It is

$$[q^i, q^j] = \sum_{k=1}^{n} \left[\frac{\partial q^i}{\partial q^k} \frac{\partial q^j}{\partial p_k} - \frac{\partial q^i}{\partial p_k} \frac{\partial q^j}{\partial q^k} \right] = 0.$$

In the same way $[p_i, p_j] = 0$. It is

$$[p_i, p_j] = \sum_{k=1}^{n} \left[\frac{\partial p_i}{\partial q^k} \frac{\partial p_j}{\partial p_k} - \frac{\partial p_i}{\partial p_k} \frac{\partial p_j}{\partial q^k} \right] = 0.$$

Finally, we prove $[q^i, p_j] = \delta_j^i$. It is

$$[q^i, p_j] = \sum_{k=1}^{n} \left[\frac{\partial q^i}{\partial q^k} \frac{\partial p_j}{\partial p_k} - \frac{\partial q^i}{\partial p_k} \frac{\partial p_j}{\partial q^k} \right] = \sum_{k} \delta_k^i \delta_j^k = \delta_j^i,$$

which is 0 for $i \neq j$ and 1 for $i = j$.

It is easy to check the same for $\frac{dF}{\partial q^i} = [F, p_i]$. It is

$$[F, p_i] = \sum_{k=1}^{n} \left[\frac{\partial F}{\partial q^k} \frac{\partial p_i}{\partial p_k} - \frac{\partial F}{\partial p_k} \frac{\partial p_i}{\partial q^k} \right] = \frac{\partial F}{\partial q^i}.$$

Let us consider a function F depending on the space coordinates $F = F(x, y.z)$. Let us take into account the x coordinate and then

$$\frac{\partial F}{\partial x} = [F, p_x],$$

that is we can express

$$F(x + dx, y, z) = F(x, y, z) + [F, p_x]dx,$$

and similar relations with respect to the other two coordinates:

$$F(x, y + dy, z) = F(x, y, z) + [F, p_y]dy,$$

$$F(x, y, z + dz) = F(x, y, z) + [F, p_z]dz.$$

Therefore the canonical moment p_x is the infinitesimal translation along the x direction, etc.

Before discussing the Quantum Mechanics, let us say that the description of physical systems requires variables called observables which can be measured. In Classical Mechanics, we can measure, by experiments, position, momentum, energy and other quantities derived from the previous ones. These quantities are completely deterministic in equations of motion like Newton's equations, Lagrange's equations or Hamilton's equations. The physical states of a system are values of observables at a given time. For a particle, we have to consider three space and three velocity coordinates. Knowing the state of the system at a given moment, we can determine any other state using the equations of motion. The Classical Mechanics determinism cannot be transferred to systems when we are working with atoms and molecules. To see why, we need to step in the territory of Quantum Mechanics and to take into account Heisenberg's Uncertainty Principle. We need to understand the meaning of observables and to look at expectation values of observables. However, in Lecture 42, Ehrenfest's theorem and its consequences will show us the possibility to restore the Classical Mechanics formalism from the Quantum Mechanics formalism. Other connections will be established via Hamilton–Jacobi theorem in Lecture 40 when we will discuss how the deterministic formalism of Classical Mechanics is replaced by the probabilistic formalism of Quantum Mechanics.

Summary of Lecture 6. Consider a real function $F = F(p_i, q_i)$, $i \in \{1, 2, \ldots, n\}$. If the Hamiltonian is $H := H(q^i, p_i)$, we can define the Poisson brackets of the functions F and H by the formula

$$[F, H] := \sum_{i=1}^{n} \left[\frac{\partial F}{\partial q^i} \frac{\partial H}{\partial p_i} - \frac{\partial F}{\partial p_i} \frac{\partial H}{\partial q^i} \right].$$

We have

$$\frac{dF}{dt} = [F, H].$$

Consequences appear in relation to the generalized coordinates:

$$[q^i, q^j] = \sum_{k=1}^{n} \left[\frac{\partial q^i}{\partial q^k} \frac{\partial q^j}{\partial p_k} - \frac{\partial q^i}{\partial p_k} \frac{\partial q^j}{\partial q^k} \right] = 0,$$

for generalized momenta $[p_i, p_j] = 0$,

$$[p_i, p_j] = \sum_{k=1}^{n} \left[\frac{\partial p_i}{\partial q^k} \frac{\partial p_j}{\partial p_k} - \frac{\partial p_i}{\partial p_k} \frac{\partial p_j}{\partial q^k} \right] = 0.$$

Finally, it is $[q^i, p_j] = \delta^i_j$ because

$$[q^i, p_j] = \sum_{k=1}^{n} \left[\frac{\partial q^i}{\partial q^k} \frac{\partial p_j}{\partial p_k} - \frac{\partial q^i}{\partial p_k} \frac{\partial p_j}{\partial q^k} \right] = \delta^i_j.$$

If $i \neq j$, we have 0; if $i = j$, the result is 1 because, in the above sum for $i = j = k$, a nonzero term appears. The partial derivatives can also be written with respect Poisson's brackets, $\frac{\partial F}{\partial q^i} = [F, p_i]$. This discussion will be continued in the framework of Quantum Mechanics.

Chapter 3
Can Light Be Described by Classical Mechanics?

Fiat Lux

Genesis

3.1 Lecture 7: The Michelson–Morley Experiment and the Principles of Special Relativity

When the Michelson–Morley experiment was performed, it was known that light is an electromagnetic wave which travels at a constant speed (denoted by c) in vacuum. Can light be treated in terms of Classical Mechanics or we need another theory to incorporate it? A quick answer is no. The reason for this statement starts from the Michelson–Morley experiment.

Suppose we have a platform of a train wagon, an open one, on a straight railway line. During the Michelson–Morley experiment, the platform is at rest or it is moving at a constant speed v.

On this platform, let us imagine two perpendicular lines which intersect at I, one, say d_1, coincident to the sense of motion, the other one, say d_2, perpendicular to the sense of motion. On d_1, the longitudinal direction, there is a source of light, denoted by S_L, an interferometer placed in I and a mirror, denoted by M_1, such that the distance between I and M_1 is l.

The interferometer is a device able to split a light-ray in the two perpendicular directions d_1 and d_2, but also to receive two light-rays from perpendicular directions and to send them separately to another given direction (Fig. 3.1).

On the line d_2, which corresponds to the transversal direction, there is another mirror, denoted by M_2, such that the distance between I and M_2 is the same l and a receiver-device R_L such that the interferometer I is between M_2 and R_L.

The receiver-device is able to capture the light-rays coming from the interferometer and to decide which one reached first the device.

© The Author(s), under exclusive license to Springer Nature Switzerland AG 2021
S. Capozziello and W.-G. Boskoff, *A Mathematical Journey to Quantum Mechanics*,
UNITEXT for Physics, https://doi.org/10.1007/978-3-030-86098-1_3

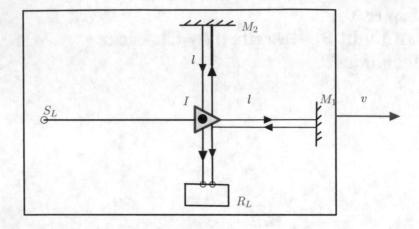

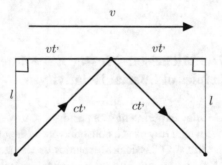

Fig. 3.1 Figure 2.1.1$_b$

The experiment is like this: when the platform is at rest or it is moving at constant speed v in the $S_L I$ longitudinal direction, a light-ray is sent by the source S_L to the interferometer I. The interferometer splits the light-ray in two light-rays. The first one is sent to the mirror M_1, it is reflected by the mirror and it is returned to the interferometer which sends it to R_L. The second one is directed to M_2, it is reflected and sent to the interferometer which sends it to R_L. Which one reaches first R_L?

In other words, we are interested in identifying the influence of the speed v on the splitted light-rays. Is there, or is there not, a difference between what is happening when the platform is at rest comparing with the case when the platform is moving at constant speed v?

Let us observe something obvious: if the platform is at rest, both light-rays reach at the same time R_L.

Now, let us try to use Classical Mechanics to describe what is happening when the platform is moving at constant speed v. First at all, let us observe that it is enough to establish only the time necessary to cover the routes $I M_1 I$ and $I M_2 I$ and to compare them.

Denote by c the speed of light. The time to cover the longitudinal route $I M_1 I$ is

$$t_1 = \frac{l}{c - v} + \frac{l}{c + v} = \frac{2lc}{c^2 - v^2},$$

because $c - v$ and $c + v$ are, in Newtonian Mechanics, the speeds for the directions $I M_1$, $M_1 I$ respectively. To be sure that the reader understands why the speeds are like this, let us focus on the first direction case. Moving at constant speed v in the sense $I M_1$, the light-ray is slowed down by the air, that is by the medium in which it is traveling, with the speed $-v$. Therefore, according to the mechanics rules, the speed of the light-ray traveling in $I M_1$ direction is $c - v$.

For the transversal direction, let us denote by t' the time necessary for the light-ray to reach the mirror M_2. During this time, the platform, therefore the mirror, travels in the longitudinal direction a $t'v$ space. The Pythagoras Theorem, in the formed rectangle triangle, is $(t'c)^2 = l^2 + (t'v)^2$, that is

$$t' = \frac{l}{\sqrt{c^2 - v^2}}.$$

It is obvious that the time necessary to the transversal ray to reach again the interferometer I is $t_2 := 2t'$, so we have

$$t_2 = \frac{2l}{\sqrt{c^2 - v^2}}.$$

Therefore

$$\frac{t_2}{t_1} = \sqrt{1 - \frac{v^2}{c^2}},$$

which implies

$$t_2 < t_1,$$

i.e. the transversal light-ray reaches earlier R_L compared to the longitudinal light-ray.

The mathematical model made with respect to the rules of Classical Mechanics has a prediction, let us repeat it: the first light-ray, arriving at R_L, is the transversal one.

If we make the experiment, the result is: the transversal and the longitudinal light-rays reach R_L at the same time. If we repeat it, the same results holds. There is no difference between what happens when the platform is at rest compared with the case when the platform is moving at constant speed v.

The error consisted in the model description: it is related to the fact that v could affect the speed of light. It seems that $c - v$ and $c + v$ are not correctly thought, therefore we cannot consider Classical Mechanics rules when we try to describe this experiment. Another rule has to be applied when we "add" velocities.

This experiment can be also seen making a parallel between the platform moving in Earth atmosphere at constant speed v and the Earth moving through the ether at constant speed v. After we establish a new theory to explain the experimental result, the main consequence is the fact that there is no ether. As a comment from a different perspective, in modern physics it has been realized that "ether" is the "physical vacuum" that is a maximally symmetric configuration of spacetime where no physical field is present. This means that matter-energy density is extremely low. In this "vacuum", electromagnetic waves propagate at the speed of light.

Essentially, Einstein formulated the Special Relativity starting from two main postulates:

The consequences of Einstein's postulates give the chance to understand how the light propagates in the context of a new physical theory, the Special Relativity, which changes the rules of Classical Mechanics when we are dealing with bodies moving at very large speeds.

Part of these results were also obtained by Henry Poincaré in his effort to explain the Michelson–Morley experiment.

The postulates are:

1. The laws of Physics are the same in all inertial reference frames.
2. The speed of light in vacuum, denoted by $c \approx 2,99 \cdot 10^8$ m/s, is the same for all the observers and it is the maximal speed reached by a moving object.

Einstein used the word *observer* with the meaning of *reference frame* from which a set of objects or events are measured. Since the measurements are generally made with respect to the center O of the frame, this special point is often called the "O observer" or we may refer to a frame with "the observer placed at O". We know that the laws of Mechanics are the same in all inertial frames. The first postulate asks for the same form of electromagnetic laws in any inertial reference frames, like in the case of mechanics laws. And in general, all laws of Physics must have the same form in all reference frames (this result will be fully achieved in General Relativity).

The second postulate plays a key role in Special Relativity being involved in the way in which we derive the Lorentz transformations.

The framework of Newton's laws of Mechanics is the 3-dimensional Euclidean space. Each object is described by a point or by a collection of points of it. Time is given by a universal clock and allows us to see the evolution of objects.

In Special Relativity, we have to work in a 4-dimensional space. Three of the dimensions are the standard dimensions used in Mechanics. We can denote them by x, y, z. The fourth dimension is related to time.

Definition 3.1.1 *A frame of coordinates (t, x, y, z) is called a spacetime.*

The geometry of a spacetime is determined by some physical postulates. They are

Definition 3.1.2 *Each point of spacetime is called "event".*

Definition 3.1.3 *A curve of the spacetime is called "world line" and represents a successions of events.*

Example 3.1.4 *Let us suppose to work in a two dimensional slice of the previous spacetime, with the coordinates* (t, z). *Consider a world line starting from the origin* $O(0, 0)$ *and suppose also that the next point is* $A(1, z_0)$. *Then the object remains* t_0 *seconds at rest with respect to* O. *This means that the world line has to be continued with the segment* AB, *where* B *has the coordinates* $B(1 + t_0, z_0)$. *Next, let us suppose the object advances in the direction* $-v_1$. *The line followed has the equation* $z - z_0 = -v_1(t - (1 + t_0))$, *etc.*

Example 3.1.5 *From the origin* $O(0, 0)$, *an object is moving for* t_1 *seconds in the direction* $-v$. *It reaches the point* $M(t_1, -vt_1)$. *Negative speed means only the direction of evolution in time.*

Example 3.1.6 *A photon is released from the origin* O. *There are two possible directions,* c *and* $-c$. *If it is released in the direction* c, *its trajectory will be the line* $z = ct$. *Or, it can be released in the direction* $-c$. *Its trajectory in this case is* $z = -ct$. *In this case, after* $t_0 > 0$ *seconds, the photon reaches the point* $L(t_0, -ct_0)$.

In order to advance into the theory, we have to consider two local frames of coordinates, one moving at constant speed v, denoted by S, and another one considered at rest, denoted by R. The letters are chosen from the words "speed" and "rest." Two observers are placed at the origins of each system denoted by $\bar{O}$ and O respectively. The first local frame S is described by the coordinates $(\tau, \bar{x}, \bar{y}, \bar{z})$, while the frame R is described by the coordinates (t, x, y, z).

Now, the reference frames of the two observers have to adapt to the second postulate of the Special Relativity. To be easier in our reasonings, let us suppose the bi-dimensional case when the frame S consists of the coordinates $(\tau, \bar{z})$ and it is moving, at constant speed v along the t-axis of R in the same plane as the one determined by R, here denoted as (t, z).

First of all, how can we express the fact that S is moving at constant speed v with respect to R? The simple mathematical answer is: the axis $\bar{O}\tau$ in R has the equation $z = vt$.

Even if later in this book, we will find out that light can also be seen as an electromagnetic wave (and the conservation of Maxwell's equations is guaranteed by the Lorentz transformations), in order to develop Special Relativity, we can consider here light-rays as trajectories of photons.[1]

What can we say about the world line of a photon in these inertial reference frames? With respect to the observers in each frame, two world lines are highlighted: a photon is moving at constant speed c with a trajectory $z = ct$ in R and $\bar{z} = c\tau$ in S, while, for a photon moving at speed $-c$, we have the lines $z = -ct$ in R and $\bar{z} = -c\tau$ in S.

The two world lines of photons at O form the *light cone* of the frame R. A similar definition holds in S.

Therefore, if we use the same diagram for both frames, that is $O = \bar{O}$, the second postulate has the following mathematical expression:

---

[1] The quantum concept of "photon" is rigorously defined in Lecture 18.

1. The lines $z = ct$ **in** R **and** $\bar{z} = c\tau$ **in** S **have the same image;**
2. The lines $z = -ct$ **in** R **and** $\bar{z} = -c\tau$ **in** S **have the same image.**

In other words, the two light cones are coincident.

Summary of Lecture 7. The set up of Michelson–Morley experiment is conceived in Classical Mechanics. This is not appropriate. If we compare the two time values t_1 and t_2 according to the mathematical description we have

$$t_1 = \frac{l}{c - v} + \frac{l}{c + v} = \frac{2lc}{c^2 - v^2},$$

$$t_2 = \frac{2l}{\sqrt{c^2 - v^2}},$$

therefore

$$\frac{t_2}{t_1} = \sqrt{1 - \frac{v^2}{c^2}},$$

implies

$$t_2 < t_1.$$

From the experimental result, we obtain $t_2 = t_1$. Therefore the mathematical model is not correct and the description is not appropriate.

The correct approach is related to another theory of space and time in which we have to consider that all the observers, at rest or at constant speed are moving in agreement with the fact that speed of light is the same in any reference system. In this picture, the additions $c + v$ and $c - v$ become senseless.

3.2　Lecture 8: Motion Among Inertial Frames. The Lorentz Transformations

Since we deal with inertial frames, as a rule, objects moving at constant speed in S move at constant speed in R, and vice versa. So, a straight line representing a world line of an object moving at constant speed in S, it is seen as a straight line representing the world line of the same object moving at (another) constant speed in R and vice versa. Transforming lines into lines, the change of coordinates between the two frames is described by a linear map; we denote it by L_v and we call it a *Lorentz transformation* corresponding to the speed v (Fig. 3.2).

Theorem 3.2.1 *In the context described before, the matrix of the Lorentz transformation corresponding to the speed* v *has the form*

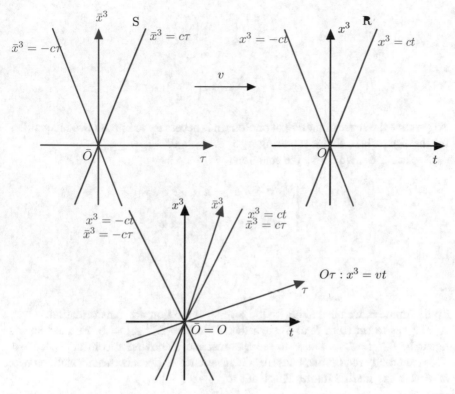

Fig. 3.2 Inertial frames and Lorentz transformation

$$L_v = \frac{1}{\sqrt{1 - v^2/c^2}} \begin{pmatrix} 1 & v/c^2 \\ v & 1 \end{pmatrix}.$$

Proof A linear map $L_v : S \to R$ has the form

$$L_v = \begin{pmatrix} a & b \\ d & e \end{pmatrix}.$$

Since $\bar{O}\tau$ axis in R has the equation $x^3 = vt$ we have

$$\begin{pmatrix} a & b \\ d & e \end{pmatrix} \begin{pmatrix} 1 \\ 0 \end{pmatrix} = \begin{pmatrix} t \\ vt \end{pmatrix},$$

that is $d = va$. In mathematical language, the second postulate is:

The eigenvectors of L_v are $\begin{pmatrix} 1 \\ c \end{pmatrix}$ and $\begin{pmatrix} 1 \\ -c \end{pmatrix}$, that is

$$L_v \cdot \begin{pmatrix} 1 \\ c \end{pmatrix} = \lambda_1 \begin{pmatrix} 1 \\ c \end{pmatrix}$$

and

$$L_v \cdot \begin{pmatrix} 1 \\ -c \end{pmatrix} = \lambda_2 \begin{pmatrix} 1 \\ -c \end{pmatrix}.$$

To preserve the sense of motion of photons, it is necessary to impose two inequalities for the eigenvalues, that is requiring $\lambda_1 > 0$, $\lambda_2 > 0$.

Replacing L_v, we derive the equations

$$\begin{cases} a\,c + b\,c^2 = a\,v + e\,c \\ -a\,c + b\,c^2 = a\,v - e\,c \end{cases}$$

i.e.

$$L_v = a \begin{pmatrix} 1 & v/c^2 \\ v & 1 \end{pmatrix}.$$

To determine a, we need to derive the inverse of the Lorentz transformation.

L_v^{-1} has to act from R to S, such that $L_v\,L_v^{-1} = L_v^{-1}\,L_v = I_2$. Furthermore, it has to be $L_v^{-1} := L_{-v}$, that is to see S at rest and R moving at constant speed $-v$. This reciprocity requirement ensures the invariance of Lorentz transformations with respect to any inertial frame. This leads to

$$I_2 = a^2 \begin{pmatrix} 1 - v^2/c^2 & 0 \\ 0 & 1 - v^2/c^2 \end{pmatrix},$$

i.e. $a^2 = \dfrac{1}{1 - v^2/c^2}$.

To determine the right sign of a, we use the Cayley Theorem. It is a simple matrix exercise: For a 2×2 real matrix B, the following equality holds

$$B^2 - 2\,TrB \cdot B + \det B \cdot I_2 = O_2.$$

In our case, $TrL_v = 2a = \lambda_1 + \lambda_2 > 0$.

The final form of the Lorentz transformation matrix is

$$L_v = \frac{1}{\sqrt{1 - v^2/c^2}} \begin{pmatrix} 1 & v/c^2 \\ v & 1 \end{pmatrix}.$$

$\square$

For v very small with respect c, that is if $v << c$, the Lorentz matrix becomes the Galilei matrix

$$S_v = \begin{pmatrix} 1 & 0 \\ v & 1 \end{pmatrix}$$

which represents how coordinates are transformed in Classical Mechanics.

Finally, we can write how the Lorentz transformation looks like in four dimensions. It is

$$\begin{cases} t = \dfrac{\tau + \bar{z}\, v/c^2}{\sqrt{1 - v^2/c^2}} \\ x = \bar{x} \\ y = \bar{y} \\ z = \dfrac{\tau v + \bar{z}}{\sqrt{1 - v^2/c^2}}. \end{cases}$$

Summary of Lecture 8. Consider two inertial frames of coordinates: S moves at constant speed v with respect to R, at rest. We can switch, from the coordinates $(\tau, \bar{z})$ of the frame S to the coordinates (t, z) of the frame R, using the matrix of the Lorentz transformation corresponding to the speed v. This transformation is obtained using the idea that speed of light is the same in both reference frames. It is

$$L_v = \frac{1}{\sqrt{1 - v^2/c^2}} \begin{pmatrix} 1 & v/c^2 \\ v & 1 \end{pmatrix}.$$

This happens in two corresponding slices of four dimensional spaces. If we consider the coordinates in each spacetime, $S : (\tau, \bar{x}, \bar{y}, \bar{z})$ and $R : (t, x, y, z)$, the transformation L_v provides the relations

$$\begin{cases} t = \dfrac{\tau + \bar{z}\, v/c^2}{\sqrt{1 - v^2/c^2}} \\ x = \bar{x} \\ y = \bar{y} \\ z = \dfrac{\tau v + \bar{z}}{\sqrt{1 - v^2/c^2}}. \end{cases}$$

3.3 Lecture 9: Addition of Velocities. The Relativistic Formula

Consider three inertial referential frames, S', S and R, such that S' is moving at constant speed w with respect to S and S is moving at constant speed v with respect to R.

The two corresponding Lorentz transformations are $L_w = \dfrac{1}{\sqrt{1 - w^2/c^2}}$

$\begin{pmatrix} 1 & w/c^2 \\ w & 1 \end{pmatrix}$ and $L_v = \dfrac{1}{\sqrt{1 - v^2/c^2}} \begin{pmatrix} 1 & v/c^2 \\ v & 1 \end{pmatrix}$.

The natural question is: which is the speed of S' with respect to R?

The answer is: we have to describe the linear map between S' and R via S, that is $L_v \cdot L_w$.

Theorem 3.3.1 $L_v \cdot L_w = L_{v \oplus w}$, where $v \oplus w = \dfrac{v + w}{1 + vw/c^2}$.

Proof After multiplying, we have

$$L_v \cdot L_w = \frac{1}{\sqrt{1 - v^2/c^2}} \frac{1}{\sqrt{1 - w^2/c^2}} \begin{pmatrix} 1 & v/c^2 \\ v & 1 \end{pmatrix} \cdot \begin{pmatrix} 1 & w/c^2 \\ w & 1 \end{pmatrix} =$$

$$= \frac{1 + vw/c^2}{\sqrt{(1 - v^2/c^2)(1 - w^2/c^2)}} \begin{pmatrix} 1 & \dfrac{v + w}{1 + vw/c^2} \cdot \dfrac{1}{c^2} \\ \dfrac{v + w}{1 + vw/c^2} & 1 \end{pmatrix} =$$

$$= \frac{1}{\sqrt{1 - \left(\dfrac{v + w}{1 + vw/c^2}\right)^2 \cdot \dfrac{1}{c^2}}} \begin{pmatrix} 1 & \dfrac{v + w}{1 + vw/c^2} \cdot \dfrac{1}{c^2} \\ \dfrac{v + w}{1 + vw/c^2} & 1 \end{pmatrix} = L_{v \oplus w},$$

where

$$v \oplus w = \frac{v + w}{1 + vw/c^2}.$$

$\square$

Definition 3.3.2 *The last formula is called the relativistic velocities addition.*

Suppose $w \to c$. The limit, in the left part of the previous equality, will be $v \oplus c$, while, in the right part, is $\dfrac{v + c}{1 + vc/c^2}$. This last ratio is c. From the relativistic point of view, we can say $v \oplus c = c$ at the limit. Therefore the classical mechanical point of view disappeared. The consequence is: light needs another treatment than the Classical Mechanics. This is the reason of the failure of Michelson–Morley experiment.

A further comment is necessary at this point. In Physics, systems of coordinates are thought with axes whose coordinates are related to physical units of time, length and so on (second, meter, etc.). The systems of coordinates corresponding to physical units are systems of *physical coordinates*. In the previous sections, we worked adopting physical coordinates. The units of measure in Physics were thought before to understand how deeply Geometry is involved in the description of the physical phenomena. If we choose an appropriate "length" (e.g. the meter) and an appropriate "time duration" (e.g. the second), the speed of light can be $c = 1$. We call these new units, *geometric units*. All formulas become simpler and the geometric images are more intuitive. The coordinates corresponding to geometric units are called geometric coordinates and are often used in Physics.

As an example, the Lorentz transformation matrix, in geometric coordinates, is

$$L_v = \frac{1}{\sqrt{1 - v^2}} \begin{pmatrix} 1 & v \\ v & 1 \end{pmatrix}.$$

and the addition of velocities becomes

$$v \oplus w = \frac{v + w}{1 + vw}.$$

Summary of Lecture 9. In this lecture, we obtain the relativistic formula for the addition of velocities

$$v \oplus w = \frac{v + w}{1 + vw/c^2}.$$

In the limit where w approaches the speed of light c, we have, from one hand, $v \oplus c$, and, from the other hand, c. Therefore, from the relativistic point of view, in the limit, we have $v \oplus c = c$. This is the reason why the Michelson–Morley experiment failed in measuring the ether velocity.

3.4 Lecture 10: The Einstein Rest Energy Formula $E = mc^2$

Let us start from an physical object at rest with a rest mass, denoted by $m_0 \neq 0$, and relativistic momentum at rest $\mathbb{P}_0 = \begin{pmatrix} m_0 \\ 0 \end{pmatrix}$.

We denote by $\mathbb{P} = \begin{pmatrix} m \\ mv \end{pmatrix}$ the relativistic momentum of a classical body moving at constant speed v. The previous considerations can be easily derived defining the generalized momentum $\mathbb{P} := m\dfrac{d\mathbf{x}}{dt}$, where $\mathbf{x} := \begin{pmatrix} t \\ x \\ y \\ z \end{pmatrix}$. In our case $\mathbf{x} := \begin{pmatrix} t \\ z \end{pmatrix}$.

Theorem 3.4.1 *If $m_0 \neq 0$ is the rest mass of a body moving at constant speed v, then*

$$m = m(v) = \frac{m_0}{\sqrt{1 - v^2/c^2}}.$$

Proof Using the Lorentz transformation L_v, we have $\mathbb{P} = L_v \cdot \mathbb{P}_0$, i.e.

$$\begin{pmatrix} m \\ mv \end{pmatrix} = \frac{1}{\sqrt{1 - v^2/c^2}} \begin{pmatrix} 1 & v/c^2 \\ v & 1 \end{pmatrix} \cdot \begin{pmatrix} m_0 \\ 0 \end{pmatrix},$$

which leads to the so called relativistic mass formula in physical coordinates,

$$m = m(v) = \frac{m_0}{\sqrt{1 - v^2/c^2}}.$$

$\square$

Denote by f', f'' the first and the second derivative of a real function f. It is easy to prove that

$$f(x) = f(0) + \frac{x}{1!} f'(0) + \frac{x^2}{2!} f''(0) + O[x^3],$$

where $O[x^3]$ contains only terms in x with powers greater than 3.

If we neglect the $O[x^3]$ terms, when we consider the real function

$$f(x) = \frac{1}{\sqrt{1 - x^2}}$$

we can write

$$\frac{1}{\sqrt{1 - x^2}} = 1 + \frac{1}{2} x^2.$$

Replacing x by v/c and multiplying by m_0 we have

$$\frac{m_0}{\sqrt{1 - v^2/c^2}} = m_0 + \frac{1}{2} m_0 v^2 / c^2.$$

Let us define the *relativistic kinetic energy* or, simply, the relativistic energy by

$$E(v) := m(v)c^2 = \frac{m_0 c^2}{\sqrt{1 - v^2/c^2}}.$$

The previous formula becomes

$$E(v) = m_0 c^2 + \frac{1}{2} m_0 v^2.$$

The *rest energy* is given by the formula

$$E_0 := m_0 c^2 .$$

It is worth stressing again that it makes sense when $m_0 \neq 0$.

Denoting the rest mass by m, we have obtained the famous Einstein formula

$$E = mc^2.$$

This formula tell us that mass means energy and energy means mass. In fact small masses can produce enormous energies. And energies mean masses, i.e. if we measure or we highlight energy there is a mass associated to it, it means we highlighted matter.

Summary of Lecture 10. The most known formula is derived starting from the Lorentz transformation and the relativistic mass concept. In the case of objects having nonzero rest mass, we have a rest energy E expressed by

$$E = mc^2.$$

This formula will be often used in this book.

3.5   Lecture 11: The Relativistic Energy Formula $E^2 = p^2 c^2 + m^2 c^4$

Theorem 3.5.1 *The relativistic energy formula is*

$$E^2 = p^2 c^2 + m^2 c^4$$

Proof We start from

$$E(v) := \frac{m_0 c^2}{\sqrt{1 - v^2/c^2}}$$

and we denote $E(v)$ by E keeping in mind the above meaning. Squaring it, we obtain

$$E^2 - E^2 \frac{v^2}{c^2} = m_0^2 c^4.$$

Now we observe that

$$E^2 \frac{v^2}{c^2} = \frac{m_0^2}{(1 - v^2/c^2)} v^2 c^2 = m^2(v) v^2 c^2 = p^2(v) c^2.$$

We denote by p the previous $p(v)$, that is the momentum corresponding to the relativistic mass. It results

$$E^2 = p^2 c^2 + m_0^2 c^4$$

and, denoting again the rest mass by m, we obtain the formula of the above statement. $\qquad\qquad\qquad\qquad\qquad\qquad\qquad\qquad\qquad\qquad\qquad\qquad\qquad\qquad\square$

Summary of Lecture 11. We obtained the relativistic energy formula

$$E^2 = p^2 c^2 + m_0^2 c^4,$$

where $E = E(v) = \dfrac{m_0 c^2}{\sqrt{1 - v^2/c^2}}$ is the relativistic energy, $m(v) = \dfrac{m_0}{\sqrt{1 - v^2/c^2}}$ is the relativistic mass, $p = p(v) = m(v)v$ is the momentum in the case of the relativistic mass $m(v)$ and $m_0 \neq 0$ is the rest mass.

We use this formula later when we study the Compton effect and the Dirac equation.

3.6   Lecture 12: Electromagnetic Waves by the Maxwell Equations

The nature of light, at the middle of the 19th century, was not known even if some progresses were made when Young understood its wave behavior. The meaning of light wave oscillations remained a mystery since James Clerk Maxwell succeeded in unifying Magnetism, Electricity and Optics. In his formulation of Electromagnetism, Maxwell presented light as a propagating electromagnetic wave. The way he used to describe the electromagnetic field is extremely important and we present it bellow. Indeed, the Maxwell equations are the "core" of Special Relativity. Essentially, this theory has been developed in view of explaining their invariance under the Lorentz transformations. We need some preliminary results and notations.

Theorem 3.6.1 *Let us consider some vector fields. If*

$$M = (M_x, M_y, M_z), \ N = (N_x, N_y, N_z), \ P = (P_x, P_y, P_z),$$

$$N \times P := \begin{vmatrix} \vec{i} & \vec{j} & \vec{k} \\ N_x & N_y & N_z \\ P_x & P_y & P_z \end{vmatrix},$$

$$M \cdot N := M_x N_x + M_y N_y + M_z N_z, \quad M \cdot P := M_x P_x + M_y P_y + M_z P_z,$$

then

$$M \times (N \times P) = (M \cdot P)N - (M \cdot N)P.$$

Proof We have

$$(M \cdot P)N - (M \cdot N)P =$$

$$= (M_x P_x + M_y P_y + M_z P_z)(N_x, N_y, N_z) - (M_x N_x + M_y N_y + M_z N_z)(P_x, P_y, P_z) =$$

$$= \begin{vmatrix} \vec{i} & \vec{j} & \vec{k} \\ M_x & M_y & M_z \\ N_y P_z - N_z P_y & -N_x P_z + N_z P_x & N_x P_y - N_y P_x \end{vmatrix} = M \times (N \times P).$$

□

Now, let us consider both the gradient operator and the Laplace operator in spatial coordinates denoted by (x, y, z), that is

$$\nabla := \left(\frac{\partial}{\partial x}, \frac{\partial}{\partial y}, \frac{\partial}{\partial z} \right),$$

$$\nabla^2 := \frac{\partial^2}{\partial x^2} + \frac{\partial^2}{\partial y^2} + \frac{\partial^2}{\partial z^2}.$$

The last formula can be also seen written in the formal way

$$\nabla^2 := \nabla \cdot \nabla$$

We formally define

$$\nabla \cdot M := \frac{\partial M_x}{\partial x} + \frac{\partial M_y}{\partial y} + \frac{\partial M_z}{\partial z}$$

and

$$\nabla \times M := \begin{vmatrix} \vec{i} & \vec{j} & \vec{k} \\ \frac{\partial}{\partial x} & \frac{\partial}{\partial y} & \frac{\partial}{\partial z} \\ M_x & M_y & M_z \end{vmatrix} = \left(\frac{\partial M_z}{\partial y} - \frac{\partial M_y}{\partial z}, \frac{\partial M_x}{\partial z} - \frac{\partial M_z}{\partial x}, \frac{\partial M_y}{\partial x} - \frac{\partial M_x}{\partial y} \right).$$

Using these operators, a consequence of the above theorem is

Corollary 3.6.2

$$\nabla \times (\nabla \times M) = (\nabla \cdot M)\nabla - (\nabla \cdot \nabla)M.$$

Another comment. We know the meaning of $\nabla^2 \phi$, where ϕ is a scalar function. The meaning of $\nabla^2 M$ is related to the fact that ∇^2 acts on each component of M, i.e.

$$\nabla^2 M := (\nabla^2 M_x, \nabla^2 M_y, \nabla^2 M_z).$$

Therefore we can write

$$\nabla \times (\nabla \times M) = (\nabla \cdot M)\nabla - \nabla^2 M.$$

If $\nabla \cdot M = 0$, the previous formula becomes

Corollary 3.6.3

$$\nabla \times (\nabla \times M) = -\nabla^2 M.$$

We will use this result later.

Let us denote by

$$E = E(t, x, y, z) := (E_x(t, x, y, z), E_y(t, x, y, z), E_z(t, x, y, z))$$

the electric force vector and by

$$B = B(t, x, y, z) := (B_x(t, x, y, z), B_y(t, x, y, z), B_z(t, x, y, z))$$

the magnetic force vector;

In geometric units and empty space, the *Maxwell equations* are

$$\begin{cases} \nabla \cdot E = 0 \\ \nabla \times E = -\dfrac{\partial B}{\partial t} \\ \nabla \cdot B = 0 \\ \nabla \times B = \dfrac{\partial E}{\partial t} \end{cases}$$

The first equation reveals the existence of an electric field in the absence of electric charge. If we are not in vacuum, the first equation is $\nabla \cdot E = \rho$, where ρ is the electric charge, therefore the first equation describes how an electric charge acts as source for the electric force, here seen as an electric field.

The second equation $\nabla \times E = -\dfrac{\partial B}{\partial t}$ shows how a time-varying magnetic field gives rise to an electric field.

The third equation $\nabla \cdot B = 0$ shows that there are no magnetic charges.

The fourth equation $\nabla \times B = \dfrac{\partial E}{\partial t}$ shows how the time variation of electric field creates the magnetic field.

Let us consider the derivative with respect t of the second equation.

$$-\frac{\partial^2 B}{\partial t^2} = \frac{\partial}{\partial t}(\nabla \times E) = \frac{\partial}{\partial t}\begin{vmatrix} \vec{i} & \vec{j} & \vec{k} \\ \frac{\partial}{\partial x} & \frac{\partial}{\partial y} & \frac{\partial}{\partial z} \\ E_x & E_y & E_z \end{vmatrix} = \begin{vmatrix} \vec{i} & \vec{j} & \vec{k} \\ \frac{\partial}{\partial x} & \frac{\partial}{\partial y} & \frac{\partial}{\partial z} \\ \frac{\partial E_x}{\partial t} & \frac{\partial E_y}{\partial t} & \frac{\partial E_z}{\partial t} \end{vmatrix} = \nabla \times \frac{\partial E}{\partial t}.$$

Using the last Maxwell equation and the above results, we find

$$-\frac{\partial^2 B}{\partial t^2} = \nabla \times \frac{\partial E}{\partial t} = \nabla \times (\nabla \times B) = -\nabla^2 B,$$

that is

$$\frac{\partial^2 B}{\partial t^2} = \nabla^2 B.$$

If we denote by

$$\Box := \frac{\partial^2}{\partial t^2} - \nabla^2$$

the d'Alembert operator, the previous equation is

$$\Box B = 0.$$

This is the wave equation corresponding to the magnetic field. Therefore, for each component B_i, $i \in \{x, y, z\}$ we have

$$\frac{\partial^2 B_i}{\partial t^2} = \nabla^2 B_i = \frac{\partial^2 B_i}{\partial x^2} + \frac{\partial^2 B_i}{\partial y^2} + \frac{\partial^2 B_i}{\partial z^2}.$$

Now, let us consider the derivative with respect t of the last equation.

$$\frac{\partial^2 E}{\partial t^2} = \frac{\partial}{\partial t}(\nabla \times B) = \frac{\partial}{\partial t}\begin{vmatrix} \vec{i} & \vec{j} & \vec{k} \\ \frac{\partial}{\partial x} & \frac{\partial}{\partial y} & \frac{\partial}{\partial z} \\ B_x & B_y & B_z \end{vmatrix} = \begin{vmatrix} \vec{i} & \vec{j} & \vec{k} \\ \frac{\partial}{\partial x} & \frac{\partial}{\partial y} & \frac{\partial}{\partial z} \\ \frac{\partial B_x}{\partial t} & \frac{\partial B_y}{\partial t} & \frac{\partial B_z}{\partial t} \end{vmatrix} = \nabla \times \frac{\partial B}{\partial t}.$$

Using the second Maxwell equation and its consequences, we find that

$$\frac{\partial^2 E}{\partial t^2} = \nabla \times \frac{\partial B}{\partial t} = -\nabla \times (\nabla \times E) = \nabla^2 E,$$

i.e.

Fig. 3.3 Two sections image
of an electromagnetic wave

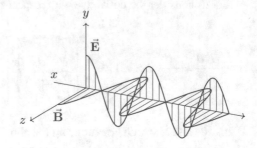

$$\Box E = 0.$$

This one is the wave equation corresponding to the electric field. We have now a picture of the electromagnetic field described by the Maxwell equations: The two waves equations of electric and magnetic field are interconnected by the four Maxwell equations. It is clear that a field cannot exist without the other in the electromagnetic description of Maxwell equations. Each one generates the other. And they travel together at the speed of light. We will discuss this fact in Lecture 15 when we will define $\Box_v$ (Fig. 3.3).

Summary of Lecture 12. Let us denote by

$$E = E(t, x, y, z) := (E_x(t, x, y, z), E_y(t, x, y, z), E_z(t, x, y, z))$$

the electric force vector and by

$$B = B(t, x, y, z) := (B_x(t, x, y, z), B_y(t, x, y, z), B_z(t, x, y, z))$$

the magnetic force vector. In geometric units, i.e. $c = 1$, the *Maxwell equations in vacuum* are

$$\begin{cases} \nabla \cdot E = 0 \\ \nabla \times E = -\dfrac{\partial B}{\partial t} \\ \nabla \cdot B = 0 \\ \nabla \times B = \dfrac{\partial E}{\partial t} \end{cases}$$

If we denote by

$$\Box := \frac{\partial^2}{\partial t^2} - \nabla^2$$

the d'Alembert operator, after some algebra, the previous equations can be written in the form

$$\begin{cases} \Box E = 0 \\ \Box B = 0. \end{cases}$$

The d'Alembert operator is used, in general, for the description of wave equations. Later, in Lecture 15, we will introduce the definition of $\Box_v$. It will result that the electromagnetic waves travel at the speed of light $c = 1$. The form obtained here can be used to show the invariance of Maxwell equations with respect to the Lorentz transformations.

3.7 Lecture 13: The Invariance of Maxwell Equations Under the Lorentz Transformations

Are these wave equations invariant under Lorentz transformations? The answer is yes, but we need to perform more steps in order to achieve these results.

In the same way as before, for each component E_i, $i \in \{x, y, z\}$, we have the equalities

$$\frac{\partial^2 E_i}{\partial t^2} = \nabla^2 E_i = \frac{\partial^2 E_i}{\partial x^2} + \frac{\partial^2 E_i}{\partial y^2} + \frac{\partial^2 E_i}{\partial z^2}.$$

To simplify, let us suppose that the electric field E depends only on the variables t and z. The previous equations can be synthesized in only one equation

$$\frac{\partial^2 E_i}{\partial t^2} - \frac{\partial^2 E_i}{\partial z^2} = 0.$$

To continue, let us choose a component only, say x. Since for the other two components, the following computations are the same, we use E_x to denote this chosen component by the letter $\mathbb{E}$. The previous equation becomes

$$\frac{\partial^2 \mathbb{E}}{\partial t^2} - \frac{\partial^2 \mathbb{E}}{\partial z^2} = 0.$$

How this simple equation looks like, in the S frame considered with coordinates $\tau, \bar{z}$, if S is supposed to move at constant speed v along the z axis in R? We have to use the Lorentz inverse transformation L_{-v}, that is

$$\begin{cases} \tau = \dfrac{t - z\,v}{\sqrt{1 - v^2}} \\ \bar{z} = \dfrac{-t\,v + z}{\sqrt{1 - v^2}}. \end{cases}$$

Denote by $\bar{\mathbb{E}}(\tau, \bar{z}) = \bar{\mathbb{E}}\left(\dfrac{t - z\,v}{\sqrt{1 - v^2}}, \dfrac{-t\,v + z}{\sqrt{1 - v^2}}\right) := \mathbb{E}(t, z)$ the corresponding component of the electric field in S, which, obviously has to be the same as in R. We have to prove the following

Theorem 3.7.1 *The Lorentz transformations preserve the Maxwell equations form.*

Proof Let us show that

$$\frac{\partial^2 \mathbb{E}}{\partial t^2} - \frac{\partial^2 \mathbb{E}}{\partial z^2} = \frac{\partial^2 \bar{\mathbb{E}}}{\partial \tau^2} - \frac{\partial^2 \bar{\mathbb{E}}}{\partial \bar{z}^2}.$$

We have

$$\frac{\partial \mathbb{E}}{\partial t} = \frac{\partial \bar{\mathbb{E}}}{\partial \tau}\frac{\partial \tau}{\partial t} + \frac{\partial \bar{\mathbb{E}}}{\partial \bar{z}}\frac{\partial \bar{z}}{\partial t} = \frac{\partial \bar{\mathbb{E}}}{\partial \tau}\frac{1}{\sqrt{1 - v^2}} + \frac{\partial \bar{\mathbb{E}}}{\partial \bar{z}}\frac{-v}{\sqrt{1 - v^2}}$$

and

$$\frac{\partial^2 \mathbb{E}}{\partial t^2} = \frac{1}{\sqrt{1 - v^2}}\left(\frac{\partial^2 \bar{\mathbb{E}}}{\partial \tau^2}\frac{\partial \tau}{\partial t} + \frac{\partial^2 \bar{\mathbb{E}}}{\partial \tau \partial \bar{z}}\frac{\partial \bar{z}}{\partial t}\right) - \frac{v}{\sqrt{1 - v^2}}\left(\frac{\partial^2 \bar{\mathbb{E}}}{\partial \bar{z}\partial \tau}\frac{\partial \tau}{\partial t} + \frac{\partial^2 \bar{\mathbb{E}}}{\partial \bar{z}^2}\frac{\partial \bar{z}}{\partial t}\right),$$

that is

$$\frac{\partial^2 \mathbb{E}}{\partial t^2} = \frac{1}{1 - v^2}\frac{\partial^2 \bar{\mathbb{E}}}{\partial \tau^2} - \frac{2v}{1 - v^2}\frac{\partial^2 \bar{\mathbb{E}}}{\partial \bar{z}\partial \tau} + \frac{v^2}{1 - v^2}\frac{\partial^2 \bar{\mathbb{E}}}{\partial \bar{z}^2}.$$

In the same way

$$\frac{\partial^2 \mathbb{E}}{\partial z^2} = \frac{v^2}{1 - v^2}\frac{\partial^2 \bar{\mathbb{E}}}{\partial \tau^2} - \frac{2v}{1 - v^2}\frac{\partial^2 \bar{\mathbb{E}}}{\partial \bar{z}\partial \tau} + \frac{1}{1 - v^2}\frac{\partial^2 \bar{\mathbb{E}}}{\partial \bar{z}^2},$$

therefore the desired relation is obtained by subtracting the two expressions. Now, from

$$\frac{\partial^2 \mathbb{E}}{\partial t^2} - \frac{\partial^2 \mathbb{E}}{\partial z^2} = 0$$

in R, we obtain

$$\frac{\partial^2 \bar{\mathbb{E}}}{\partial \tau^2} - \frac{\partial^2 \bar{\mathbb{E}}}{\partial \bar{z}^2} = 0$$

in S, that is the corresponding equation is the same as it has to be. Therefore, in a moving inertial frame, the Maxwell equations have the same form as in a frame at rest.   □

Theorem 3.7.2 *The Galilei transformations do not preserve the Maxwell equations form.*

Proof The proof is similar to the one seen before. We ask if the equality

$$\frac{\partial^2 \mathbb{E}}{\partial t^2} - \frac{\partial^2 \mathbb{E}}{\partial z^2} = \frac{\partial^2 \bar{\mathbb{E}}}{\partial \tau^2} - \frac{\partial^2 \bar{\mathbb{E}}}{\partial \bar{z}^2}$$

holds for the inverse of Galilean transformations $\bar{\mathbb{E}}(\tau, \bar{z}) = \bar{\mathbb{E}}\,(t, -vt + z) := \mathbb{E}(t, z)$. The answer is no. This can be easily shown performing the following computations

$$\frac{\partial \mathbb{E}}{\partial t} = \frac{\partial \bar{\mathbb{E}}}{\partial \tau}\frac{\partial \tau}{\partial t} + \frac{\partial \bar{\mathbb{E}}}{\partial \bar{x}^3}\frac{\partial \bar{x}^3}{\partial t} = \frac{\partial \bar{\mathbb{E}}}{\partial \tau} - v\frac{\partial \bar{\mathbb{E}}}{\partial \bar{x}^3}$$

and

$$\frac{\partial^2 \mathbb{E}}{\partial t^2} = \left(\frac{\partial^2 \bar{\mathbb{E}}}{\partial \tau^2}\frac{\partial \tau}{\partial t} + \frac{\partial^2 \bar{\mathbb{E}}}{\partial \tau \partial \bar{x}^3}\frac{\partial \bar{x}^3}{\partial t} \right) - v\left(\frac{\partial^2 \bar{\mathbb{E}}}{\partial \bar{x}^3 \partial \tau}\frac{\partial \tau}{\partial t} + \frac{\partial^2 \bar{\mathbb{E}}}{(\partial \bar{x}^3)^2}\frac{\partial \bar{x}^3}{\partial t} \right) =$$

$$= \frac{\partial^2 \bar{\mathbb{E}}}{\partial \tau^2} - 2v\frac{\partial^2 \bar{\mathbb{E}}}{\partial \bar{x}^3 \partial \tau} + v^2\frac{\partial^2 \bar{\mathbb{E}}}{(\partial \bar{x}^3)^2}.$$

Then,

$$\frac{\partial \mathbb{E}}{\partial x^3} = \frac{\partial \bar{\mathbb{E}}}{\partial \bar{x}^3}$$

and

$$\frac{\partial^2 \mathbb{E}}{(\partial x^3)^2} = \frac{\partial^2 \bar{\mathbb{E}}}{(\partial \bar{x}^3)^2},$$

that is

$$\frac{\partial^2 \mathbb{E}}{\partial t^2} - \frac{\partial^2 \mathbb{E}}{(\partial x^3)^2} = \frac{\partial^2 \bar{\mathbb{E}}}{\partial \tau^2} - \frac{\partial^2 \bar{\mathbb{E}}}{(\partial \bar{x}^3)^2} - 2v\frac{\partial^2 \bar{\mathbb{E}}}{\partial \bar{x}^3 \partial \tau} + v^2\frac{\partial^2 \bar{\mathbb{E}}}{(\partial \bar{x}^3)^2} \neq \frac{\partial^2 \bar{\mathbb{E}}}{\partial \tau^2} - \frac{\partial^2 \bar{\mathbb{E}}}{(\partial \bar{x}^3)^2}.$$

$\square$

The final conclusion is: Classical Mechanics, through the Galilei transformations, does not preserve the Maxwell equations form while the Special Relativity, through the Lorentz transformations, does it.

Summary of Lecture 13. The simplest way to express the invariance of Maxwell equations with respect to the Lorentz transformations is:

1. First, we write the Maxwell equations in a simplified form and we look only at a component of the electric field, say

$$\frac{\partial^2 \mathbb{E}}{\partial t^2} - \frac{\partial^2 \mathbb{E}}{\partial z^2} = 0.$$

2. This equation depends on the coordinates (t, z) in the frame R at rest. How this equation looks like in S frame considered with coordinates $(\tau, \bar{z})$, if S is supposed to move at constant speed v along the z axis in R?

To answer, we have to use the Lorentz inverse transformation L_{-v}, that is

$$\begin{cases} \tau = \dfrac{t - z\,v}{\sqrt{1 - v^2}} \\ \bar{z} = \dfrac{-t\,v + z}{\sqrt{1 - v^2}}. \end{cases}$$

Let us denote by $\bar{\mathbb{E}}(\tau, \bar{z}) = \bar{\mathbb{E}}\left(\dfrac{t - z\,v}{\sqrt{1 - v^2}}, \dfrac{-t\,v + z}{\sqrt{1 - v^2}} \right) := \mathbb{E}(t, z)$ the corresponding component of the electric field in S, which, obviously, has to be the same as in R. It remains to prove

$$\frac{\partial^2 \mathbb{E}}{\partial t^2} - \frac{\partial^2 \mathbb{E}}{\partial z^2} = \frac{\partial^2 \bar{\mathbb{E}}}{\partial \tau^2} - \frac{\partial^2 \bar{\mathbb{E}}}{\partial \bar{z}^2}.$$

After simple computations the equality is proved. It is important to say that if we try to achieve the same result by the Galilei transformations, the equality does not hold, therefore light, seen as an electromagnetic phenomenon, cannot be represented in Classical Mechanics.

Chapter 4
Why Quantum Mechanics?

Anyone who is not shocked by Quantum Theory has not understood it.

Niels Bohr

4.1 Lecture 14: The Problem of the Nature of Matter

When we are talking about matter we can think at the following thought experiment: consider a rectangular parallelepiped block of iron and suppose we can perfectly cut it in two equal pieces without loosing material. One of its dimensions, say the height L becomes $L/2$. One of the two smaller equal rectangular parallelepipeds is also cut in two equal pieces. Its height becomes now $L/2^2$. We continue this process and, at each step, the power of 2 from denominator increases, so at the n step the height length is $L/2^n$.

Can we continue indefinitely this process, or the matter is such that at a certain moment we must stop?

Aristotle's opinion was: yes, we can continue indefinitely because at each step we have something material to cut.

Contrary, Democritus thought there is no possible to indefinitely do it because matter has to be composed by small pieces putted together and this small pieces cannot more be divided. They are the constitutive blocks of matter, and he gave them the name of "atoms". Democritus cannot explain how these blocks look like or how they can stay sticked together and in ancient epochs the problem remained one without an acceptable answer.

As you can see later in this book, between these two schools of thinking, the one which supposed the basic blocks of matter gave a more accurate picture of what we might call reality.

Of course, the Greek philosophers Aristotle and Democritus did not think about a rectangular parallelepiped of iron, may be it was one of stone or a wooden bar, but we choose an element as iron because the next step was made in the Middle

© The Author(s), under exclusive license to Springer Nature Switzerland AG 2021
S. Capozziello and W.-G. Boskoff, *A Mathematical Journey to Quantum Mechanics*,
UNITEXT for Physics, https://doi.org/10.1007/978-3-030-86098-1_4

Ages when the progress of chemistry made known a lot of elements like gold, silver, iron, etc. People started to think these elements as atoms. For most part of scientists and natural philosophers, when they said one iron element, they precisely thought at one atom of iron. It was the moment when chemists described matter as compounds of different atoms and it was understood that some different types of atoms could combine producing substances. This is how the ordinary salt is made; salt is sodium chloride. Its chemical formula is NaCl. Here the basic elements are Sodium whose chemical symbol is Na and the chloride whose chemical symbol is Cl. At the same time, chemists understood that from iron you cannot obtain gold or vice versa, so the golden dream of alchemists to obtain gold from some other elements failed.

John Dalton's model of an atom. Only in 1804 someone succeeded to synthesize the previous ideas in a set of rules, say axioms, that can be considered at the chemistry foundations. Influenced by some ideas of Bryan Higgins, the scientist (meteorologist, physicist and chemist) John Dalton thought of atoms as pure philosophical concepts needed to describe the combinations of different gases from atmosphere. He developed a refined theory to explain what happen with these gases, whose axioms are:

1. Elements are made of small particles called atoms.
2. Atoms of a given element are identical in all their properties concerning size, mass, etc. Atoms of different elements differ in size, mass, and other properties.
3. Atoms cannot be subdivided, created and destroyed.
4. Atoms of different elements combine to form chemical compounds.
5. In chemical reactions, atoms are combined, separated or rearranged.

Even if the first axiom seems to be a little beat confusing and saying that now we have the correct description, the third axiom is not true, these statements are an important step in understanding the basic chemical concepts and how they rule the description of matter. In the following, it is possible to demonstrate why the third axiom is not true.

Joseph J. Thomson's model of the atom. The next step was made after the electricity was used by scientists in experiments with "glass bottles" having different gases inside and electric circuits attached to the bottle. In fact when we are saying "glass bottles", we are talking about cathode ray tubes in which most of the air was evacuated and a ray originates by a high voltage electric circuit at the cathode flows to the anode. The ray can be detected after the anode. To simplify, let us consider a gas, say hydrogen in the tube after the anode. The hydrogen glows when the ray passes through it. Two opposite charged plates are installed near the tube.

When someone is looking at what is happening, she/he can see a shadow flow attracted by the positive charged plate, when this one is charged and the other plate is not. Another shadow flow bends into the direction of the negative plate when this plate is charged and the other one is not. There is a difference between the two shadow flows. The second shadow flow is not so bended as the first one. It is only slightly bended. So, let us draw some conclusions. The hydrogen is neutral from electric point of view. If not, all the atoms were attracted to the same electrically charged plate. It seems that, in the atoms of hydrogen, there is "something" having

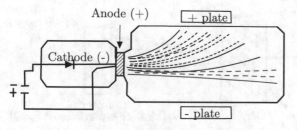 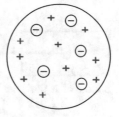

Fig. 4.1 Thomson's model of atom

a negative charge which is attracted by the positive plate. And there is "something" having positive charge slightly attracted by the negative plate. Let us call electron the "something" having negative charge. The positively charged part will be called proton. The proton is massive with respect to the electron. Its mass is much more heavier than the electron mass and we concluded this by looking at how much the two shadow flows are bended. Something remarkable happens. The atom of hydrogen has constituents. Such an experiment was made in 1897 when the physicist J. J. Thomson was credited with the discovery of the electron as a part of the hydrogen atom. Let us add that the chemical symbol of the hydrogen is H. And let us underline two important things.

1. Dalton's atomic theory was disproved, at least at the level of the third axiom.
2. A model of atom has to be developed.

So, Thomson thought at a model starting from the facts that all atoms of elements are electrically neutral. He proposed the plum pudding model in which the negative charged particles are floating in a soup of positive charge such that the atom remains electrically neutral (Fig. 4.1).

Ernest Rutherford's model of the atom. The Thomson model of the atom is not the only one which has been formulated. The next step was done by the physicist Ernest Rutherford in 1911 after the analysis of *Geiger–Marsden experiment* made in 1909 (Fig. 4.2).

At that time, radioactive materials were known and also the fact that they emit rays was known. There are three kinds of rays denoted after the first letters of the Greek alphabet. These rays have different electrical charges.

- α rays are positively charged because they are attracted by negative charged plates.
- β rays are negatively charged because they are attracted by positive charged plates.
- γ rays are not charged at all because they are not attracted neither by negative charged plates nor by positive charged plates.

Geiger–Marsden experiment is simple. Suppose a device capable of emitting α rays, a sort of tube containing a substance called radium bromide which is known emittings α rays. Another device detects α rays. Between the two there is a golden foil.

The α rays go through the golden foil to the detector. Moving the detector in the proximity of the source, we can find some deflected α rays even there. According to Thomson model of gold atom, this cannot happen. The α rays have to pass through

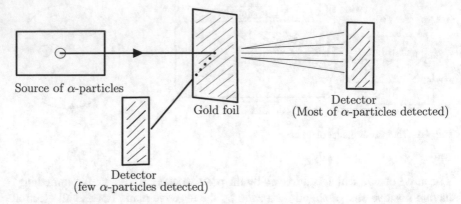

Fig. 4.2 Geiger–Marsden experiment

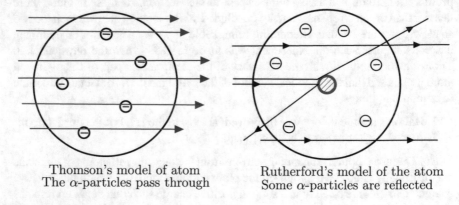

Thomson's model of atom   Rutherford's model of the atom
The α-particles pass through   Some α-particles are reflected

Fig. 4.3 Rutherford's view on experimental results

the gold atoms. The fact that there are some reflected rays might be related to the existence of something they meet in the interior of atoms (Fig. 4.3).

Therefore, Rutherford proposed a model having a nucleus where all positive charges are concentrated and electrons orbiting around it. In the case of the hydrogen, the nucleus contains one proton and the atom has only one electron.

This model has two inconsistencies.

According to the first one, let us think about the hydrogen atom model having one proton, that is a positive charge, and one electron, that is a negative charge, moving around it. The opposite charges attract. Why the electron do not spiral on the nucleus and so, the atom can no longer exist? What is the nature of the electron? Is it a particle or something else? As we will see, we need first to know the fundamental nature of light.

The second question is related to atoms having more than one proton in the nucleus. The same sign charges have to repel. Why does not the nucleus of the atom fall apart and so the atom no longer exists? How are they kept together?

Before to continue let us say how small is a nucleus with respect to the dimension of the atom. If we increase the atom at the size of a football field, the nucleus is exactly a grain of sand in the middle of it. The atomic mass is almost done by the nucleus. In nucleus there are protons and neutrons, too. The neutrons are electrically neutral and their mass is almost like the proton mass. Now we know that electrons are building block of matter, but protons and neutrons are not. They are made by quarks and about this topic, we refer the reader to more advanced texts on the subjects.

In a given atom, the number of protons coincides with the number of electrons, this fact makes the atoms electrically neutral.

The stability of the atom depends on the nature of light. At the end of this chapter, we will see that light has a dual nature. It is wave and particle, at the same time.

Summary of Lecture 14. This lecture is a quick summary on ideas and experimental facts which led to Quantum Mechanics. Starting from the different visions by Aristotle and Democritus about the existence or not of indivisible parts of matter, the atoms, the lecture continues with the perspective of Middle Ages scientists who discovered chemical elements and compounds. In 1804, John Dalton considered atoms, chemical elements, and compounds in a sort of axiomatic way according to some experimental facts. Among these axioms, there is one related to the elements which are made by atoms and atoms cannot be subdivided, created or destroyed. This theory was influenced by his meteorological studies about the gases from the Earth atmosphere.

After, Joseph J. Thomson studied gases as hydrogen and succeeded to show that the hydrogen atoms contain an electric negative charge, called electron, and a positive charge with a grater mass. Thomson imagined an atom as a plum pudding, where the positive charge part (the pudding) contains in its interior the small particles (plums) with negative charge.

His student, Ernst Rutherford changed the perspective about the atom considering the positive part in the middle, that is all positive charges staying together in an nucleus and the electrons orbiting around it.

Thinking only at the model of the hydrogen atom where only a proton (a positive charge) and an electron (a negative charge) exist, two questions appeared:

1. The opposite charges attract. Why the electrons do not spiral on the nucleus and so, the atom can no longer exist? What is the nature of the electron? Is it a particle or something else?

2. The second question is related to the atoms having more than one proton in the nucleus. Objects with the same sign charges have to repel. Why does not the nucleus of the atom fall apart and so the atom no longer exist? How these positive charges are kept together? Starting from these fundamental questions on matter has been possible to develop Quantum Mechanics.

4.2   Lecture 15: Monochromatic Plane Waves—The One Dimensional Case

To have an intuitive view on the wave behavior of light and matter, we can start describing a *wave* as a function

$$\psi(t, x) = A \cos(kx - \omega t)$$

depending on t and x which are real valued. The meaning of k and ω will be clear by the mathematical description of monochromatic plane waves.

All propagating dynamic perturbations and disturbances are in fact waves. A wave has crests and troughs and can be mathematically described by the wave equation. Crests of a wave can move in one direction. Such waves are called traveling waves. They travel at a given speed and they carry energy. An example of such waves are the electromagnetic waves. Their equation, in the case of magnetic field, is

$$\frac{\partial^2 B}{\partial t^2} - \nabla^2 B = 0,$$

where ∇^2 is the Laplace operator,

$$\nabla^2 := \frac{\partial^2}{(\partial x^1)^2} + \frac{\partial^2}{(\partial x^2)^2} + \frac{\partial^2}{(\partial x^3)^2}.$$

If we denote by

$$\square := \frac{\partial^2}{\partial t^2} - \nabla^2$$

the d'Alembert operator, the previous equation is

$$\square B = 0.$$

The same equation holds in the case of electric field, that is

$$\square E = 0,$$

where E is the electric field. Here we write the equations of magnetic and electric fields for $t \in \mathbb{R}$, $\mathbf{x} \in \mathbb{R}^3$. In order to reduce them to the planar case, we consider only one space variable and the Laplace operator reduced to one term only, say $\dfrac{\partial^2}{\partial (x^1)^2}$ in the case of the electric field.

For general waves, terms involved in the d'Alembert operator can have coefficients.

Let us consider now the Euler formula

$$e^{i\alpha} = \cos\alpha + i\sin\alpha,$$

where i is the immaginary unit having the property $i^2 = -1$. If t is the time and x is the position, a monochromatic plane wave can be imagined as

$$\Psi(t,x) = Ae^{i(kx-\omega t)} = A[\cos(kx - \omega t) + i\sin(kx - \omega t)].$$

The positive number A is called an *amplitude*. To have a picture of the situation, one has to think that sin and cos functions oscillate between -1 and 1, while A sin or A cos functions oscillate between $-A$ and A.

The same holds for the function $\cos(kx - \omega t)$. The crest is reached for $kx - \omega t = 0$ and at any 2π translations. From $kx - \omega t = 0$, it results

$$\frac{x}{t} = \frac{\omega}{k}.$$

From adimensional point of view, this ratio is space over time, that is a velocity. So, we have obtained the formula

$$v := \frac{\omega}{k}.$$

If we write

$$kx - \omega t = k\left[x - \frac{\omega}{k}t\right] = k[x - vt]$$

at a given $t_0 > 0$, we have $k[x - vt_0]$, that is $k(x - x_0)$. Such a formula shows that the origin $O(0,0)$ switches to $(x_0, 0)$ moving on the right.

Let us denote by T the smallest positive number such

$$e^{i(kx-\omega T)} = e^{ikx}, \text{ that is } e^{i\omega T} = 1.$$

It results that $T = \dfrac{2\pi}{\omega}$. Since the *period* T is related to the *frequency* ν by the formula $T = \dfrac{1}{\nu}$, it results $\omega = \dfrac{2\pi}{T} = 2\pi\nu$. Therefore ω is the *angular velocity* of the wave.

The *wavelength* is the distance between two consecutive crests. It is denoted by the Greek letter λ. Since $\lambda = vT$, it results

$$\lambda = \frac{\omega}{k}T = \frac{\omega}{k}\frac{2\pi}{\omega} = \frac{2\pi}{k}.$$

We have obtained

$$k := \frac{2\pi}{\lambda}.$$

If we have the function $\sin\alpha$ and the distance between two consecutive crests is $\lambda = 2\pi$, we have $k = 1$. If we have $\sin 2\alpha$ and the distance between two consecutive crests is $\lambda = \pi$, therefore we have $k = 2$ and so on. According to this definition, we call k the *wavenumber* of Ψ.

It results now the meaning of all letters which appears in the describing of the planar right-traveling wave Ψ:

- t is the time parameter,
- x is the space parameter,
- ω is the angular velocity derived from the frequency ν,
- k is the wavenumber derived from the period T.

It remains to prove that $\Psi(t, x)$ is indeed a wave. It is easy to compute $\dfrac{\partial^2 \Psi}{\partial t^2}$ and $\dfrac{\partial^2 \Psi}{\partial x^2}$. Calculations lead to

$$\frac{\partial^2 \Psi}{\partial t^2} = -\omega^2 \Psi; \quad \frac{\partial^2 \Psi}{\partial x^2} = -k^2 \Psi,$$

therefore

$$\frac{\partial^2 \Psi}{\partial t^2} - \frac{k^2}{\omega^2} \frac{\partial^2 \Psi}{\partial x^2} = 0.$$

The last equation can be written with respect to the speed v of the traveling wave, so

$$\frac{\partial^2 \Psi}{\partial t^2} - \frac{1}{v^2} \frac{\partial^2 \Psi}{\partial x^2} = 0.$$

The two dimensional d'Alembert operator can be written as

$$\Box_v := \frac{\partial^2}{\partial t^2} - \frac{1}{v^2} \frac{\partial^2}{\partial x^2}.$$

At this point, a short discussion about waves is necessary. They can be reflected, refracted, diffracted and they can interfere. When they interfere, we can see an interference pattern. This interference pattern can be imagined using the following observations. Let us take into account two waves with the same amplitude. There are crests and troughs. When two crests come together they determine a double crest. We say that this is a constructive interference. The same for two troughs which determine a double trough. When a trough come together a crest they cancel, so we are talking about a destructive interference. The wave interference appears when two waves traveling in the same medium meet.

The interference of waves can produce a special type of waves, called standing waves. They have null points and the wave oscillates keeping these points fixed. An example of such waves are string vibrations. Let us look at standing waves in the 1-dimensional case. Consider an infinite length string along the x-axis which can oscillate only in a plane determined by the previous x-axis and also consider an y-axis orthogonal to it. The wave described by the formula

$$R(t, x) = A \sin(kx - \omega t)$$

is, accordingly to the above facts on planar waves, a traveling wave to the right along the string, while

$$L(t, x) = A \sin(kx + \omega t)$$

is a traveling wave moving to the left along the string. Let us remember the well known trigonometry formula

$$\sin \alpha + \sin \beta = 2 \sin \frac{\alpha + \beta}{2} \sin \frac{\alpha - \beta}{2}.$$

Let us apply it for the interference of the two previous waves in the form

$$y(t, x) := A \sin(kx - \omega t) + A \sin(kx + \omega t) = 2A \sin kx \cos \omega t.$$

What we have obtained is not a traveling wave, but a *standing wave* having nodes at each $x = \dfrac{n\pi}{k}$, $n \in \mathbb{N}$. The wave oscillates up and down preserving the nodes as fixed points.

The length of the string has not to be infinite. It can be finite with two ends, one fixed point at $(0, 0)$, the other one at $(L, 0)$, where L is the length of the string. The standing wave has the same form

$$y(t, x) := 2A \sin kx \cos \omega t$$

but we have to impose boundary conditions to fix the endpoints of the string. We observe that the chosen form has the property $y(0, t) = 0$, therefore the boundary condition at the left is satisfied accordingly to the way we defined $y(t, x)$. At the right, we have to impose the condition

$$\sin kL = 0,$$

i.e. the wavenumber k has to satisfy $kL = n\pi$, $n \in \mathbb{N}$. Since $k = \dfrac{2\pi}{\lambda}$, we obtain the wavelength condition imposed by the two fixed ends of the string: $\lambda = \dfrac{2L}{n}$. The associated frequencies of this standing wave are

$$\nu = \frac{1}{T} = \frac{v}{\lambda} = \frac{nv}{2L}.$$

By definition, $n = 1$ gives the fundamental frequency of the string, while higher integers than 1 correspond to the so called, harmonics or overtones.

An important observation is necessary at this point: let us consider that a standing wave can be modeled around a given circle of length L. In such a case, the mandatory condition to obtain a standing wave is $kL = 2n\pi$, that is the wave oscillates with an integer multiple of the period of the function sin. Since the circle has length L, its radius is $r = \dfrac{L}{2\pi}$. The formula we obtain is $rk = n$. It will be used later when we study the electron seen as a standing wave around the nucleus. An exercise for the reader is to obtain the same results using the wave description $\Psi(t, x) = A(t)e^{ikx}$.

Summary of Lecture 15. If t is the time and x is the position, a planar wave traveling to the right can be imagined as

$$\Psi(t, x) = Ae^{i(kx - \omega t)} = A[\cos(kx - \omega t) + i \sin(kx - \omega t)].$$

The positive number A is called amplitude. The wavelength λ is the distance between two consecutive crests.

k is the wavenumber and it is connected to the wavelength λ through the formula

$$k := \frac{2\pi}{\lambda}.$$

ω is the angular velocity related to the period T through the formula

$$T := \frac{2\pi}{\omega}.$$

The frequency of the wave, denoted ν is defined by

$$\nu := \frac{1}{T}.$$

The speed of traveling wave is given by the formula

$$v := \frac{x}{t} = \frac{\omega}{k}.$$

The traveling wave Ψ satisfies the equation

$$\frac{\partial^2 \Psi}{\partial t^2} - \frac{1}{v^2} \frac{\partial^2 \Psi}{\partial x^2} = 0$$

or, written with respect to the d'Alembert operator

$$\Box_v := \frac{\partial^2}{\partial t^2} - \frac{1}{v^2} \frac{\partial^2}{\partial x^2}$$

it has the form

$$\Box_v \Psi = 0.$$

In the one dimensional case, a standing wave can be seen as the interference between two traveling waves along the x-axis, that is

$$y(t, x) := A \sin(kx - \omega t) + A \sin(kx + \omega t) = 2A \sin kx \cos \omega t.$$

The nodes are at each $x = \dfrac{n\pi}{k}$, $n \in \mathbb{N}$. The wave oscillates up and down preserving the nodes as fixed points.

In this case the length of the string is infinite. But we can model standing waves whose initial string has a finite length. If the two ends are at $(0, 0)$ and at $(L, 0)$, where L is the length of the string, the standing wave has the same form

$$y(t, x) := 2A \sin kx \cos \omega t,$$

but we have to impose boundary conditions to fix the endpoints of the string. We observe that the chosen form has the property $y(0, t) = 0$, therefore the boundary condition to the left is satisfied accordingly to the way we defined $y(t, x)$. To the right, we have to impose the condition

$$\sin kL = 0,$$

i.e. the wavenumber k has to satisfy $kL = n\pi$, $n \in \mathbb{N}$. Since $k = \dfrac{2\pi}{\lambda}$, we obtain the wavelength condition imposed by the two fixed ends of the string, $\lambda = \dfrac{2L}{n}$.

The associated frequencies of this standing wave are

$$\nu = \frac{1}{T} = \frac{v}{\lambda} = \frac{nv}{2L}.$$

Now, a very important remark: let us observe that a standing wave can be modeled around a given circle of length L. In such a case, the mandatory condition to obtain a standing wave is

$$kL = 2n\pi,$$

that is the wave oscillates an integer multiple of the period of the function sin. Since the circle has length L, its radius is $r = \dfrac{L}{2\pi}$. The formula we obtain, that is $rk = n$, will be used later when we will study the electron considered as a standing wave around the nucleus.

4.3   Lecture 16: The Young Double Split Experiment. Light Seen as a Wave

It seems René Descartes was the first who states that light is made by particles traveling at a finite speed in straight line. This description can be used to understand reflection of light. Light acts as "small bullets" bouncing when they meet a surface at a given angle (Fig. 4.4).

This point of view was kept also by Isaac Newton in his treatise concerning Optics. The light is explained by two rules:

1. Every source of light emits a large number of tiny particles called corpuscles in the medium which surround the source.

2. These corpuscles are perfectly elastic, rigid and weightless.

These allowed Newton to see corpuscles as a sort of material points which can follow the basic rules of Mechanics. The trajectories, being lines, lead to a law for the reflection: the incidence angle equals the reflected angle. But not all of the known properties of light can be explained in a satisfactory way following this point of view. A century later, scientists still believed Newton, even if Christian Huygens represented light as a wave because Newton's theory cannot explain phenomena like polarization or interference. However, in 1801, Thomas Young made an experiment which showed, without any doubt, that light can be seen as a wave.

If we have a one-slit plate parallel to the plate of a table and we leave sand to go through, on the plate of the table we obtain a pile of sands. If we have a double-slit plate parallel to the plate of a table and we do the same "experiment", two piles of sand will form. The small grains of sand will create a pattern corresponding to all other grains of something which pass through the double-slit.

Let us consider now the same experiment, this time using, say, blue light. The *double-slit experiment* can be described as follows. Consider a plate ΔS with two

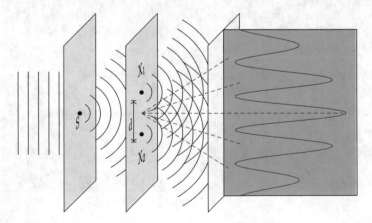

Fig. 4.4   Young's double-slit experiment (1)

Fig. 4.5 Young's double-slit experiment (2)

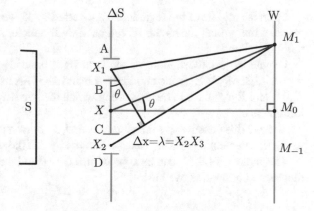

parallel slits, a source of monochromatic blue light S, that is only one optical frequency ν exists corresponding to the blue light, and a screen W placed such that the order of objects is S, ΔS and W.

If we cover one slit, we see a blue light strip on the screen.

Let us consider now light passing through both slits. The light illuminates the screen. In fact, on the screen, we can see many vertical blue light and dark bands. Physicist call all these bands, fringes.

Of course, if the light is made up of particles like the little bullets we discussed earlier, only two blue light fringes would appear on the screen. So, in this experiment, light cannot be thought as "made of particles".

Here we see an interference pattern as happen when two small stones fall in the water at the same time. The constructive interference produces the blue light fringes, while the destructive interference produces the dark fringes.

The main conclusion of this experiment is that light acts as a wave. The wave nature of light is the only possible explanation of the interference pattern we observe. The waves passing through the two slits interfere producing blue light and dark fringes.

In this experiment, it is important the magnitude of slits, the distance between the two slits and the distance between the double-slit plate to the screen. The magnitude of slits and the distance between them have to be almost as the wavelength of the monochromatic light produced by the source. If the slits are too large, the chance to produce the interference pattern decreases. The same happens if the screen is too close to the double-slit plate. If the two slits are too narrow, the same, we cannot see the interference pattern. The experiment has to be well "prepared".

Now, we have to understand why the blue light and dark fringes appear (Fig. 4.5).

Consider a perpendicular section, that is a plane perpendicular to the plate containing the two slits. Let us denote with AB the first slit and with CD the second slit, and A, B, C, D are collinear points. Therefore BC is the distance between the two slits of length d. The sizes of the two slits are equal and they are very tight as we explained above.

From the middle of the segment BC, denoted by X, we draw, in the perpendicular plane, a line which meets the screen at M_0. Of course, M_0 is at the middle of the main light band.

Denote by M_1 the middle of the next light band. Now we consider the angle $\theta := \angle M_1 X M_0$. If X_1 is the middle of AB and X_2 is the middle of CD, then the lines $X_1 M_1$ and $X_2 M_1$ have to be axes along which the two waves travel in a constructive interference.

To have this constructive interference, first we draw a perpendicular line from X_1 to $X_2 M_1$ intersecting at X_3. Then, we denote by Δx the segment $X_2 X_3$.

The angle $\angle X_2 X_1 X_3$ can be approximated by θ and the length $X_1 X_2$ by d, so, for these small quantities, we have

$$\sin \theta = \frac{\Delta x}{d}.$$

The constructive interference condition is $\Delta x = n\lambda$. That is a correspondence crest to crest and trough to trough appears for waves having axes $X_1 M_1$ and $X_3 M_1$, respectively. Therefore the condition of constructive interference in the double-slit experiment is

$$n\lambda = d \sin \theta,$$

for those integers n such that $\dfrac{n\lambda}{d} \in [-1, 1]$.

In the same way we find the condition for destructive interference. The corresponding figure is easy to construct and the condition is:

$$\frac{2n+1}{2}\lambda = d \sin \theta,$$

for those integers n such that $\dfrac{2n+1}{2}\dfrac{\lambda}{d} \in [-1, 1]$.

If we denote by y the length of $M_0 M_1$ and by D the length of $X M_0$, for small angles θ we have

$$\frac{\Delta x}{d} = \sin \theta \approx \tan \theta = \frac{y}{D},$$

therefore y can be computed with the formula $y = \dfrac{Dn\lambda}{d}$. Of course we can compute the distance between the middle of two consecutive light or dark fringes, that is we can mathematically model in a satisfactory way the double-slit experiment results (Fig. 4.6).

Fig. 4.6 Young's double-slit experiment (3)

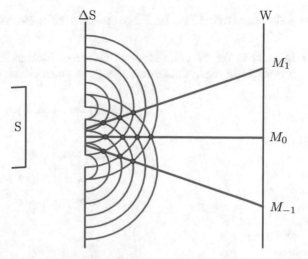

Summary of Lecture 16. The double-slit experiment can be described as follows. Consider a plate ΔS with two parallel slits, a source of monochromatic blue light S, that is a defined optical frequency ν exists corresponding to the blue light, and a screen W placed such that the order of the components is S, ΔS and W.

If we cover one slit, we see a blue light strip on the screen.

Let us consider now light passing through both slits. The light illuminates the screen. In fact, on the screen we can see many vertical blue light and dark bands. Physicists call these bands, fringes.

Of course, if the light made up of particles like the Newton little bullets we discussed before, only two fringes would appeared on the screen. So, in this experiment, light cannot be thought as "made of particles".

Here we see an interference pattern like that happens when two small stones fall in the water at the same time. The constructive interference produces the blue light fringes, while the destructive interference produces the dark fringes.

The main conclusion of this experiment is that light acts as a wave. The result of the experiment is described by the equation

$$\frac{\Delta x}{d} = \sin \theta \approx \tan \theta = \frac{y}{D},$$

which is derived above.

4.4   Lecture 17: The Planck–Einstein Formula $E = h\nu$

In order to fix the basic ideas of Quantum Mechanics, let us write some useful formulas whose physical meaning will be given below. It is

$$R(\lambda, T) = \frac{c}{4} \, \rho(\lambda, T)$$

$$\rho(\lambda, T) = \frac{8\pi}{\lambda^4} \, \bar{\varepsilon}$$

$$\beta = \frac{1}{k\,T}$$

and finally

$$\bar{\varepsilon} = \frac{\int_0^\infty \varepsilon \, e^{-\beta\varepsilon} d\varepsilon}{\int_0^\infty e^{-\beta\varepsilon} d\varepsilon}.$$

The last fraction with integrals can be easily computed. Many readers will prefer to separately compute the integrals from nominator and denominator. It is:

$$\int_0^\infty \varepsilon \, e^{-\beta\varepsilon} d\varepsilon = -\frac{1}{\beta} \, \varepsilon \, e^{-\beta\varepsilon} \, |_0^\infty + \frac{1}{\beta} \int_0^\infty e^{-\beta\varepsilon} d\varepsilon = \frac{1}{\beta^2},$$

$$\int_0^\infty e^{-\beta\varepsilon} d\varepsilon = -\frac{1}{\beta} \, e^{-\beta} |_0^\infty = \frac{1}{\beta},$$

therefore the final result is

$$\bar{\varepsilon} = \frac{1}{\beta} = kT.$$

Another way is to observe this is

$$\bar{\varepsilon} = \frac{\int_0^\infty \varepsilon \, e^{-\beta\varepsilon} d\varepsilon}{\int_0^\infty e^{-\beta\varepsilon} d\varepsilon} = -\frac{d}{d\beta} \left(\ln \left[\int_0^\infty e^{-\beta\varepsilon} d\varepsilon \right] \right) = \frac{1}{\beta} = kT,$$

and the same result is obtained. Replacing in the formula of $R(\lambda, T)$, we have

$$R(\lambda, T) = \frac{c}{4} \, \rho(\lambda, T) = \frac{2\pi c}{\lambda^4} \, kT.$$

Now let us observe that

$$\lim_{\lambda \to 0_+} R(\lambda, T) = +\infty.$$

This is a mathematical model formulated in a given context which gives rise to some predictions. The approach is like in the case of Michelson–Morley experiment when, in a given Classical Mechanics context, we obtained some predictions. In the Michelson–Morley context, the experimental results contradict the mathematical prediction obtained from Classical Mechanics. In that case, a new theory, the Special relativity, must be formulated to fit the observations.

Here the situation is similar. Something in the set of formulas presented above leads to results which do not fit experiments. We have to change something, the infinite obtained in the above formula cannot exist in reality.

The only place where we can act seems to be when the two integrals are considered.

What happens if we change the continuous variable ε, belonging to the real interval $[0, +\infty)$, with $n\varepsilon_0$, where n belongs to the natural number set $\mathbb{N}$?

We get a modified formula for $\bar{\varepsilon}$, i.e.

$$\bar{\varepsilon} = \frac{\sum_{n=0}^{\infty} n\varepsilon_0\, e^{-\beta n\varepsilon_0}}{\sum_{n=0}^{\infty} e^{-\beta n\varepsilon_0}} = -\frac{d}{d\beta}\left[\ln\left(\sum_{n=0}^{\infty} e^{-\beta n\varepsilon_0}\right)\right] = -\frac{d}{d\beta}\left[\ln\frac{1}{1 - e^{-\beta\varepsilon_0}}\right] = \frac{\varepsilon_0}{e^{\beta\varepsilon_0} - 1}.$$

It results

$$R(\lambda, T) = \frac{2\pi c}{\lambda^4}\left(\frac{\varepsilon_0}{e^{\beta\varepsilon_0} - 1}\right) = \frac{2\pi c}{\lambda^4}\left(\frac{\varepsilon_0}{e^{\frac{\varepsilon_0}{kT}} - 1}\right).$$

According to the physical meaning, it is

$$\varepsilon_0 = h\nu = h\frac{c}{\lambda},$$

where h is a constant, the *Planck constant*, ν is a frequency, c is the speed of light, and λ is a wavelength. The above formula is now

$$R(\lambda, T) = \frac{2\pi hc^2}{\lambda^5}\left(\frac{1}{e^{\frac{hc}{kT}\frac{1}{\lambda}} - 1}\right)$$

and

$$\lim_{\lambda \to 0_+} \frac{2\pi hc^2}{\lambda^5}\left(\frac{1}{e^{\frac{hc}{kT}\frac{1}{\lambda}} - 1}\right) = 0$$

which fits the experiment which we are going to describe below.

The main change, made in order to fit the experiment, is related to the fact that ε cannot have a continuous set of values, but only natural multiples of a given energy ε_0. This energy depends on a constant, called Planck constant, and on the frequency

ν. *This means that the energy cannot have a continuous set os values but only discrete values which depends on a finite quantity of energy, called "quanta". Therefore, the corresponding formula is*

$$E := h\nu.$$

This approach was postulated by Max Planck in order to explain the black-body radiation. It is an electromagnetic radiation emitted by a black body. A black body is an idealized physical body that absorbs all incident electromagnetic radiation, regardless of frequency or angle of incidence. The name "black body" is given because it absorbs all colors of light. A black body also emits black-body radiation.

According to the above notations, let us say that λ is the wavelength of the black-body radiation. T is the absolute temperature of black body. The Planck constant h can be experimentally evaluated being about 6.626×10^{-34} joule· second. For the experiments leading to this result, we refer the reader to the Bibliography.

The function $R(\lambda, T)$ is the spectral radiance produced by the surface of the black body having the absolute temperature T and wavelength λ. Here k is the Boltzmann constant. . The formula

$$R(\lambda, T) = \frac{2\pi c}{\lambda^4}\left(\frac{\varepsilon_0}{e^{\beta \varepsilon_0} - 1}\right) = \frac{2\pi c}{\lambda^4}\left(\frac{\varepsilon_0}{e^{\frac{\varepsilon_0}{kT}} - 1}\right) = \frac{2\pi hc^2}{\lambda^5}\left(\frac{1}{e^{\frac{hc}{kT}\frac{1}{\lambda}} - 1}\right).$$

is now called the Planck spectral radiance formula. To obtain this, Max Planck hypothesized that $\bar{\varepsilon}$, the average energy produced by all the possible frequencies, is produced by energies of type $n\varepsilon_0$ where the quantum energy is $\varepsilon_0 = h\nu = h\frac{c}{\lambda}$. In this way, the spectral radiance does not diverge and is coherent with experimental results. This means that a black body cannot accumulate infinite energy.

Specifically, the Planck spectral radiance formula allows to calculate

$$R(T) := \int_0^\infty R(\lambda, T)d\lambda$$

which is a finite number depending on T. Indeed, if we compute the integral

$$\int_0^\infty \frac{\frac{1}{\lambda^5}}{e^{\alpha\frac{1}{\lambda}} - 1}d\lambda$$

and substitute $y := \frac{1}{\lambda}$, that is $dy := -\frac{1}{\lambda^2}d\lambda$, the previous integral becomes

$$\int_0^\infty \frac{y^3}{e^{\alpha y} - 1}dy.$$

Another obvious change of variable is $x := \alpha y$, $dx = \alpha dy$, and then we can recast the integral as

$$K(\alpha) \int_0^\infty \frac{x^3}{e^x - 1} dx,$$

where $K(\alpha)$ is a constant depending on T^4, being $\alpha = \dfrac{hc}{kT}$. It is

$$\int_0^\infty \frac{x^3}{e^x - 1} dx = \int_0^\infty \frac{x^3 e^{-x}}{1 - e^{-x}} dx = \int_0^\infty x^3 \sum_{n=1}^\infty e^{-nx} dx = \sum_{n=1}^\infty \int_0^\infty x^3 e^{-nx} dx.$$

Using $u := nx$ and $du = ndx$, it results that

$$\int_0^\infty \frac{x^3}{e^x - 1} dx = 6 \sum_{n=1}^\infty \frac{1}{n^4}$$

which is a finite number as we are going to demonstrate.

For this purpose, let us use the Fourier series which we will reconsider in details in Lecture 37. In particular, let us consider the Plancharel-Parseval theorem. We start from $f(x) = x^2$ for $x \in [-\pi, \pi]$, which is a continuous function on the given interval. The same is its square, f^2. So, both f and f^2 are integrable. We compute the Fourier coefficients:

$$a_n := \frac{1}{\pi} \int_{-\pi}^\pi x^2 e^{inx} dx = \frac{2\cos(n\pi)}{n^2}, \ n \in \mathbb{Z}, \ n \neq 0,$$

$$a_0 := \frac{1}{2\pi} \int_{-\pi}^\pi x^2 dx = \frac{\pi^2}{3}.$$

It results $|a_n|^2 = \dfrac{4}{n^4}$, $|a_0|^2 = \dfrac{\pi^4}{9}$, that is, according to the Plancharel-Parseval theorem, we have

$$\frac{1}{2\pi} \int_{-\pi}^\pi f^2(x) dx = \sum_{n \in \mathbb{Z}} a_n^2 = a_0^2 + 2 \sum_{n=1}^\infty a_n^2.$$

The last line can be written in the form

$$\frac{\pi^4}{5} = \frac{\pi^4}{9} + 8 \sum_{n=1}^\infty \frac{1}{n^4}.$$

Therefore

$$\sum_{n=1}^\infty \frac{6}{n^4} = \frac{\pi^4}{15}.$$

We have the entire landscape view on Planck's work after we observe that $R(T)$, the total spectral radiance, is a constant multiplied by T^4, exactly as it appears in the Stefan–Boltzmann law. In other words, the Planck idea to consider a quantized energy is the key for a correct physical description of the spectral radiance. Specifically, the formula

$$R(\lambda, T) = \frac{2\pi h c^2}{\lambda^5} \left(\frac{1}{e^{\frac{hc}{kT}\frac{1}{\lambda}} - 1} \right),$$

is in agreement with experiments and allows, after computations, to highlight the correct proportionality between the total spectral radiance $R(T)$ and T^4 as it appears in Stefan–Boltzmann formula.

Summary of Lecture 17. The following formula

$$\bar{\varepsilon} = \frac{\int_0^\infty \varepsilon\, e^{-\beta\varepsilon} d\varepsilon}{\int_0^\infty e^{-\beta\varepsilon} d\varepsilon} = \frac{1}{\beta},$$

is the key ingredient which leads to the concept of quantized energy introduced by Planck. If

$$R(\lambda, T) = \frac{2\pi c}{\lambda^4} \bar{\varepsilon}$$

and $\beta = \dfrac{1}{kT}$, it results

$$R(\lambda, T) = \frac{2\pi c}{\lambda^4} kT$$

and

$$\lim_{\lambda \to 0_+} R(\lambda, T) = +\infty.$$

These formulas have to be related to a experiments whose result cannot be $+\infty$. What is wrong in this description? Let us change the continuous variable ε, which belongs to the real interval $[0, +\infty)$, with $n\varepsilon_0$, where n belongs to the natural number set $\mathbb{N}$. Therefore we have a modified formula for $\bar{\varepsilon}$, which is

$$\bar{\varepsilon} = \frac{\sum_0^\infty n\varepsilon_0\, e^{-\beta n\varepsilon_0}}{\sum_0^\infty e^{-\beta n\varepsilon_0}} = -\frac{d}{d\beta}\left[\ln\left(\sum_0^\infty e^{-\beta\varepsilon_0} \right) \right] = -\frac{d}{d\beta}\left[\ln \frac{1}{1 - e^{-\beta\varepsilon_0}} \right] = \frac{\varepsilon_0}{e^{\beta\varepsilon_0} - 1}.$$

It results

$$R(\lambda, T) = \frac{2\pi c}{\lambda^4}\left(\frac{\varepsilon_0}{e^{\beta\varepsilon_0} - 1} \right) = \frac{2\pi c}{\lambda^4}\left(\frac{\varepsilon_0}{e^{\frac{\varepsilon_0}{kT}} - 1} \right).$$

According to the physical meaning of the experiment, we have

$$\varepsilon_0 = h\nu = h\frac{c}{\lambda},$$

where h is a constant, ν is the frequency, c is the speed of light and λ the wavelength. Therefore

$$R(\lambda, T) = \frac{2\pi hc^2}{\lambda^5} \left(\frac{1}{e^{\frac{hc}{kT}\frac{1}{\lambda}} - 1} \right),$$

and

$$\lim_{\lambda \to 0_+} \frac{2\pi hc^2}{\lambda^5} \left(\frac{1}{e^{\frac{hc}{kT}\frac{1}{\lambda}} - 1} \right) = 0$$

which fits the experiment. The main change made, in order to fit the experiments, is related to the fact that ε cannot have a continuous set of values, but only natural multiples of a given energy ε_0. This energy depends on a constant, called the Planck constant, and on the frequency ν. According to this result, the energy cannot have a continuous set of values but only discrete values which depends on finite quantities of energy, called quanta. Therefore, the corresponding Planck–Einstein formula $E := h\nu$ must be considered as an axiom which allows the correct mathematical description and the physical interpretation of the experiment.

4.5 Lecture 18: Light as Particles. The Einstein Photoelectric Effect

In an experiment performed in 1887, Heinrich Hertz detected the photoelectric effect by observing that ultraviolet light hiting a metal produces sparks. The sparks are actually electrons released from the metal used. The ultraviolet light can be replaced by any light beam.

A theoretical explanation of this effect was given by Einstein using Planck's idea related to energy conceived as packets. Einstein proposed to consider light formed by photons which are tiny packets of energy, each packet carrying the energy $E = h\nu$. Einstein proved that the extraction of electrons happens if the energy of photons, which hit the metal, is grater than the energy necessary to electrons to escape from the metal. The photon energy is proportional to the frequency of the light, so increasing the frequency we increase the energy and we can see the "photoemission" happening.

So, when a photon from the light beam hits the surface, it uses all its energy $E = h\nu$ to extract an electron. How we can model this? In 1905, Einstein's idea was to observe that if electrons are released at a speed, denoted v, and if m is the mass of the electron, the following equality holds

$$\frac{1}{2}mv^2 = h\nu - W.$$

In this formula the minimum amount of energy necessary to extract the electron is denoted by W. The meaning of the previous formula is the following. If $v = 0$, this extraction does not happen. In this case, $E = h\nu = W$. But, if $E - W > 0$, that is $h\nu - W > 0$, the electron is released at speed v according to the above formula.

Einstein's idea to see light made of particles (the photons) carrying energy oblige us to develop another point of view when we are dealing with it: The light can be described as a wave and a particle at the same time. In some experiments, light has to be considered as a wave, while in other experiments it must be seen as a particle. Therefore, light has a dual nature. This point of view is incompatible with Classical Mechanics. In this perspective, we need another approach to microscopic phenomena leading to Quantum Mechanics.

Summary of Lecture 18. In the long story related to the question "What is the nature of light?" two possible answers are possible. The wave nature of light is highlighted by the two slit experiment. A description of light as electromagnetic waves was given by Maxwell. Above we presented the Mathematics behind this wave nature.

Heinrich Hertz detected the photoelectric effect, when, in an experiment, he observed that ultraviolet light, hiting a metal, produces sparks. The sparks are actually electrons released from the metal used in this experiment. Einstein gave the explanation of the phenomenon considering light as at a collection of particles called photons and having quantized energies, $E = h\nu$. The equality

$$\frac{1}{2}mv^2 = h\nu - W$$

makes sense if the energy of light is grater than the energy of extraction, W. If so, electrons are expelled from the metal according to the above formula. According to this explanation for the photoelectric effect, he concluded about the dual nature of the light: light can be wave and particle at the same time.

4.6 Lecture 19: Atomic Spectra and Bohr's Model of Hydrogen Atom

In this lecture, we are going to present an atomic model based on the above facts and results. In 1913, when this model was presented the physicists still did not know about the possibility to detect electrons as waves. Let us start from the experiments discussed above on hydrogen in a glass bottle. Excited atoms of hydrogen emit light. If this light passes through a prism, we can observe a series of four colored lines with dark spaces in between. We obtain the same experimental result by considering other elements. Each time, we observe more or less colored lines with dark spaces in between. These series of colored lines are called atomic spectra. Each element has its own atomic spectrum, therefore elements can be identified by their line spectrum. Johann Jakob Balmer was the first who, in 1885, observed a regularity in spectral lines of atoms. In fact he obtained a formula related to the nine known lines of hydrogen in terms of their wavelengths:

$$\lambda = C \frac{n^2}{n^2 - 4}, \ n = 3, 4, 5, \ldots, 11,$$

where $C = 3646 \text{Å}$ is a constant.

In 1889, Johannes Rydberg described the wavelengths of photons emitted by changes in energy level of an electron in a hydrogen atom. The considered formula is in fact the Balmer one written in a different way, that is:

$$\frac{1}{\lambda} = R_H \left(\frac{1}{2^2} - \frac{1}{n^2} \right), \ n = 3, 4, 5, \ldots$$

where R_H is the Rydberg constant, $R_H = 109677, 58 \ cm^{-1}$. Rydberg's formula can be generalized for spectral series of hydrogen seen in ultraviolet or infrared light as

$$\frac{1}{\lambda} = R_H \left(\frac{1}{n_1^2} - \frac{1}{n_2^2} \right), n_1 < n_2, \ n_1 = 1, 2, 3, 4, 5, \ldots, \ , n_2 = 2, 3, 4, 5, \ldots,$$

where n_1 is called the principal quantum number of the upper energy level and n_2 is called the principal quantum number of the lower energy level for the atomic electron transition. The existence of spectral series cannot be explained by Classical Physics. Another model of an atom should be created to explain these observations.

It was the Danish physicist Niels Bohr, based on some quantum postulates, who attempted to describe the hydrogen atom. He created a mixed classical-quantum description based on Rutherford's planetary model with a positive charged massive nucleus at center and a negative charged electron around it. He used Einstein–Planck's relation i.e. photons as carriers of energy, as we have seen in Einstein's photoelectric effect. The postulates are:

1. The electron moves around the nucleus on a circular orbit imposed by the electrostatic force who acts between the two electric charges.

The second Bohr's postulate is related to the fact that not all circular orbits are possible:

2. Among all orbits, only those corresponding to given energy levels are possible.

3. The electron preserves its energy while moving on circular trajectories related to these given levels.

Let us first explain the idea before enunciating the fourth postulate. Energy can be lost or gained only if the electron switches between two allowed energy levels. A photon having the energy $E = h\nu$ interacts with the electron found on an orbit corresponding to an energy E_k if there exists a possible orbit corresponding to the energy $E_k + h\nu$. In this case the electron will jump to this orbit and the atom gains energy. An electron of an orbit corresponding to an energy E_m can release a photon with an energy $E = h\nu$ only if this energy of the photon allows the electron to jump to another allowed level, E_p.

4. Therefore, a photon can make the electron jump between two allowed energy levels E_l and E_j, $E_j < E_l$ if and only if its energy $h\nu$ satisfies the relation

$$E_l - E_j = h\nu.$$

5. There is a first fundamental level E_1 corresponding to a possible minimum energy such that the electron cannot spiral to the nucleus.

The mathematical description of the model starts with the definitions:

– ε_0 which is now the vacuum permittivity,
– $|e|$, the absolute value of the charge for electron and proton,
– Z, the number of protons in nucleus, in the case of hydrogen atom $Z = 1$,
– the Coulomb force,

$$F = \frac{1}{4\pi\varepsilon_0} \frac{Z|q_1||q_2|}{r^2},$$

in this case

$$F = \frac{1}{4\pi\varepsilon_0} \frac{e^2}{r^2},$$

– Bohr first postulate is

$$mvr = n\hbar, \ n = 1, 2, 3, \ldots$$

where $\hbar := \dfrac{h}{2\pi}$ is the *reduced Planck constant*. It means the stationary orbits are at distances r such that the angular momentum is quantized (m is the mass of the electron). The lowest value $n = 1$ corresponds to the smallest possible orbit, the one stated in the fifth postulate.

– the centripetal force corresponding to the orbit(s) is

$$F = \frac{mv^2}{r},$$

– the electron potential energy is

$$V = -\frac{e^2}{4\pi\varepsilon_0}\frac{1}{r},$$

– the electron kinetic energy is

$$T = \frac{1}{2}mv^2.$$

Bohr started from the Classical Mechanics energy condition

$$\frac{mv^2}{r} = \frac{1}{4\pi\varepsilon_0}\frac{e^2}{r^2},$$

i.e.

$$v = \frac{e^2}{4\pi\varepsilon_0}\frac{1}{mvr} = \frac{e^2}{4\pi\varepsilon_0 n\hbar}.$$

It results

$$r = \frac{n\hbar}{mv} = \frac{4\pi\varepsilon_0\hbar^2}{me^2}n^2, \quad n = 1, 2, 3, \ldots$$

Replacing in the potential and kinetic energy formulas, we obtain

$$V = -\frac{m}{\hbar^2}\left(\frac{e^2}{4\pi\varepsilon_0}\right)^2\frac{1}{n^2}, \quad T = \frac{m}{2\hbar^2}\left(\frac{e^2}{4\pi\varepsilon_0}\right)^2\frac{1}{n^2}.$$

The hydrogen atom total energy formula is

$$E_n = -\frac{m}{2\hbar^2}\left(\frac{e^2}{4\pi\varepsilon_0}\right)^2\frac{1}{n^2}, \quad n = 1, 2, 3, \ldots.$$

If we compute the energy difference corresponding to two different orbits, we find a Rydberg type formula

$$E = E_l - E_j = R_E\left(\frac{1}{n_l} - \frac{1}{n_j}\right),$$

where R_E is a constant with the obvious physical dimensions of energy.

This model, which mixes Quantum and Classical Mechanics, is today obsolete. The main objection is related to the statement of first postulate; de Broglie gave a consistent explanation for it, as we will discuss in the next lecture. Another major objection is related to the concept of circular orbits. It means that the electron exhibits a deterministic behavior, that is, we know at any moment, its position and its velocity. When we will study the Heisenberg Uncertainty Principle, we will see that it is impossible to determine, at the same time, the electron trajectory and momentum. However, it is worth studying the Bohr model of atom because it is both an important step towards a self-consistent formulation of Quantum Mechanics and because it gives good predictions for the Rydberg series.

Summary of Lecture 19. The spectral series of hydrogen atom, the Balmer's one

$$\frac{1}{\lambda_{vac}} = R_H \left(\frac{1}{2^2} - \frac{1}{n^2} \right), \ n = 3, 4, 5, \ldots$$

and the Rydberg's one,

$$\frac{1}{\lambda_{vac}} = R_H \left(\frac{1}{n_1^2} - \frac{1}{n_2^2} \right), n_1 < n_2, \ n_1 = 1, 2, 3, 4, 5, \ldots, \ , n_2 = 2, 3, 4, 5, \ldots,$$

cannot be explained by Classical Physics. In 1913, Niels Bohr proposed an atomic model which mixed Rutherford's model, Planck's quanta and Einstein's photoelectric effect. Among the postulates used by Bohr, one is very important: the electron moves on a circular orbit around the nucleus. This one is written in terms of quantized angular momentum,

$$mvr = n\hbar.$$

The mathematical description of the model starts with the Coulomb force written in the case of hydrogen atom as

$$F = \frac{1}{4\pi\varepsilon_0} \frac{e^2}{r^2}$$

which equals the centripetal force

$$F = \frac{mv^2}{r}.$$

Therefore

$$\frac{mv^2}{r} = \frac{1}{4\pi\varepsilon_0} \frac{e^2}{r^2},$$

i.e.

$$r = \frac{n\hbar}{mv} = \frac{4\pi\varepsilon_0\hbar^2}{me^2} n^2, \ n = 1, 2, 3, \ldots$$

Replacing in the potential and kinetic energy formulas, we obtain the hydrogen atom total energy formula

$$E_n = -\frac{m}{2\hbar^2}\left(\frac{e^2}{4\pi\varepsilon_0}\right)^2\frac{1}{n^2}, \; n = 1, 2, 3, \ldots.$$

If we compute the energy difference between two orbits, we find a Rydberg type formula

$$E = E_l - E_j = R_E\left(\frac{1}{n_l} - \frac{1}{n_j}\right).$$

The model was firstly corrected by de Broglie which obtained the Bohr's first postulate from another hypothesis. However, another major objection remains. The circular orbits contradicts the Heisenberg Uncertainty Principle we will study later.

4.7 Lecture 20: De Broglie's Hypothesis. Material Objects as Waves

To translate the title, de Broglie thought that all what we consider matter acts as wave, too; that is any material object has an associated wavelength. This means that the wave-corpuscle duality principle can be extended to particles like electrons and, in general, to any form of matter. How did de Broglie come to this hypothesis?

Let us start from the formula $E = mc^2$. It does not work for photons, because we know that their rest mass is assumed to be 0. Let us write the formula with respect the momentum p.

Then $E = pc$. And let us consider the other formula we obtained for energy, $E = h\nu$. Therefore we obtain

$$p = \frac{h}{c}\nu = \frac{h}{c}\frac{1}{T} = \frac{h}{\lambda}.$$

According to this formula, we can assume that an electron (and any particle) has a proper wavelength. Clearly the wavenumber can be given in all its possible expressions like

$$k = \frac{2\pi}{\lambda} = \frac{2\pi}{cT} = \frac{2\pi}{c}\nu.$$

Multiplying both members by h, and considering again $\hbar$, it results $\frac{h}{2\pi}k = \frac{E}{c} = p$, that is

$$p = \hbar k.$$

The last formula can be seen as a quantization of the momentum. Any time h or $\hbar$ appear in a formula, we are dealing with a *quantized quantity*.

These considerations can be summarized in the de Broglie formula

$$\lambda = \frac{h}{p} = \frac{h}{mv}$$

according to which an electron (or any particle) can be considered as a wave.

In order to test this hypothesis, we can take into account the Davisson-Germer double slit experiment where the particle beam is made of electrons. The same phenomenology holds also for other elementary particles, atoms and even molecules. It is possible to confirm the wave behavior, justifying de Broglie hypothesis, in any case. We refer the interested reader to the Bibliography for detailed descriptions of these experiments.

According to this picture, we can guess why electrons do not spiral into the atomic nucleus.

The electron in an atom is a standing wave. In other words, to correct the hydrogen atom model by Bohr, de Broglie proposed to look at the electron as at a standing wave. Therefore the electron is a wave and an integer number of wavelengths represent the circumference of the orbit, i.e.

$$n\lambda = 2\pi r.$$

From this relation, it is clear that, for a given wavelength λ, only some radius are possible. According to the de Broglie formula, it results

$$\frac{nh}{mv} = 2\pi r$$

which is equivalent to the Bohr postulate $mvr = n\hbar$.

The next step to understand correctly the atom model will be realized after studying the Heisenberg Uncertainty Principle.

To conclude, matter has in general a wave behavior. But why we cannot see this wave behavior for a macroscopic object having, for example, a mass of 1 Kg which moves at speed 1 m s^{-1}? The answer is very simple. If we compute λ, we find out it is h, that is of the order of magnitude 10^{-34}Kg m^{2}s^{-1}. Therefore we cannot observe such a wave having this extremely small wavelength. Summing up: All these facts cannot be framed in the standard scheme of Classical Mechanics. This means that a new paradigm is necessary. It is the Quantum Mechanics.

Summary of Lecture 20. The de Broglie hypothesis can be summarized in following way. We start from the rest energy formula written in the form $E = pc$. Consider now the other formula for energy, $E = h\nu$. Therefore we obtain

$$p = \frac{E}{c} = \frac{h}{c}\nu = \frac{h}{c}\frac{1}{T} = \frac{h}{\lambda}.$$

It can be arranged in the form

$$\lambda = \frac{h}{p} = \frac{h}{mv}.$$

This formula can describe the wavelength of an electron. Starting from the formula of the wavenumber

$$k = \frac{2\pi}{c}\nu,$$

we can derive

$$p = \hbar k.$$

The last formula can be seen as a quantization of the momentum, which is the next step after the quantization of energy. The Davisson-Germer double slit experiment, made with electrons, confirmed the de Broglie hypothesis. The standing wave behavior of the electron in the hydrogen atom is a consequence of this reasoning. The Bohr postulate about circular orbits, $mvr = n\hbar$, can be deduced by de Broglie hypothesis starting from the electron wavelength formula, which is

$$\lambda = \frac{h}{mv}.$$

4.8 Lecture 21: Strengthening the Einstein Idea of Photons. The Compton Effect

This very important experiment proves definitely that light cannot be seen only as an electromagnetic wave but also as a particle. We already know, from the photoelectric effect, that light is described as made of particles, called photons, carrying tiny packets of energy. The remarkable Einstein explanation has been demonstrated to be experimentally valid, but: Is there an experiment where light starts as a wave but it shows particle characteristics? This will strengthen Einstein's point of view about light seen as made of particles. Measuring something predicted gives the proof of the statement. Such an experiment was performed by Arthur Compton in 1922. He highlighted what we call now the *Compton effect* described by a formula which predicts a change of wavelength which can be measured. From a classical point of view, there is no possibility to explain why the initial wavelength can change after interactions with charged particles as electrons. The fact that light wave, after interaction, has a different wavelength obliges us to consider light as a flux of particles whose energy changes after interaction.

Suppose a source which emits X-rays. The X-rays hit a substance and act on the electrons in a way described by Compton using the following model. An incident photon, whose initially momentum is p_0, hits an electron e supposed at rest. After the collision, we denote by p_1, the momentum of the photon, and by p_2 the momentum of the electron. The photon and the electron are spread in two different directions by angles θ and ϕ with respect to the axis determined by the direction of the incoming photon. After the collision with the photon, the speed of the electron is high enough, in comparison with the speed of light, therefore we can use the Special Relativity.

Let us denote by m the mass of the electron. The kinetic energy of the electron is defined as

$$T := E - mc^2,$$

where E is the total relativistic energy and mc^2 is the rest energy of the electron. In the case of photon, its energy is

$$E = h\nu = h\frac{c}{\lambda}$$

and its momentum is

$$p = \frac{E}{c} = \frac{h}{\lambda}.$$

The total relativistic energy is described by the formula

$$E^2 = m^2c^4 + p^2c^2,$$

where the impulse of the electron is

$$p = \frac{mv}{\sqrt{1 - v^2/c^2}}.$$

Let us apply these formulas to the case of the experiment described above. The photon has the initial energy $E_0 = h\dfrac{c}{\lambda_0}$ and the initial momentum $p_0 = \dfrac{E_0}{c} = \dfrac{h}{\lambda_0}$.

After the impact with the electron e, the previous quantities switched into $E_1 = h\dfrac{c}{\lambda_1}$ and $p_1 = \dfrac{E_1}{c} = \dfrac{h}{\lambda_1}$. As we previously asserted, θ is the angle made by the incidence direction with the new direction taken by the photon.

We denote by p_2 the momentum of electron after the collision and by ϕ the angle between the incidence direction and the direction taken by the electron e. The momentum conservation leads to the following equalities:

$$\begin{cases} p_1 \cos\theta + p_2 \cos\phi = p_0 \\ p_1 \sin\theta - p_2 \sin\phi = 0 \end{cases}$$

that is

$$\begin{cases} p_0 - p_1 \cos\theta = p_2 \cos\phi \\ \quad p_1 \sin\theta = p_2 \sin\phi. \end{cases}$$

We square each line and we add the two equations. Finally, we obtain

$$p_2^2 = p_0^2 + p_1^2 - 2p_0p_1 \cos\theta.$$

Now let us focus on the conservation of energy. We have an energy before the collision, and the energy after the collision. Before the collision the energy is $E_0 + mc^2$. After the collision the energy is $E_1 + \sqrt{m^2c^4 + p_2^2c^2}$. The energy is conserved, therefore

$$E_0 + mc^2 = E_1 + \sqrt{m^2c^4 + p_2^2c^2}.$$

We arrange in the form

$$E_0 - E_1 = \sqrt{m^2c^4 + p_2^2c^2} - mc^2,$$

where the left hand side represents the difference between the energies of the photon before and after the collision while the right hand is the difference between the energies of electron after and before the collision. It results

$$\sqrt{m^2c^4 + p_2^2c^2} - mc^2 = E_0 - E_1 = c(p_0 - p_1).$$

Canceling c we obtain

$$\sqrt{m^2c^2 + p_2^2} = (p_0 - p_1) + mc,$$

that is

$$(p_0 - p_1)^2 + 2mc(p_0 - p_1) = p_2^2.$$

Since we have also

$$p_0^2 + p_1^2 - 2p_0p_1 \cos\theta = p_2^2$$

it results

$$mc(p_0 - p_1) = 2p_0p_1 \sin^2\frac{\theta}{2}.$$

It can be written as

$$mc\left(\frac{h}{\lambda_0} - \frac{h}{\lambda_1}\right) = 2\frac{h}{\lambda_0}\frac{h}{\lambda_1} \sin^2\frac{\theta}{2}.$$

The final relation obtained is the Compton effect formula

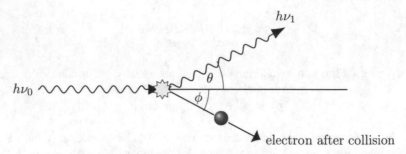

Fig. 4.7 Compton's effect. The basic picture of the collision

$$\lambda_1 - \lambda_0 = 2\frac{h}{mc}\sin^2\frac{\theta}{2}$$

which shows that light has not only an electromagnetic wave behavior but also a corpuscular behavior (Fig. 4.7).

Summary of Lecture 21. Suppose a source which emits X-rays. The X-rays hit a substance and act on the electrons in a way described by Compton using the following model: an incident photon, whose initial momentum is p_0, hits an electron e supposed at rest. After the collision, we denote by p_1 the momentum of the photon and by p_2 the momentum of the electron. The photon and the electron are spread in two different directions forming angles θ and ϕ respectively with respect to the axis determined by the initial direction of the photon. If m is the mass of the electron, E is the total relativistic energy then

$$T := E - mc^2,$$

is the kinetic energy of the electron. The photon has the initial energy $E_0 = h\frac{c}{\lambda_0}$ and the initial momentum $p_0 = \frac{E_0}{c} = \frac{h}{\lambda_0}$. After the impact with the electron e, the previous quantities switch into $E_1 = h\frac{c}{\lambda_1}$ and $p_1 = \frac{E_1}{c} = \frac{h}{\lambda_1}$. The total relativistic energy is described by the formula

$$E^2 = m^2c^4 + p^2c^2,$$

where the impulse of the electron is

$$p = \frac{mv}{\sqrt{1 - v^2/c^2}}.$$

Imposing the conservation of energy, we have

$$E_0 + mc^2 = E_1 + \sqrt{m^2 c^4 + p_2^2 c^2}.$$

The conservation of momentum leads to

$$\begin{cases} p_0 - p_1 \cos \theta = p_2 \cos \phi \\ \quad p_1 \sin \theta = p_2 \sin \phi \end{cases}$$

Manipulating the formulas, we finally obtain the Compton effect formula

$$\lambda_1 - \lambda_0 = 2 \frac{h}{mc} \sin^2 \frac{\theta}{2}.$$

What we have obtained can be measured and shows that light, considered only as an electromagnetic wave, cannot have such behavior. Its wavelength cannot change without having a corpuscular behavior which manifests during the experiment.

Chapter 5
The Schrödinger Equations and Their Consequences

Art and science have their meeting-point in method.

Edward George Buwler-Lytton

5.1 Lecture 22: The Schrödinger Equations. The One Dimensional Case

As the title suggests we have to discuss about more than one equation. These Schrödinger equations can be considered as axioms and principles of Quantum Mechanics due to the practical and conceptual consequences of their formulation. The starting point for our considerations is asking how Erwin Schrödinger conceived these equations. According to the results of previous chapter, he took into account the fact that objects of Quantum Mechanics are both particles and waves at the same time. This means that we have to describe these objects according to the wave definition we choose. Therefore an appropriate wave equation has to be constructed. In the one-dimensional case, the solution should be a monochromatic wave described as

$$\Psi(t, x) := \Psi(x) = Ae^{i(kx - \omega t)}, \ t \in \mathbb{R}, \ x \in \mathbb{R}.$$

We observe

$$\frac{d\Psi}{dx} = ik\Psi,$$

and

$$\frac{d^2\Psi}{dx^2} = -k^2\Psi.$$

Let us take into account that $p = \hbar k$, i.e. $k = \dfrac{p}{\hbar}$. It results

© The Author(s), under exclusive license to Springer Nature Switzerland AG 2021
S. Capozziello and W.-G. Boskoff, *A Mathematical Journey to Quantum Mechanics*,
UNITEXT for Physics, https://doi.org/10.1007/978-3-030-86098-1_5

$$-\hbar^2\frac{d^2\Psi}{dx^2} = p^2\Psi,$$

that is

$$-\frac{\hbar^2}{2m}\frac{d^2\Psi}{dx^2} = \frac{p^2}{2m}\Psi.$$

We know the formula

$$\frac{p^2}{2m} = H - V(x),$$

where H is the total energy, i.e. the Hamiltonian H. Multiplying it by Ψ, we obtain

$$-\frac{\hbar^2}{2m}\frac{d^2\Psi}{dx^2} = \frac{p^2}{2m}\Psi = H\Psi - V\Psi.$$

It results the *Schrödinger time-independent equation* in the form

$$H\Psi = -\frac{\hbar^2}{2m}\frac{d^2\Psi}{dx^2} + V(x)\Psi.$$

The *time-dependent Schrödinger's equation* uses the derivative of the wave function with respect to time and the multiplication by $\hbar$:

$$\hbar\frac{\partial\Psi}{\partial t} = -i\omega\hbar\Psi = -iH\Psi.$$

Therefore

$$i\hbar\frac{\partial\Psi}{\partial t} = H\Psi.$$

We have used the energy formula, here in the form $H = \hbar\omega$, because we are dealing with the total energy which may contain a potential energy term.

Replacing $H\Psi$ with $i\hbar\dfrac{\partial\Psi}{\partial t}$ in the time-independent Schrödinger equation, we get the time dependent Schrödinger equation

$$i\hbar\frac{\partial\Psi}{\partial t}(t, x) = -\frac{\hbar^2}{2m}\frac{\partial^2\Psi}{\partial x^2}(t, x) + V(x)\Psi(t, x).$$

If the potential is dependent on t and x, we have $V(t, x)$.

These quick considerations point out the fact that there is not a deduction of Schrödinger's equations, because they were postulated. Let us now specifically take into account the one dimensional case.

The time-independent Schrödinger equation refers to a wave function Ψ dependent on the position x only. The postulated equation is

$$H\Psi(x) = \left(-\frac{\hbar^2}{2m}\frac{d^2}{dx^2} + V(x)\right)\Psi(x).$$

For the time-dependent Schrödinger equation both Ψ and the potential V depends on (t, x), therefore the postulated equation is

$$i\hbar\frac{d}{dt}\Psi(t, x) = \left(-\frac{\hbar^2}{2m}\frac{d^2}{dx^2} + V(t, x)\right)\Psi(t, x).$$

This form is related to the operator form we develop later. In the case of an electric field, a particle is subject of a time-dependent potential, so it makes sense the last form.

The time-dependent Schrödinger equation is used in the theory of the harmonic oscillator we will present later.

Despite of these considerations, in Lecture 32 we find out that something is wrong: particles as electron cannot be described by the monochromatic planar wave

$$\Psi(t, x) = Ae^{i(kx-wt)}, \ t \in \mathbb{R}, \ x \in \mathbb{R},$$

so some further ingredient is necessary in our discussion. However the Schrödinger equation can have solutions expressed with respect to the wave function form written above and in the same Lecture 32 we understand the meaning of such a solution. Furthermore, let us observe the linearity of the Schrödinger equation with implications in the existence of solutions written as linear combination of solutions as we will see in the next lectures.

Summary of Lecture 22. In the one-dimensional case, let us consider a wave described as

$$\Psi(t, x) = Ae^{i(kx-\omega t)}, \ t \in \mathbb{R}, \ x \in \mathbb{R}.$$

Using the Hamiltonian H

$$H(t, x) = \frac{p^2}{2m} + V(x)$$

we can "derive" both Schrödinger's time-independent equation

$$H\Psi = -\frac{\hbar^2}{2m}\frac{d^2\Psi}{dx^2} + V(x)\Psi$$

and Schrödinger's time-dependent equation

$$i\hbar\frac{\partial\Psi}{\partial t}(t, x) = -\frac{\hbar^2}{2m}\frac{\partial^2\Psi}{\partial x^2}(t, x) + V(t, x)\Psi(t, x).$$

Even if these equations were postulated in Quantum Mechanics, a problem remains. Is

$$\Psi(t, x) = A e^{i(kx - wt)}, \ t \in \mathbb{R}, \ x \in \mathbb{R}$$

an appropriate description for waves related to particles as, for example, the electron? The answer is no. As a consequence, we have to replace the wave of the above form with wave packets.

5.2   Lecture 23: Solving the Schrödinger Equation for the Free Particle

Let us start from the time independent Schrödinger equation

$$H\Psi = -\frac{\hbar^2}{2m} \frac{d^2\Psi}{dx^2} + V(x)\Psi.$$

For a free particle, it is $V(x) = 0$. Suppose that the particle moves along a line segment of length L; the corresponding Schrödinger equation becomes

$$\frac{d^2\Psi}{dx^2} = -\frac{2m}{\hbar^2} H\Psi,$$

which can be written as

$$\frac{d^2\Psi}{dx^2} = -k^2\Psi.$$

This formula comes from $H = T = \dfrac{p^2}{2m}$ and $p = \hbar k$. It results

$$k^2 = \frac{2m}{\hbar^2} H.$$

The general solution is

$$\Psi(x) = A_1 e^{ikx} + A_2 e^{-ikx}$$

as you can immediately verify. From $\lambda = \dfrac{h}{mv}$ and $n\lambda = L$, the momentum is described by the formula

$$p = \frac{nh}{L},$$

therefore, if we replace in the energy formula, we obtain

$$H = H_n = \frac{h^2}{2mL^2}n^2, \ n = 1, 2, 3, \ldots$$

It means that the energy of the free particle is quantized. If the particle moves between two energy levels, the role of photons is highlighted because

$$\Delta H_{nk} = H_n - H_k = h\nu,$$

where $\nu = \dfrac{h}{2mL^2}(n^2 - k^2)$ as it is easy to show.

Summary of Lecture 23. We start from the time independent Schrödinger equation for a free particle (i.e. $V(x) = 0$) which moves along a line segment of length L. The equation can be written with respect to the wavenumber k:

$$\frac{d^2\Psi}{dx^2} = -k^2\Psi.$$

The general solution is

$$\Psi(x) = A_1 e^{ikx} + A_2 e^{-ikx}.$$

The related momentum is described by the formula

$$p = \frac{nh}{L}$$

therefore, if we replace in the energy formula, we obtain

$$H = H_n = \frac{h^2}{2mL^2}n^2, \ n = 1, 2, 3, \ldots.$$

The energy of the free particle is quantized and, if the particle moves between two energy levels, photons are highlighted because

$$\Delta H_{nk} = H_n - H_k = h\nu,$$

where $\nu = \dfrac{h}{2mL^2}(n^2 - k^2).$

5.3 Lecture 24: Solving the Schrödinger Equation for a Particle in a Box

In a plane of coordinates (x, V), let us imagine a box determined by the points $U_1(0, V_0)$, $O(0, 0)$, $P(L, 0)$, $U_2(L, V_0)$ and the segments U_1O, OP, PU_2. The box interior is determined by the interior of the rectangle U_1OPU_2. We start from the same time independent Schrödinger equation,

$$H\Psi = -\frac{\hbar^2}{2m}\frac{d^2\Psi}{dx^2} + V(x)\Psi,$$

which we can write in the form

$$\frac{d^2\Psi}{dx^2} = \frac{2m}{\hbar^2}(V - H)\Psi.$$

$V - H$ is positive quantity because we supposed the particle in the box. The previous equation can be written as

$$\frac{d^2\Psi}{dx^2} = K^2\Psi,$$

where

$$K^2 = \frac{2m}{\hbar^2}(V - H)$$

and the solution now is

$$\Psi(x) = A_1e^{Kx} + A_2e^{-Kx}.$$

The solution has to describe a particle in the box, that is $A_1 = 0$. If not, as $x \to +\infty$ and the solution Ψ approaches $+\infty$. Therefore, inside the box, the solution remains only

$$\Psi(x) = A_2e^{-Kx}.$$

We can imagine the particle being in a given moment at the point U_2 and falling outside the box with the same type of exponential function. The particle is somewhere at a height less than the height V_0 of the box. Unless the box is not an infinite rectangle, this situation is possible. The lateral wall PU_2 acts as barrier and the particle inside the box, which cannot have enough energy to jump the barrier, can however be outside the box. There is a small probability to happen, but it can happen. In Quantum Mechanics this is called the quantum tunneling effect.

Let us now suppose that $V(x) = 0$ for the particle inside the box. The corresponding Schrödinger equation is again

$$\frac{d^2\Psi}{dx^2} = -k^2\Psi,$$

because

$$k^2 = \frac{2m}{\hbar^2} H.$$

The solution is similar to the case of free particle but some conditions has to exist to preserve the particle inside the box. This can happen when the box is infinitely deep. We write

$$\Psi(x) = A_1 e^{ikx} + A_2 e^{-ikx} = C_1 \cos kx + C_2 \sin kx.$$

If the box is infinitely deep, the particle will fall at the corners of the box. Therefore $\Psi(0) = C_1 = 0$. When $x = L$, we have $\Psi(L) = C_2 \sin kL = 0$ and this happens for $kL = n\pi$, i.e.

$$\Psi(x) = C_2 \sin \frac{n\pi}{L} x.$$

Now looking at the natural values of n and at the energy

$$H = T = \frac{p^2}{2m} = \frac{h^2 \pi^2}{2mL^2} n^2,$$

we observe that energy is quantized. Let us try to introduce from now a possible interpretation for Ψ. We can consider $|\Psi(x)|^2 dx$ related to the probability to have the particle in the box. Since the particle has always to be in the box, the constant C_2 is determined from the condition

$$\int_0^L |\Psi(x)|^2 dx = 1.$$

It results

$$\int_0^L C_2^2 \sin^2 \frac{n\pi}{L} x dx = 1.$$

Using

$$\sin^2 \alpha = \frac{1 - \cos 2\alpha}{2}$$

we obtain

$$C_2 = \sqrt{\frac{2}{L}}.$$

The form of the Ψ function, in the case of an infinitely deep box, is

$$\Psi(x) = \sqrt{\frac{2}{L}} \sin \frac{n\pi}{L} x.$$

Summary of Lecture 24. In a plane of coordinates (x, V) consider a box, as described in the present lecture, and the time independent Schrödinger equation

$$\frac{d^2\Psi}{dx^2} = \frac{2m}{\hbar^2}(V - H)\Psi,$$

where $V - H$ is a positive quantity assuming the particle inside the box. The previous equation can be written as

$$\frac{d^2\Psi}{dx^2} = K^2\Psi,$$

where

$$K^2 = \frac{2m}{\hbar^2}(V - H)$$

and the solution is now

$$\Psi(x) = A_1 e^{Kx} + A_2 e^{-Kx}.$$

The solution describes a particle in the box if $A_1 = 0$. If not, as $x \to +\infty$, the solution Ψ approaches to $+\infty$. Therefore inside the box the solution remains only

$$\Psi(x) = A_2 e^{-Kx}.$$

Let us now suppose that $V(x) = 0$ for the particle inside the box. We have again

$$\frac{d^2\Psi}{dx^2} = -k^2\Psi, \ k^2 = \frac{2m}{\hbar^2}H.$$

The solution is similar to the case of free particle, but some conditions has to exist to preserve the particle inside the box.

This can happen when the box is infinitely deep. We write

$$\Psi(x) = A_1 e^{ikx} + A_2 e^{-ikx} = C_1 \cos kx + C_2 \sin kx.$$

The constants are identified imposing some physical conditions and we find

$$\Psi(x) = C_2 \sin \frac{n\pi}{L}x.$$

The energy of the particle,

$$H = T = \frac{p^2}{2m} = \frac{h^2\pi^2}{2mL^2}n^2,$$

is quantized. Since the particle has always to be inside the box, the constant C_2 is determined from the condition

$$\int_0^L |\Psi(x)|^2 dx = 1.$$

It results

$$\Psi(x) = \sqrt{\frac{2}{L}} \sin \frac{n\pi}{L}x.$$

5.4   Lecture 25: Solving the Schrödinger Equation of Harmonic Oscillator. The Quantized Energies

The first lectures of this book were related to Classical Mechanics and we presented there the classical harmonic oscillator: it can be represented as a massive point at the end of a spring which oscillates around the equilibrium position. The classical image was presented using three possible pictures: Newton's one, Lagrange's one and Hamilton's one, all three being related to a wave equation which describes the phenomenon. Hamilton's language is related to the total energy of the harmonic oscillator, $H(q, p) = \frac{p^2}{2m} + \frac{1}{2}kq^2$, and it is closer to the Quantum Mechanics language where the wave equation is the Schrödinger one. In the quantum world, we found out that an electron can be seen as a standing wave which vibrates, a restoring force $-kx$ being implied. This restoring force depends on a potential such that $F = -\frac{dV}{dx}$. We intend to use the time-dependent Schrödinger equation to describe the possible values of the harmonic oscillator energy, thinking at the electron described as a wave function in the form

$$\Psi(t, x) = e^{-iHt/\hbar}\Psi(x),$$

that is with the spatial part separated by the time evolution. In this way, the amplitude of the wave is described by its spatial part. This form of solution can represent a particle as the reader will understand in Lecture 32. We can replace this form into the Schrödinger time-dependent equation

$$i\hbar\frac{\partial\Psi}{\partial t}(t, x) = -\frac{\hbar^2}{2m}\frac{\partial^2\Psi}{\partial x^2}(t, x) + V(t, x)\Psi(t, x)$$

corresponding to the potential which generates the restoring force of the classical harmonic oscillator, that is

$$V(t, x) = V(x) = \frac{1}{2}m\omega^2 x^2.$$

Developing the computations, we have

$$i\hbar\frac{-iH}{\hbar}e^{-iHt/\hbar}\Psi(x) = -\frac{\hbar^2}{2m}e^{-iHt/\hbar}\frac{d^2\Psi}{dx^2}(x) + V(x)e^{-iHt/\hbar}\Psi(x).$$

It results

$$-\frac{\hbar^2}{2m}\frac{d^2\Psi}{dx^2}(x) + \frac{1}{2}m\omega^2 x^2\Psi(x) = H\Psi(x).$$

If we denote $u = \sqrt{\frac{m\omega}{\hbar}}x$ and $\varepsilon = \frac{2H}{\hbar\omega}$ we obtain the equation in the variable u, that is

$$\frac{d^2\Psi}{du^2}(u) + (\varepsilon - u^2)\Psi(u) = 0.$$

Let us anticipate the result. We will obtain solutions if $\varepsilon = 2n + 1$, $n \in \mathbb{N}$. They are

$$\Psi_n(u) = A_n H_n(u)e^{-u^2/2},$$

where $H_n(u)$ are the Hermite polynomials and A_n is a constant we find according to the rule

$$\int_{-\infty}^{\infty} |\Psi_n(u)|^2 du = 1$$

considered also in the previous examples. Why do we need Hermite polynomials to describe the stationary solutions? The next lecture will tell us more about the mathematics behind the harmonic oscillator. Is this mathematics also involved in finding A_n? Yes, we will prove that it is. Since

$$\varepsilon = \varepsilon_n = 2n + 1 = \frac{2H}{\hbar\omega}$$

we find that the total energy H is quantized and depends on n, being in fact H_n. We have to point out that there is a notation risk with respect to Hermite polynomials. Therefore, we denote, in this example, the total energy by the letter E, and we obtain the n-level quantized energy as

$$E_n = \left(n + \frac{1}{2}\right)\hbar\omega.$$

If we return to variable x, after all computations, we find

$$\Psi_n(x) = \frac{1}{\sqrt{2^n n!}} \left(\frac{m\omega}{\pi\hbar}\right)^{1/4} H_n\left(\sqrt{\frac{m\omega}{\hbar}}x\right) e^{-m\omega x^2/2\hbar}$$

and

$$\Psi_n(t, x) = e^{-iE_n t/\hbar} \frac{1}{\sqrt{2^n n!}} \left(\frac{m\omega}{\pi\hbar}\right)^{1/4} H_n\left(\sqrt{\frac{m\omega}{\hbar}}x\right) e^{-m\omega x^2/2\hbar}.$$

Now, the last step. Since the Schrödinger equation is linear, the general solution is described by the sum of all modes of n-oscillations, that is

$$\Psi(t, x) = \sum_n c_n e^{-iE_n t/\hbar} \Psi_n(x).$$

The meaning of this sum is related to the mathematics of Hermite polynomials, used to describe the harmonic oscillator. We develop in details these considerations in the next chapter.

Let us now define the *Hermite polynomials* in view of their forthcoming applications. The first two polynomials are $H_0(u) = 1$; $H_1(u) = 2u$ and the recurrence relation is

$$H_{n+1}(u) = 2u H_n(u) - 2n H_{n-1}(u).$$

Therefore $H_2(u) = 4u^2 - 2$, $H_3(u) = 8u^3 - 12u$, etc. Let us first observe that

$$\Psi_0(u) = A_0 H_0(u)e^{-u^2/2} = A_0 e^{-u^2/2}$$

verifies the equation

$$\frac{d^2\Psi}{du^2}(u) + (1 - u^2)\Psi(u) = 0.$$

The constant does not play a role in the equation. However the constant A_0 is determined from the

$$\int_{-\infty}^{\infty} |\Psi_0(u)|^2 du = A_0^2 \int_{-\infty}^{\infty} e^{-u^2} du = 1$$

as we did it in the previous example. As we will see below, this result has a prominent role in many concepts of Quantum Mechanics. In a forthcoming lecture, when we study the Gauss wave packets, we will find out the value of this integral as $\sqrt{\pi}$, therefore $A_0 = \frac{1}{\sqrt[4]{\pi}}$. Then, let us show that

$$\Psi_1(u) = A_1 H_1(u)e^{-u^2/2} = 2A_1 u e^{-u^2/2}$$

verifies the equation

$$\frac{d^2\Psi}{du^2}(u) + (3 - u^2)\Psi(u) = 0.$$

In this case, we give the details of calculations. As before, we can cancel the constant A_1 and check the solution:

$$\frac{d^2}{du^2}\left(2ue^{-u^2/2}\right) + (3 - u^2)2ue^{-u^2/2} = \frac{d}{du}\left(2e^{-u^2/2} - 2u^2e^{-u^2/2}\right) + (3 - u^2)2ue^{-u^2/2} =$$

$$= \left(-2u - 4u - 2u^3 + 2u(3 - u^2)\right)e^{-u^2/2} = 0.$$

The constant A_1 is computed by the same condition

$$\int_{-\infty}^{\infty} |\Psi_1(u)|^2 du = 1,$$

where

$$4A_1^2 \int_{-\infty}^{\infty} u^2 e^{-u^2} du = 1.$$

We do not intend to continue the determination of any A_n here. In the next lectures, we will offer a general solution for the form of these coefficients.

We show that Ψ_n, corresponding to the n-Hermite polynomial, i.e.

$$\Psi_n(u) = A_n H_n(u)e^{-u^2/2},$$

verifies the equation

$$\frac{d^2\Psi_n}{du^2}(u) + (2n + 1 - u^2)\Psi_n(u) = 0.$$

To check this, we act by mathematical induction in the following way: Let us suppose that $\Psi_{n-1}(u) = A_{n-1}H_{n-1}(u)e^{-u^2/2}$ and $\Psi_{n-2}(u) = A_{n-2}H_{n-2}(u)e^{-u^2/2}$ verify the equations

$$\frac{d^2\Psi_{n-1}}{du^2}(u) + (2n - 1 - u^2)\Psi_{n-1}(u) = 0,$$

$$\frac{d^2\Psi_{n-2}}{du^2}(u) + (2n - 3 - u^2)\Psi_{n-2}(u) = 0.$$

Now we perform the computations. Let us denote $H'(u) = \frac{dH}{du}(u)$ and $H''(u) = \frac{d^2H}{du^2}(u)$.

The last two relations, after canceling the constant and writing without the variable u, are

$$H_{n-1}'' - 2u H_{n-1}' + 2(n-1) H_{n-1} = 0,$$

$$H_{n-2}'' - 2u H_{n-2}' + 2(n-2) H_{n-2} = 0.$$

Supposing true this two relations, it remains to prove

$$H_n'' - 2u H_n' + 2n H_n = 0$$

which correspond to the equation

$$\frac{d^2 \Psi_n}{du^2}(u) + (2n + 1 - u^2) \Psi_n(u) = 0$$

for

$$\Psi_n(u) = A_n H_n(u) e^{-u^2/2}.$$

Let us start by deriving two times the basic recurrence of Hermite polynomials

$$H_n = 2u H_{n-1} - 2(n-1) H_{n-2}.$$

It results

$$H_n' = 2 H_{n-1} + 2u H_{n-1}' - 2(n-1) H_{n-2}'$$

and

$$H_n'' = 4 H_{n-1}' + 2u 2 H_{n-1}'' - 2(n-1) 2 H_{n-2}''.$$

Replacing all these three equalities in the equation

$$H_n'' - 2u H_n' + 2n H_n = 0$$

which we have to verify, it results that we have to check

$$\left(4H_{n-1}' + 2u H_{n-1}'' - 2(n-1) H_{n-2}'' \right) - 2u \left(2 H_{n-1} + 2u H_{n-1}' - 2(n-1) H_{n-2}' \right) +$$

$$+ 2n \left(2u H_{n-1} - 2(n-1) H_{n-2} \right) = 0.$$

Separating the terms containing H_{n-1} and H_{n-2}, the last relation can be written in the form

$$2u \left[H_{n-1}'' - 2u H_{n-1}' + 2(n-1) H_{n-1} \right] + 4 H_{n-1}' -$$

$$-2(n-1) \left[H_{n-2}'' - 2u H_{n-2}' + 2(n-2) H_{n-2} \right] - 8(n-1) H_{n-2} = 0.$$

The terms in squared brackets are null according to our induction hypothesis, therefore we have to prove the equality

$$H'_{n-1} - 2(n-1)H_{n-2} = 0.$$

At the same time, let us observe that the following two equalities are consequences of our induction hypothesis, that is

$$H'_{n-2} - 2(n-2)H_{n-3} = 0$$

and

$$H'_{n-3} - 2(n-3)H_{n-4} = 0.$$

Replacing them into the relation resulted from the recurrence

$$H'_{n-1} = 2H_{n-2} + 2uH'_{n-2} - 2(n-2)H'_{n-3},$$

we obtain

$$H'_{n-1} = 2H_{n-2} + 4(n-2)uH_{n-3} - 2(n-2)(n-3)H_{n-4} = 2(n-1)H_{n-2},$$

which had to be proved. The replacements are easy to do. It remains to trace the Mathematics behind the Hermite polynomials and to point out their deep physical meaning.

Summary of Lecture 25. In the quantum world, we found out that an electron can be seen as a standing wave which vibrates. We intend to use the Schrödinger time-dependent equation to describe the values of its quantized energy. To do this, the electron must be described by a wave function in the form

$$\Psi(t, x) = e^{-iEt/\hbar}\Psi(x).$$

We used the letter E instead of H to describe the total energy because the solution Ψ_n corresponding to the energy level E_n is related to the Hermite polynomials which are denoted by H_n in their recurrence relation,

$$H_{n+1}(u) = 2uH_n(u) - 2nH_{n-1}(u)$$

with $H_0(u) = 1$; $H_1(u) = 2u$. Replacing in the Schrödinger time-dependent equation

$$i\hbar\frac{d\Psi}{dt}(t, x) = -\frac{\hbar^2}{2m}\frac{d^2\Psi}{dx^2}(t, x) + V(t, x)\Psi(t, x)$$

corresponding to the case when $V(t, x) = V(x) = \frac{1}{2}m\omega^2 x^2$, that is when the vibration is modeled by a harmonic oscillator, we obtain

$$-\frac{\hbar^2}{2m}\frac{d^2\Psi}{dx^2}(x) + \frac{1}{2}m\omega^2 x^2 \Psi(x) = E\Psi(x).$$

If we denote $u = \sqrt{\frac{m\omega}{\hbar}}x$ and $\varepsilon = \frac{2E}{\hbar\omega}$, we obtain the equation in the variable u, that is

$$\frac{d^2\Psi}{du^2}(u) + (\varepsilon - u^2)\Psi(u) = 0.$$

This equation can be solved if $\varepsilon = \varepsilon_n = 2n + 1$. Therefore we obtain the n-level quantized energy as

$$E_n = \left(n + \frac{1}{2}\right)\hbar\omega$$

for

$$\Psi_n(x) = \frac{1}{\sqrt{2^n n!}}\left(\frac{m\omega}{\pi\hbar}\right)^{1/4} H_n\left(\sqrt{\frac{m\omega}{\hbar}}x\right)e^{-m\omega x^2/2\hbar}$$

and

$$\Psi_n(t, x) = e^{-iE_n t/\hbar}\frac{1}{\sqrt{2^n n!}}\left(\frac{m\omega}{\pi\hbar}\right)^{1/4} H_n\left(\sqrt{\frac{m\omega}{\hbar}}x\right)e^{-m\omega x^2/2\hbar}.$$

Now, the last step: since the Schrödinger equation is linear, the general solution is described by a sum of all n-oscillation modes,

$$\Psi(t, x) = \sum_n c_n e^{-iE_n t/\hbar}\Psi_n(x).$$

Chapter 6
The Mathematics Behind the Harmonic Oscillator

If the facts don't fit the theory,
change the facts!

Albert Einstein

6.1 Lecture 26: The Hermite Polynomials

Originally, the so-called Hermite polynomials were introduced by Pierre-Simon de Laplace in 1810 and studied with respect to different applications both by Pafnuty Chebyshev and Charles Hermite. They were named after Hermite because his works, in 1865, were more visible than those of the others. We start from *Hermite's differential equation*

$$\frac{d^2y}{dx^2} - 2x\frac{dy}{dx} + 2ny = 0$$

which is a second-order differential equation in $y = y(x)$. Let us try to find a solution in the polynomial form

$$y(x) = \sum_{m=0} B_m x^{k-m}.$$

We have

$$\frac{dy}{dx} = \sum_{m=0}(k-m)B_m x^{k-m-1}$$

and

$$\frac{d^2y}{dx^2} = \sum_{m=0}(k-m)(k-m-1)B_m x^{k-m-2}.$$

If we replace in Hermite's equation, we obtain

$$\sum_{m=0}(k-m)(k-m-1)B_m x^{k-m-2} - 2\sum_{m=0}(k-m)B_m x^{k-m} + 2n\sum_{m=0}B_m x^{k-m} = 0$$

which can be written in the simplified form

$$2\sum_{m=0}(k-m-n)B_m x^{k-m} - \sum_{m=0}(k-m)(k-m-1)B_m x^{k-m-2} = 0.$$

Let us look at the coefficient of x^k, that is if $m = 0$. It results

$$2(k-n)B_0 = 0,$$

that is $B_0 \neq 0$ if $k = n$. We keep in mind the possibility to have a nonzero coefficient B_0, that is $k = n$. In the same way, the coefficient of x^{k-1}, which is obtained for $m = 1$, leads to

$$2(k-1-n)B_1 = 0,$$

that is $B_1 = 0$, under the assumption $k = n$. If we look at the coefficient of x^{k-m}, we obtain the relation

$$2(k-m-n)B_m - (k-m+2)(k-m+1)B_{m-2} = 0,$$

that is

$$B_m = \frac{(k-m+2)(k-m+1)}{2(k-m-n)}B_{m-2}.$$

Therefore $B_1 = B_3 = B_5 = \ldots$ are all 0 if $k = n$ and, in this $k = n$ case, for even numbers we have nonzero B_{2n}. Indeed, from

$$B_m = -\frac{(n-m+2)(n-m+1)}{2m}B_{m-2}$$

we obtain

$$B_2 = -\frac{n(n-1)}{2 \cdot 2}B_0,$$

$$B_4 = \frac{n(n-1)(n-2)(n-3)}{2^2 \cdot 2 \cdot 4}B_0,$$

etc.,

$$y(x) = B_0 x^n + B_2 x^{n-2} + B_4 x^{n-4} + \ldots = B_0\left[x^n - \frac{n(n-1)}{2 \cdot 2}x^{n-2} + \frac{n(n-1)(n-2)(n-3)}{2^2 \cdot 2 \cdot 4}x^{n-4} - \ldots\right]$$

If we choose the constant $B_0 = 2^n$ the solution becomes the Hermite polynomial of degree n

$$H_n(x) = 2^n \left[x^n - \frac{n(n-1)}{2 \cdot 2} x^{n-2} + \frac{n(n-1)(n-2)(n-3)}{2^2 \cdot 2 \cdot 4} x^{n-4} - .. \right].$$

Let us observe that, if n is even, the Hermite polynomial has a constant term, if not, the last term has degree 1. Therefore if $n - 2r \geq 0$ in the two previous cases, we can write

$$H_n(x) = \sum_{r=0}^{\left[\frac{n}{2}\right]} \frac{(-1)^r}{r!} \frac{n!}{(n-2r)!} (2x)^{n-2r}.$$

Exercise 6.1.1 *Prove that $H_n(-x) = (-1)^n H_n(x)$.*

To advance in the theory let us prove the *Rodrigues formula*.

Exercise 6.1.2 *Prove*

$$e^{2tx-t^2} = \sum_{n=0}^{\infty} \frac{t^n}{n!} H_n(x).$$

Hint. We can use

$$e^x = \sum_{n=0}^{\infty} \frac{x^n}{n!}.$$

$$e^{2tx-t^2} = e^{2tx} e^{-t^2} = \sum_{r=0}^{\infty} \frac{(-t^2)^r}{r!} \sum_{s=0}^{\infty} \frac{(2tx)^s}{s!} = \sum_{r=0}^{\infty} \sum_{s=0}^{\infty} (-1)^r \frac{t^{s+2r}}{r!s!} (2x)^s.$$

Now, for $n := s + 2r$, we can write

$$e^{2tx-t^2} = \sum_{s=0}^{\infty} \left(\sum_{r=0}^{\left[\frac{n}{2}\right]} \frac{(-1)^r}{r!(n-2r)!} (2x)^{n-2r} \right) \frac{t^n}{n!} = \sum_{n=0}^{\infty} \frac{t^n}{n!} H_n(x).$$

The next exercise is part of the mathematical language of Quantum Mechanics. The road we are establishing, step by step, in this book started from Classical Mechanics seen in its different formulations. Lagrangian Mechanics is used when Newton's formulation, especially in dynamics, is not convenient. Hamiltonian Mechanics is more implied in establishing a language for Quantum Mechanics. As we have seen, the Schrödinger equation is based on it. The electromagnetic waves, as light, derived through Maxwell's equations led us to Special Relativity where these equations are invariant via the Lorentz transformations, while the Galilei transformations of Classical Mechanics do not preserve their form. The various experiments, that led to the paradigm shift on what we called elementary particles, also led to the development and adaption of the mathematical apparatus necessary for their description.

Planar waves needed complex numbers in their representation, d'Alembert operator in describing wave equations being now replaced by the Schrödinger equation to describe planar waves evolution. The last examples show us that the wave function Ψ could have another interpretation. This will be clear later, after we study the Gauss wave packets and we see that the momentum and the position of an electron cannot be known simultaneously. We will identify operators, eigenfunctions, eigenvectors and structures as Hilbert spaces where another viewpoint on Quantum Mechanics is offered. All these concepts and results can be related to the Hermite polynomials. Therefore, the next exercise will offer to the reader two perspectives. One is related to the number we wrote in the front of the wave function Ψ_n. The other one is related to the orthogonal basis in the Hilbert space as we will highlight in our next lectures.

Exercise 6.1.3 *Prove that*

$$\int_{-\infty}^{\infty} H_n(x) H_m(x) e^{-x^2} dx = \begin{cases} 0, & m \neq n \\ 2^n \sqrt{\pi}\, n!, & m = n. \end{cases}$$

Hint. We have

$$e^{2tx-t^2} = \sum_{n=0}^{\infty} \frac{t^n}{n!} H_n(x); \quad e^{2sx-s^2} = \sum_{m=0}^{\infty} \frac{s^m}{m!} H_m(x).$$

It results

$$e^{2tx-t^2+2sx-s^2} = \sum_{n=0}^{\infty} \sum_{m=0}^{+\infty} \frac{t^n s^m}{n!m!} H_n(x) H_m(x)$$

and

$$\int_{-\infty}^{\infty} e^{-x^2+2tx-t^2+2sx-s^2} dx = \sum_{n=0}^{\infty} \sum_{m=0}^{\infty} \frac{t^n s^m}{n!m!} \int_{-\infty}^{\infty} H_n(x) H_m(x) e^{-x^2} dx.$$

We continue with the left side denoted as LS,

$$LS = \int_{-\infty}^{\infty} e^{-[x^2+t^2+s^2-2tx-2sx+2ts]} e^{2ts} dx = e^{2ts} \int_{-\infty}^{\infty} e^{-[x-(t+s)^2]} dx.$$

Denoting $u := x - (t + s)$, we obtain

$$LS = e^{2ts} \int_{-\infty}^{\infty} e^{-u^2} du = e^{2ts} \sqrt{\pi},$$

where the last integral will be solved in Gauss wave packets lecture. We identify the corresponding coefficients of the following equality as

$$e^{2ts}\sqrt{\pi} = \sqrt{\pi}\left(1 + \frac{(2ts)}{1!} + \frac{(2ts)^2}{2!} + \frac{(2ts)^3}{3!} + ...\right) = \sum_{n=0}^{\infty}\sum_{m=0}^{\infty}\frac{t^n s^m}{n!m!}\int_{-\infty}^{\infty} H_n(x)H_m(x)e^{-x^2}dx$$

and we obtain the desired result.

In conclusion, we have obtained more results about the polynomials which appeared as solutions of the Hermite differential equation but we still do not know they are the polynomials we used to solve the quantum harmonic oscillator problem. Therefore we have to check that the solutions of differential equation satisfy the recurrence met there. There is one more step to do.

Exercise 6.1.4 *Prove the Rodrigues formula:*

$$H_n(x) = (-1)^n e^{x^2}\frac{d^n}{dx^n}\left(e^{-x^2}\right).$$

Hint. We consider the n-derivative with respect to t at t = 0 in the following formula

$$e^{2tx-t^2} = \sum_{n=0}^{\infty}\frac{t^n}{n!}H_n(x) = H_0(x) + \frac{t}{1!}H_1(x) + \frac{t^2}{2!}H_2(x) + ... + \frac{t^n}{n!}H_n(x) + \frac{t^{n+1}}{(n+1)!}H_{n+1}(x) + ...$$

and we observe that all terms containing powers of t greater than $n+1$ are 0, therefore we obtain

$$H_n(x) = \frac{\partial^n}{\partial t^n}\left[e^{2tx-t^2}\right]_{t=0} = e^{x^2}\frac{\partial^n}{\partial t^n}\left[e^{-x^2+2tx-t^2}\right]_{t=0} = e^{x^2}\frac{\partial^n}{\partial t^n}\left[e^{-(x-t)^2}\right]_{t=0}.$$

Now using $y := x - t$, we have both $y|_{t=0} = x$ and $\frac{\partial}{\partial t} = \frac{\partial}{\partial y}\frac{\partial y}{\partial t} = -\frac{\partial}{\partial y}$, therefore

$$H_n(x) = e^{x^2}(-1)^n\frac{\partial^n}{\partial y^n}\left[e^{-y^2}\right]_{t=0} = (-1)^n e^{x^2}\frac{d^n}{dx^n}\left(e^{-x^2}\right).$$

Let us observe that now, using the Rodrigues formula, we can compute $H_0(x) = 1$, $H_1(x) = 2x$, $H_2(x) = 4x^2 - 2$, etc. The next step is

Exercise 6.1.5 *Prove $H_n' = 2nH_{n-1}$.*
Hint. The derivative with respect to x of the equality

$$e^{2tx-t^2} = \sum_{n=0}^{+\infty}\frac{t^n}{n!}H_n(x)$$

leads to

$$2te^{2tx-t^2} = \sum_{n=0}^{+\infty}\frac{t^n}{n!}H_n'(x),$$

therefore

$$2t \sum_{n=0}^{+\infty} \frac{t^n}{n!} H_n(x) = \sum_{n=0}^{+\infty} \frac{t^n}{n!} H_n'(x).$$

Comparing the coefficients of t^n in both members, we obtain

$$\frac{2H_{n-1}(x)}{(n-1)!} = \frac{H_n'(x)}{n!},$$

that is the desired result.

Exercise 6.1.6 *Prove that the solutions of the Hermite differential equation satisfy the recurrence relation $H_{n+1}(x) = 2x H_n(x) - 2n H_{n-1}(x)$ starting from the derivative with respect t of the relation*

$$e^{2tx-t^2} = \sum_{n=0}^{+\infty} \frac{t^n}{n!} H_n(x).$$

Hint. We have

$$2(x - t)e^{2tx-t^2} = \sum_{n=0}^{+\infty} \frac{t^{n-1}}{(n-1)!} H_n(x),$$

i.e.

$$2(x - t) \sum_{n=0}^{+\infty} \frac{t^n}{n!} H_n(x) = \sum_{n=0}^{+\infty} \frac{t^{n-1}}{(n-1)!} H_n(x).$$

If we look at the coefficients of t^n in both members we obtain

$$\frac{1}{n!} 2x H_n(x) - \frac{2H_{n-1}(x)}{(n-1)!} = \frac{1}{n!} H_{n+1}(x),$$

that is the desired relation.

Let us make another observation. The Hermite differential equation

$$\frac{d^2 y}{dx^2} - 2x \frac{dy}{dx} + 2ny = 0$$

can be written as

$$\frac{d}{dx}\left(e^{-x^2} \frac{dy}{dx}\right) + 2ne^{-x^2} y = 0$$

being, in this way, related with another important equation,

$$\frac{d}{dx}\left(e^{-x^2/2} \frac{dy}{dx}\right) + ne^{-x^2/2} y = 0,$$

called the *probabilistic Hermite equation*. It is easy to see that its solutions are Hermite polynomials with changed coefficients, i.e. $H^0(x) = 1$, then $H^1(x) = x$ instead of $H_1(x) = 2x$, $H^2(x) = x^2 - 1$ instead $H_2(x) = 4x^2 - 2$, etc., because their differential equation

$$\frac{d^2 y}{dx^2} - x\frac{dy}{dx} + ny = 0$$

is similar to the Hermite one and its general solution expressed in terms of Rodrigues formula is

$$H^n(x) = (-1)^n e^{-x^2/2} \frac{d^n}{dx^n}\left(e^{-x^2/2}\right).$$

More discussion about the mathematics behind Hermite polynomials will be given in the next lecture.

Summary of Lecture 26. Starting from Hermite's differential equation

$$\frac{d^2 y}{dx^2} - 2x\frac{dy}{dx} + 2ny = 0$$

we find the solutions $y(x)$ as depending on n and having a polynomial form

$$H_n(x) = \sum_{r=0}^{\left[\frac{n}{2}\right]} \frac{(-1)^r}{r!}\frac{n!}{(n-2r)!}(2x)^{n-2r}.$$

Step by step, we discover the properties of these solutions

1. $H_n(-x) = (-1)^n H_n(x)$

2. $e^{2tx - t^2} = \sum_{n=0}^{+\infty} \frac{t^n}{n!}H_n(x)$

3. $\int_{-\infty}^{\infty} H_n(x)H_m(x)e^{-x^2}dx = \begin{cases} 0, & m \neq n \\ 2^n\sqrt{\pi}\,n!, & m = n \end{cases}$

4. $H_n' = 2nH_{n-1}$

until we find the property used in the quantum description of harmonic oscillator, that is

5. $H_{n+1}(x) = 2xH_n(x) - 2nH_{n-1}(x)$.

In the same description, we have the wave functions

$$\Psi_n(x) = A_n H_n(x) e^{-x^2/2}.$$

Looking at the third property, it is possible to derive the orthogonality for such wave functions, i.e.

$$\int_{-\infty}^{\infty} \Psi_n(x)\Psi_m(x)e^{-x^2}dx = A_n A_m \int_{-\infty}^{\infty} H_n(x)H_m(x)e^{-x^2}dx = \begin{cases} 0, & m \neq n \\ 2^n \sqrt{\pi}\, n! A_n^2, & m = n. \end{cases}$$

According to previous results, we can impose $2^n \sqrt{\pi}\, n! A_n^2 = 1$, that is

$$A_n = \frac{1}{\sqrt[4]{\pi}\sqrt{2^n\, n!}}.$$

Now, since

$$\Psi_n(t, x) = e^{-iEt/\hbar}\Psi_n(x),$$

the previous formula can be written as

$$\int_{-\infty}^{\infty} \Psi_n(t, x)\Psi_m^*(t, x)dx = \begin{cases} 0, & m \neq n \\ 1, & m = n, \end{cases}$$

where $\Psi_m^*(t, x)$ is the complex conjugate of $\Psi_m(t, x)$, therefore the orthogonality is extended to another type of wave functions.

6.2 Lecture 27: Real and Complex Vector Structures

Let us start from the final result of the previous summary:

$$\int_{-\infty}^{\infty} \Psi_n(t, x)\Psi_m^*(t, x)dx = \begin{cases} 0, & m \neq n \\ 1, & m = n. \end{cases}$$

It comes from

$$\int_{-\infty}^{\infty} \Psi_n(x)\Psi_m(x)e^{-x^2}dx = A_n A_m \int_{-\infty}^{\infty} H_n(x)H_m(x)e^{-x^2}dx = \begin{cases} 0, & m \neq n \\ 1, & m = n. \end{cases}$$

if

$$\Psi_n(t, x) = e^{-iEt/\hbar}\Psi_n(x),$$

where

$$\Psi_n(x) = \frac{1}{\sqrt[4]{\pi}\sqrt{2^n\, n!}} H_n(x) e^{-x^2/2}.$$

What sets and structures can we highlight here?

6.2.1 Finite Dimensional Real and Complex Vector Spaces, Inner Product, Norm, Distance, Completeness

Let us recall what an *inner product* for a vector space $\mathbb{V}$ over the field $\mathbb{R}$ is: by definition, the function $\langle , \rangle : \mathbb{V} \times \mathbb{V} \to \mathbb{R}$ is an inner product, if $\langle , \rangle$ is a bilinear real function, i.e.

$$\langle \alpha_1 u_1 + \alpha_2 u_2, v \rangle = \alpha_1 \langle u_1, v \rangle + \alpha_2 \langle u_2, v \rangle, \quad \forall\, u_1, u_2, v \in \mathbb{V}, \ \alpha_1, \alpha_2 \in \mathbb{R}$$

$$\langle u, \beta_1 v_1 + \beta_2 v_2 \rangle = \beta_1 \langle u, v_1 \rangle + \beta_2 \langle u, v_2 \rangle, \quad \forall\, v_1, v_2, u \in \mathbb{V}, \ \beta_1, \beta_2 \in \mathbb{R},$$

fulfilling the two properties:

$$\langle u, v \rangle = \langle v, u \rangle, \ \forall\, u, v \in \mathbb{V},$$

$$\langle u, u \rangle \geq 0 \ \forall\, u \in \mathbb{V}; \ \langle u, u \rangle = 0 \iff u = 0.$$

The first property is called symmetry, the second one, positivity.
If we work in a finite dimensional real vector space $\mathbb{V}$, endowed with an inner product $\langle , \rangle$ and we have a basis $e_1, e_2, e_3, ..., e_n$ such that

$$\langle e_m, e_n \rangle = \begin{cases} 0, & m \neq n \\ 1, & m = n \end{cases}$$

we call this basis an orthonormal one. Each vector $u \in \mathbb{V}$ can be written as a finite combination with real coefficients a_k,

$$u = \sum_{k=1}^{n} a_k e_k.$$

Considering

$$u = \sum_{k=1}^{n} a_k e_k, \quad v = \sum_{k=1}^{n} b_k e_k,$$

it is a simple exercise to see that the inner product with respect the orthonormal basis has the form

$$\langle u, v \rangle = \sum_{k=1}^{n} a_k b_k.$$

Now, if we work in a finite dimensional complex vector space $\mathbb{V}$ endowed with an inner product $\langle, \rangle$ and we have a basis $e_1, e_2, e_3, \ldots, e_n$ such that

$$\langle e_m, e_n \rangle = \begin{cases} 0, & m \neq n \\ 1, & m = n \end{cases}$$

we call as previously this basis an orthogonal one. Each vector $u \in \mathbb{V}$ can be written as a finite combination of complex coefficients a_k,

$$u = \sum_{k=1}^{n} a_k e_k.$$

For the complex number $z = x + iy$ we denote by z^* its complex conjugate, i.e. $z^* = x - iy$. By definition, in this complex vector space case, the inner product $\langle \cdot, \cdot \rangle : \mathbb{V} \times \mathbb{V} \to \mathbb{C}$ is a bilinear complex function, i.e.

$$\langle \alpha_1 u_1 + \alpha_2 u_2, v \rangle = \alpha_1^* \langle u_1, v \rangle + \alpha_2^* \langle u_2, v \rangle, \quad \forall u_1, u_2, v \in \mathbb{V}, \ \alpha_1, \alpha_2 \in \mathbb{C}$$

$$\langle u, \beta_1 v_1 + \beta_2 v_2 \rangle = \beta_1 \langle u, v_1 \rangle + \beta_2 \langle u, v_2 \rangle, \quad \forall v_1, v_2, u \in \mathbb{V}, \ \beta_1, \beta_2 \in \mathbb{C},$$

fulfilling two properties:

$$\langle v, u \rangle = \langle u, v \rangle^*, \ \forall u, v \in \mathbb{V},$$

$$\langle u, u \rangle \geq 0 \ \forall u \in \mathbb{V}; \ \langle u, u \rangle = 0 \iff u = 0.$$

Considering

$$u = \sum_{k=1}^{n} a_k e_k, \quad v = \sum_{k=1}^{n} b_k e_k,$$

it is a simple exercise to see that the inner product with respect the the orthonormal basis has the form

$$\langle u, v \rangle = \sum_{k=1}^{n} a_k^* b_k.$$

Let us think at the first example of vector spaces endowed with an inner product. It is $\mathbb{V} = \mathbb{R}^n$ over the field $\mathbb{R}$ endowed with the inner product

$$\langle u, v \rangle = \sum_{k=1}^{n} a_k b_k,$$

where the vectors u and v have the components a_k, respectively b_k. The Euclidean norm $|| \ ||$ and the Euclidean distance $d(,)$ are defined by the formulas

$$||u|| = \sqrt{\langle u, u \rangle},$$

$$d(u, v) = ||u - v|| = \sqrt{\langle u - v, u - v \rangle}.$$

Both the *Cauchy-Buniakowski-Schwarz inequality*

$$\langle u, v \rangle^2 \leq ||u||^2 ||v||^2$$

and the triangle inequality

$$d(u, v) \leq d(u, w) + d(w, v)$$

hold. The angle θ between the vectors u and v is defined by the formula

$$\cos \theta = \frac{\langle u, v \rangle}{||u|| \cdot ||v||}$$

which is equivalent to the generalized Pythagoras theorem.
We leave to the reader the demonstration of the following exercises:

Exercise 6.2.1 *Pythagoras theorem: If $\langle u, v \rangle = 0$ then $||u + v||^2 = ||u||^2 + ||v||^2$. Hint: Use $||u + v||^2 = \langle u + v, u + v \rangle$, the properties of inner product and the orthogonality.*

Exercise 6.2.2 *Parallelogram law: If $\langle u, v \rangle = 0$ then $||u + v||^2 + ||u - v||^2 = 2||u||^2 + 2||v||^2$. Hint: Use $||u + v||^2 = \langle u + v, u + v \rangle$, $||u - v||^2 = \langle u - v, u - v \rangle$ and the inner product properties.*

Each component of the Euclidean n-dimensional space is a real number and we know how to define the limit of a set of real numbers. If we think at the set of vectors $\{x_k\} \subset \mathbb{R}^n$, we say that the set converges to $x \in \mathbb{R}^n$ if

$$||x_k - x|| \to 0.$$

This means that each component of x_k converges to its correspondent component of x.

Let us now consider each component of a set of vectors x_k in $\mathbb{R}^n$ as a Cauchy set of real numbers.
Denote the set corresponding to one component, say the second one, of the set x_k by

$x_{k,2}$. That is, from the vector x_1, we select the component $x_{1,2}$, from the vector x_2, we select the component $x_{2,2}$, etc. The set $(x_{k,2})_{k\in\mathbb{N}}$ is a set of real numbers and suppose $(x_{k,2})_{k\in\mathbb{N}}$ is a Cauchy set: given $\varepsilon > 0$, there exists $n_0 \in \mathbb{N}$ such that $|x_{m,2} - x_{n,2}| < \varepsilon$ for all $m, n \geq n_0$. Each *Cauchy set of real numbers* is a convergent set in $\mathbb{R}$. This property means that $\mathbb{R}$ is complete. The vector formed by all these component limits, denoted by x, is the limit of the vectors sequence x_k, that is

$$||x_k - x|| \to 0.$$

This implies $\mathbb{R}^n$ is a complete space.

Summary of Lecture 27. In this lecture, we reviewed the basic facts related to finite dimensional real and complex vector spaces. All these vector spaces are examples of Hilbert spaces. Several aspects of our geometric intuition of the Hilbert structure is Euclidean. Some of these geometric aspects will be transferred from finite to infinite dimensional vector spaces.

The Pythagoras theorem:
"if $\langle u, v \rangle = 0$ then $||u + v||^2 = ||u||^2 + ||v||^2$"
and the Parallelogram law:
"if $\langle u, v \rangle = 0$ then $||u + v||^2 + ||u - v||^2 = 2||u||^2 + 2||v||^2$" are part of the Euclidean intuition.

We also look at the basic calculus facts written with respect to the norm induced by the inner product and we observe that these spaces are complete: any Cauchy sequence becomes a convergent sequence with respect the norm.

We are now prepared to introduce the concept of Hilbert space.

6.3   Lecture 28: Pre-Hilbert and Hilbert Spaces

In the development of Quantum Mechanics, the previous mathematical notations were replaced by the Dirac notation which allows a better description of operators and their actions. A straightforward description, in a finite dimensional complex vector space of *Dirac notation*, is offered in Lecture 44 before presenting its utility when we study the light polarization.

For a vector space $\mathbb{V}$ over the field of complex numbers $\mathbb{C}$, the elements are denoted by $|u\rangle$ instead of u and they are named ket vectors. For any two ket vectors, the addition is made by the rule

$$|u\rangle + |v\rangle = |r\rangle$$

and the result is a ket vector, here denoted by $|r\rangle \in \mathbb{V}$. The summation of ket vectors must determine a commutative group structure $(\mathbb{V}, +)$, that is the following properties hold:

$$(|u\rangle + |v\rangle) + |w\rangle = |u\rangle + (|v\rangle + |w\rangle) \ \text{(associativity)}$$

$$|u\rangle + |0\rangle = |0\rangle + |u\rangle = |u\rangle \ \text{(existence of zero vector for any } |u\rangle \in \mathbb{V})$$

$$\forall |u\rangle \in \mathbb{V} \ \exists | - u\rangle \in \mathbb{V} \text{ such that } |u\rangle + | - u\rangle = | - u\rangle + |u\rangle = |0\rangle \ \text{(the additive inverse)}$$

$$|u\rangle + |v\rangle = |v\rangle + |u\rangle \ \text{(commutativity)}$$

The multiplication by $\mathbb{C}$ scalars follows the same type of definition:

$$\forall |u\rangle \in \mathbb{V}, \ \forall a \in \mathbb{C} \ \exists \, a|u\rangle \in \mathbb{V}.$$

Multiplication by scalar has the properties

$$a(b|u\rangle) = (ab)|u\rangle$$

$$1 \cdot |u\rangle = |u\rangle$$

$$a(|u\rangle + |v\rangle) = a|u\rangle + a|v\rangle$$

$$(a + b)|u\rangle = a|u\rangle + b|u\rangle$$

Some consequences of the vector space structure exist and we left as exercises for readers the following statements

$$0 \cdot |u\rangle = |0\rangle$$

$$(-1) \cdot |u\rangle = | - u\rangle$$

$$|u\rangle - |v\rangle = |u\rangle + | - v\rangle \ .$$

Consider u and v as column vectors, that is vectors in the same space $\mathbb{V}$. When we write the inner product as a sum of components, we have

$$\langle u, v\rangle = (u_1^*, u_2^*, .., u_n^*) \begin{pmatrix} v_1 \\ v_2 \\ . \\ . \\ v_n \end{pmatrix} = \sum_{k=1}^{n} u_k^* v_k.$$

We need some definitions for these objects. A *ket vector* $|u\rangle$ is a column vector. Its transpose is denoted by $\langle u|$ and it is called a *bra vector*, and the components will be the complex conjugate components of $|u\rangle$. The inner product become a multiplication between a bra vector $\langle u|$ and a ket vector $|v\rangle$, that is a "bracket" $\langle u|v\rangle$. The result of a such an inner product can be a real or a complex number.

We use the convenient notations $|u\rangle + |v\rangle = |u + v\rangle$; $a|u\rangle = |au\rangle$ and z^* for the complex conjugate of z.

The intuitive explanation of the above formulas allow us to define the formal inner product for two abstract ket vectors $|u\rangle$ and $|v\rangle$ in a vector space with an infinite number of dimensions through the formula $\langle u|v\rangle$, that is $\langle \cdot | \cdot \rangle : \mathbb{V} \times \mathbb{V} \to \mathbb{C}$ fulfilling the conditions

$$\langle \alpha_1 u_1 + \alpha_2 u_2 | v \rangle = \alpha_1^* \langle u_1 | v \rangle + \alpha_2^* \langle u_2 | v \rangle, \quad \forall |u_1\rangle, |u_2\rangle, |v\rangle \in \mathbb{V}, \ \alpha_1, \alpha_2 \in \mathbb{C}$$

$$\langle u | \beta_1 v_1 + \beta_2 v_2 \rangle = \beta_1 \langle u | v_1 \rangle + \beta_2 \langle u | v_2 \rangle, \quad \forall |v_1\rangle, |v_2\rangle, |u\rangle \in \mathbb{V}, \ \beta_1, \beta_2 \in \mathbb{C},$$

$$\langle v | u \rangle = \langle u | v \rangle^*, \ \forall |u\rangle, |v\rangle \in \mathbb{V},$$

$$\langle u | u \rangle \geq 0 \ \forall |u\rangle \in \mathbb{V}; \ \langle u, u \rangle = 0 \iff |u\rangle = |0\rangle.$$

The inner product allows to define a norm $|| \ ||$ and a distance $d(,)$:

$$||u|| = \sqrt{\langle u | u \rangle},$$

$$d(u, v) = ||u - v|| = \sqrt{\langle u - v | u - v \rangle}.$$

By definition, a norm is a function from a vector space $\mathbb{V}$ on $\mathbb{R}$ satisfying the following conditions
$$||u|| = 0 \ \text{implies} \ u = 0,$$

$$||\alpha u|| = |\alpha| ||u|| \ \text{for} \ u \in \mathbb{V}, \ \alpha \in \mathbb{C},$$

$$||u + v|| \leq ||u|| + ||v|| \ \text{for every} \ u, v \in \mathbb{V}.$$

The formal definition of the distance is: A function from $\mathbb{V} \times \mathbb{V}$ to $\mathbb{R}_+$ such that the following three axioms are fulfilled

$$d(u, v) = 0 \ \equiv u = v,$$

$$d(u, v) = d(v, u) \ \text{for all} \ u, v \in \mathbb{V},$$

$$d(u, v) \leq d(u, w) + d(w, v) \ \text{for all} \ u, v, w \in \mathbb{V}.$$

$(|x_n\rangle)_{n \in \mathbb{N}}$ is a Cauchy sequence in $\mathbb{V}$ if, given any $\varepsilon > 0$, there exists a natural number n_0 such that
$$|| \ |x_m - x_n\rangle \ || < \varepsilon$$

for all $m, n \in \mathbb{N}, \ m, n \geq n_0$.

A set $(|x_n\rangle)_{n\in\mathbb{N}}$ is called "convergent" in the norm of $\mathbb{V}$ to $x \in \mathbb{V}$, if, given any $\varepsilon > 0$, it exists $n_\varepsilon \in \mathbb{N}$ such that $||\,|x_n - x\rangle\,|| < \varepsilon$ for all $n \geq n_\varepsilon$.

The vector space $\mathbb{V}$ is called "complete" if any Cauchy sequence, in the norm of $\mathbb{V}$, is convergent in the norm of $\mathbb{V}$. Therefore we can formally define the Hilbert space as we did before when we discussed the Hilbert structure of the n-dimensional Euclidean space.

A pre-Hilbert space is a formal vector space $\mathbb{V}$ over the field of complex numbers $\mathbb{C}$ endowed with an abstract inner product $\langle\cdot|\cdot\rangle$. A Hilbert space is a complete pre-Hilbert space.

Again, the Cauchy-Buniakowski-Schwarz inequality

$$\langle u, v\rangle^2 \leq ||u||^2 ||v||^2$$

and the triangle inequality

$$d(u, v) \leq d(u, w) + d(w, v)$$

hold.

Summary of Lecture 28. We introduced the definition of the Hilbert space. First we considered the definition of a vector space in Dirac notation: a set endowed with two operations, an internal one denoted by $+$ and an external operation called multiplication by scalars, here in $\mathbb{C}$ such that the following axioms hold

$$(|u\rangle + |v\rangle) + |w\rangle = |u\rangle + (|v\rangle + |w\rangle) \ \text{(associativity)}$$

$$|u\rangle + |0\rangle = |0\rangle + |u\rangle = |u\rangle \ \text{(existence of zero vector for any } |u\rangle \in \mathbb{V})$$

$$\forall |u\rangle \in \mathbb{V} \ \exists |-u\rangle \in \mathbb{V} \text{such that} |u\rangle + |-u\rangle = |-u\rangle + |u\rangle = |0\rangle \ \text{(the additive inverse)}$$

$$|u\rangle + |v\rangle = |v\rangle + |u\rangle \ \text{(commutativity)}$$

$$a(b|u\rangle) = (ab)|u\rangle$$

$$1 \cdot |u\rangle = |u\rangle$$

$$(a(|u\rangle + |v\rangle)) = a|u\rangle + a|v\rangle$$

$$(a + b)|u\rangle = a|u\rangle + b|u\rangle$$

The inner product is a complex valued function having the following properties

$$\langle u|\beta_1 v_1 + \beta_2 v_2\rangle = \beta_1\langle u|v_1\rangle + \beta_2\langle u|v_2\rangle, \quad \forall \, |v_1\rangle, |v_2\rangle, |u\rangle \in \mathbb{V}, \; \beta_1, \beta_2 \in \mathbb{C},$$

$$\langle v|u\rangle = \langle u|v\rangle^*, \; \forall \, |u\rangle, |v\rangle \in \mathbb{V},$$

$$\langle u|u\rangle \geq 0 \; \forall \, |u\rangle \in \mathbb{V}; \; \langle u, u\rangle = 0 \iff |u\rangle = |0\rangle.$$

The inner product allows to define a norm $|| \; ||$ and a distance $d(,)$:

$$||u|| = \sqrt{\langle u|u\rangle},$$

$$d(u, v) = ||u - v|| = \sqrt{\langle u - v|u - v\rangle}.$$

$(|x_n\rangle)_{n\in\mathbb{N}}$ is a Cauchy sequence in $\mathbb{V}$ if, given any $\varepsilon > 0$, there exists a natural number n_0 such that

$$|| \, |x_m - x_n\rangle \, || < \varepsilon$$

for all $m, n \in \mathbb{N}, \; m, n \geq n_0$.

A set $(|x_n\rangle)_{n\in\mathbb{N}}$ is called convergent in the norm of $\mathbb{V}$ to $x \in \mathbb{V}$ if, given any $\varepsilon > 0$, it exists $n_\varepsilon \in \mathbb{N}$ such that $|| \, |x_n - x\rangle \, || < \varepsilon$ for all $n \geq n_\varepsilon$.

The vector space $\mathbb{V}$ is called complete if any Cauchy sequence is convergent in the norm of $\mathbb{V}$. Therefore we can formally define the Hilbert space as we did before when we discussed the Hilbert structure of the n-dimensional Euclidean space.

A pre-Hilbert space is a formal vector space $\mathbb{V}$ over the field of complex numbers $\mathbb{C}$ endowed with an abstract inner product $\langle\cdot|\cdot\rangle$. A Hilbert space is a complete pre-Hilbert space.

6.4 Lecture 29: Examples of Hilbert Spaces

The next four exercises are needed in our mathematical construction to allow the presentation of two important infinite dimensional examples of Hilbert spaces. The vectors are written in bra notation. It is important to mention that, if a given vector in ket notation $|x\rangle$ has the components $x_k \in \mathbb{C}$ and it is represented as a column vector, the corresponding bra vector, $\langle x|$, is written as a row vector with complex conjugate elements

$$\langle x| = (x_1^*, x_2^*, \dots\dots).$$

Exercise 6.4.1 *Consider the set denoted by* l^2 *of all sequences* $\langle a| = (a_1, a_2, ...,$ $a_n, ...)$, $a_k \in \mathbb{C}$, $k \in \mathbb{N} - \{0\}$ *such that*

$$\sum_n |a_k|^2 < \infty,$$

where $|a_k|^2 = a_k a_k^*$. *Consider two operations which generalize the operations for finite dimensional vectors,*

$$\langle a| + \langle b| := (a_1 + b_1, a_2 + b_2, ..., a_n + b_n, ...),$$

and

$$\alpha \langle a| := (\alpha a_1, \alpha a_2, ..., \alpha a_n, ...),$$

which define $\langle a + b|$ *and* $\langle \alpha a|$ *respectively. Show that* l^2 *endowed by the two operations is a vector space over the field* $\mathbb{C}$.
Hint. Verify all properties of operations presented before in the case of ket vectors notation.
However, we have to explain why, for two bra vectors $\langle a|$ *and* $\langle b|$ *in* l^2, *we have* $\langle a| + \langle b|$ *as a bra vector of* l^2. *The explanation is the Minkowski inequality*

$$\sqrt{\sum_n |a_n + b_n|^2} \leq \sqrt{\sum_n |a_n|^2} + \sqrt{\sum_n |b_n|^2}$$

which can be easily proved.

Exercise 6.4.2 *For two bra vectors* $\langle a|$ *and* $\langle b|$, *let us define*

$$\langle a|b \rangle := \sum_n a_n^* b_n.$$

Show that the previous formula is an inner product in the sense of the above definition. First, we have to understand why the previous formula which defined the inner product makes sense. We have the Cauchy-Buniakowski-Schwarz inequality involved

$$\left(\sum_n |a_n^* b_n| \right)^2 \leq \sum_n |a_n^*|^2 \sum_n |b_n|^2.$$

It remains to the reader checking the other properties of the inner product for the proposed formula. Until now we have proved that l^2 *is a pre-Hilbert space.*

Exercise 6.4.3 *Prove that the set of vectors*

$$\langle e_1| = (1, 0, 0, ..., 0, ...)$$

$$\langle e_2| = (0, 1, 0, ..., 0, ...)$$

$$\langle e_3| = (0, 0, 1, ..., 0, ...)$$

$$..............................$$

$$\langle e_n| = (0, 0, 0, ..., 1, ...)$$

$$..............................$$

satisfies

$$\langle e_k|e_m\rangle := \delta_{kn},$$

i.e. is an orthonormal basis for l^2. Hint. Use the definition of the inner product seen above.

Exercise 6.4.4 *The completeness of l^2. The inner product in l^2 allows us to define a norm under the usual formula*

$$||a|| = \sqrt{\langle a|a\rangle}.$$

We can define a set of bra vectors in l^2 using the notation $(\langle a_n|)_{n\in\mathbb{N}}$ considering the following notation

$$\langle a_n| = (a_{n,1}, a_{n,2}, ..., a_{n,n}, ...).$$

If $(\langle a_n|)_{n\in\mathbb{N}}$ is a Cauchy sequence, that is, given any $\varepsilon > 0$, there exists a natural number n_0 such that

$$|| \langle a_m - a_n| || < \varepsilon$$

for all $m, n \in \mathbb{N}$, $m, n \geq n_0$, we intend to show that the set $(\langle a_n|)_{n\in\mathbb{N}}$ is convergent. From

$$|a_{m,p} - a_{n,p}|^2 \leq \sum_k |a_{m,k} - a_{n,k}|^2 < \varepsilon^2,$$

each component is a Cauchy set in $\mathbb{R}$, that is a convergent set. Denoting each component limit by

$$a_p = \lim_{n\to\infty} a_{n,p}, \ p = 1, 2, ...$$

we can generate $\langle a| = (a_1, a_2, ...)$. First, we have to prove that $\langle a|$ is an element of l^2. If $n, m \geq n_0$, we can write

$$\sum_k |a_k - a_{n,k}|^2 = \lim_{m\to\infty} \sum_k |a_{m,k} - a_{n,k}|^2 < \varepsilon^2$$

The finite number of missing terms leads to a finite sum of squares of components of $\langle a|$, that is

$$\sqrt{\sum_{k=1} |a_k|^2} = \sqrt{\sum_{k=1}(|a_k| - |a_{n_0,k}| + |a_{n_0,k}|)^2} \leq \sqrt{\sum_{k=1}(|a_k| - |a_{n_0,k}|)^2} + \sqrt{\sum_{k=1}|a_{n_0,k}|^2} < \infty.$$

At the same time

$$\lim_{n \to \infty} \| (a - a_n)\| = \lim_{n \to \infty} \sum_k |a_k - a_{n,k}|^2 = 0,$$

i.e. ($\langle a_n \rangle|$ is convergent to ($\langle a \rangle|$ in l^2. Being complete, the pre-Hilbert space l^2 becomes an infinite dimensional Hilbert space.

The next examples are more complicated. The completeness is difficult to be proved. In fact, a lot of mathematical techniques are involved together with other mathematical structures.

Consider the set of complex valued functions $f : \mathbb{R} \to \mathbb{C}$ such that $|f|$ is integrable on $\mathbb{R}$, denoted by $L^1(\mathbb{R})$. Among these functions, we denote by $L^2(\mathbb{R})$ the set of complex valued functions such that $|f|^2$ is integrable on $\mathbb{R}$. We define their norm by

$$\|f\|_2 := \left(\int_{-\infty}^{\infty} |f(x)|^2 dx \right)^{\frac{1}{2}}.$$

According to the first structure of functions, it exists another possible norm,

$$\|f\|_1 := \int_{-\infty}^{\infty} |f(x)| dx.$$

By definition, the null functions are those which fulfill the property

$$\int_{-\infty}^{\infty} |f(x)| dx = 0.$$

Let us observe that a null function is not necessarily $f = 0$. If we consider $f(x) = 0$, for $x \in \mathbb{R}$ except x_0 where the function is, for example, 1, we have

$$\int_{-\infty}^{\infty} |f(x)| dx = 0$$

without having $f = 0$ everywhere.

If it is easy to show that $\| \cdot \|_1$ is a norm, it is not easy to check that $\| \cdot \|_2$ is a norm.

The third condition needs some technicalities which we explain below.

The following inequality

$$\|fg\|_1 \leq \|f\|_2 \|g\|_2,$$

that is

$$\left(\int_{-\infty}^{\infty} |f(x)g(x)|dx\right)^2 \le \left(\int_{-\infty}^{\infty} |f(x)|^2 dx\right)\left(\int_{-\infty}^{+\infty} |g(x)|^2 dx\right),$$

is known as the *integral form of Cauchy-Buniakowski-Schwarz inequality*.

Exercise 6.4.5 *Prove the previous inequality.*
Hint. If $u, v \ge 0$, we have

$$uv \le \frac{u^2 + v^2}{2}.$$

Replacing

$$u := \frac{|f(x)|}{\|f\|_2} \quad and \quad v := \frac{|g(x)|}{\|g\|_2}$$

in the previous formula, we obtain

$$\frac{|f(x)|}{\|f\|_2}\frac{|g(x)|}{\|g\|_2} \le \frac{|f(x)|^2}{2\|f\|_2^2} + \frac{|g(x)|^2}{2\|g\|_2^2}.$$

Then we integrate both members and it remains

$$\int_{-\infty}^{+\infty} \frac{|f(x)|}{\|f\|_2}\frac{|g(x)|}{\|g\|_2}dx \le \frac{1}{2} + \frac{1}{2} = 1$$

which proves the statement.

Exercise 6.4.6 *The following exercise is known as Minkowski's inequality: For any $f, g \in L^2(\mathbb{R})$, we have*

$$\|f + g\|_2 \le \|f\|_2 + \|g\|_2.$$

Hint. We start from $u, v \ge 0$ and

$$(u + v)^2 \le 2(u^2 + v^2).$$

Therefore

$$|f + g|^2 \le |f||f + g| + |g||f + g|,$$

that is

$$\|f + g\|_2^2 = \int_{-\infty}^{\infty} |f(x) + g(x)|^2 dx \le \int_{-\infty}^{\infty} |f(x)||f(x) + g(x)|dx + \int_{-\infty}^{\infty} |g(x)||f(x) + g(x)|dx \le$$

$$\le \|f\|_2\|f + g\|_2 + \|g\|_2\|f + g\|_2$$

which ends the proof of the statement.

Exercise 6.4.7 *Check the axioms of vector spaces for $L^2(\mathbb{R})$ endowed with the addition and multiplication by scalars.*
Even if checking the axioms of this vector space seems to be obvious, it is useful to better understand the properties of $L^2(\mathbb{R})$ functions.

Exercise 6.4.8 *Show that the space $L^2(\mathbb{R})$ is complete.*

It is not simple to offer a detailed proof of this statement, if one is not familiar with the concepts of Lebesgue measure, measurable sets, "almost everywhere" equality, convergence almost everywhere, Lebesgue integrable functions (which in fact are subject of $L^1(\mathbb{R})$ and $L^2(\mathbb{R})$ sets), Fatou's lemma, the fact that $L^1(\mathbb{R})$ is complete with respect to the norm $|| \cdot ||_1$, etc.

Let us only give an overview of the proof. We start from a Cauchy sequence $(f_n)_{n \in \mathbb{N}}$ in the norm of $L^2(\mathbb{R})$, that is

$$||f_n - f_m||_2 \to 0 \text{ as } m, n \to +\infty$$

and we have to construct $f \in L^2(\mathbb{R})$ such that

$$||f_n - f||_2 \to 0.$$

Consider $a > 0$ a fixed number. The Cauchy-Buniakovsky-Schwarz inequality leads to

$$\left(\int_{-a}^{a} |f_n(x) - f_m(x)| dx \right)^2 \leq \int_{-a}^{a} |f_m(x) - f_n(x)|^2 dx \int_{-a}^{a} 1 \cdot dx = 2a||f_n - f_m||_2^2 \to 0$$

that is the set $(f_n)_{n \in \mathbb{N}}$ is a Cauchy set with respect to $|| \cdot ||_1$ in $L^1([-a, a])$. Since $L^1([-a, a])$ is complete, there exists $f \subset L^1([-a, a])$ such that

$$||f_n - f||_1 = \int_{-a}^{a} |f_n(x) - f(x)| dx \to 0 \text{ as } n \to +\infty.$$

Now, with all techniques developed by studying the subjects mentioned above and assuming $a = 1, a = 2, a = 3, ..., a = n, ...$, it can be proved that this f belongs to $L^2(\mathbb{R})$ and

$$||f_n - f||_2 \to 0.$$

Let us observe that the completeness of $L^2(\mathbb{R})$ is based on the completeness of $L^2([-a, a])$ which is a Hilbert space whose pre-Hilbert structure is given by the inner product

$$\langle f | g \rangle = \int_{-a}^{a} f^*(x) g(x) dx$$

where $f^*(x)$ is the complex conjugate of $f(x)$, i.e. the norm is

$$\|f\|_2 := \left(\int_{-a}^{a} |f(x)|^2 dx \right)^{\frac{1}{2}}.$$

Therefore, if we look at $L^2(\mathbb{R})$, we have the norm

$$\|f\|_2 := \left(\int_{-\infty}^{\infty} |f(x)|^2 dx \right)^{\frac{1}{2}}$$

coming from the inner product

$$\langle f|g \rangle = \int_{-\infty}^{\infty} f^*(x) g(x) dx$$

where $f^*(x)$ is the complex conjugate of $f(x)$. $L^2(\mathbb{R})$ becomes another example of Hilbert space.

An important remark is necessary at this point. Using the Cauchy-Buniakowski-Schwarz inequality, if we consider the functions $f \in L^2([-a, a])$ and $g = 1$, it results that any $f \in L^2([-a, a])$ belongs to $L^1([-a, a])$, that is $L^2([-a, a]) \subset L^1([-a, a])$. This is not the case for $L^2(\mathbb{R})$ and $L^1(\mathbb{R})$ and, in this case, one has to look at the function

$$f(x) = \begin{cases} -\dfrac{1}{x}, & x \in (-\infty, -1) \\ 1, & x \in [-1, 1] \\ \dfrac{1}{x}, & x \in (1, +\infty). \end{cases}$$

which belongs to $L^2(\mathbb{R})$ but not to $L^1(\mathbb{R})$.

Summary of Lecture 29. The first example: Let us consider the set denoted by l^2 of all sequences $\langle a| = (a_1, a_2, ..., a_n, ...)$, $a_k \in \mathbb{C}$, $k \in \mathbb{N} - \{0\}$ such that

$$\sum_{n} |a_k|^2 < \infty,$$

where $|a_k|^2 = a_k a_k^*$. Consider two operations which generalize the operations for finite dimensional vectors,

$$\langle a| + \langle b| := (a_1 + b_1, a_2 + b_2, ..., a_n + b_n, ...),$$

and

$$\alpha \langle a| := (\alpha a_1, \alpha a_2, ..., \alpha a_n, ...),$$

which define $\langle a + b|$ and $\langle \alpha a|$ respectively. We endow this vector space with the inner product

$$\langle a|b \rangle := \sum_n a_n^* b_n$$

and consequently with the norm

$$||a|| = \sqrt{\langle a|a \rangle}.$$

A Hilbert space structure is highlighted on l^2.

The second example gives another Hilbert space. Let us consider the set of complex valued functions $f : \mathbb{R} \to \mathbb{C}$ such that $|f|$ is integrable on $\mathbb{R}$ and denoted by $L^1(\mathbb{R})$. Among these functions, we denote by $L^2(\mathbb{R})$ the set of complex valued functions such that $|f|^2$ is integrable on $\mathbb{R}$. We define their norm by

$$||f||_2 := \left(\int_{-\infty}^{\infty} |f(x)|^2 dx \right)^{\frac{1}{2}}.$$

According to the first structure of these functions, another possible norm exists. It is

$$||f||_1 := \int_{-\infty}^{\infty} |f(x)| dx.$$

By definition, the null functions are those which fulfill the property

$$\int_{-\infty}^{\infty} |f(x)| dx = 0.$$

The inner product is

$$\langle f|g \rangle = \int_{-\infty}^{\infty} f^*(x) g(x) dx$$

where $f^*(x)$ is the complex conjugate of $f(x)$. $L^2(\mathbb{R})$ becomes another example of Hilbert space. According to the completeness of $L^2(\mathbb{R})$, it is possible to highlight another Hilbert space, $L^2([-a, a])$, where a is a positive real number.

6.5   Lecture 30: Orthogonal and Orthonormal Systems in Hilbert Spaces

By definition a set (finite or infinite) of elements of a Hilbert space is called an *orthogonal system* if any two distinct elements x and y of the system have the property $x \perp y$. This means that, with respect to the inner product involved in the pre-Hilbert structure, it is $\langle x | y \rangle = 0$.

Example 6.5.1   Consider the set

$$\{f_n \in L^2([-\pi, \pi]) | \, n \in \mathbb{Z}\} \text{ where } f_n(x) = \frac{1}{\sqrt{2\pi}} e^{inx}.$$

We have

$$\langle f_n | f_n \rangle = \frac{1}{2\pi} \int_{-\pi}^{\pi} f_n^*(x) f_n(x) dx = \frac{1}{2\pi} \int_{-\pi}^{\pi} e^{-inx} e^{inx} dx = 1,$$

and for $n \neq m$,

$$\langle f_n | f_m \rangle = \frac{1}{2\pi} \int_{-\pi}^{\pi} f_n^*(x) f_m(x) dx = \frac{1}{2\pi} \int_{-\pi}^{\pi} e^{-inx} e^{imx} dx = \frac{1}{2\pi} \int_{-\pi}^{\pi} e^{i(m-n)x} dx =$$

$$= \frac{1}{2\pi} \frac{e^{i(m-n)x}}{i(m-n)} \Big|_{-\pi}^{\pi} = \frac{1}{2\pi} \frac{e^{i(m-n)\pi} - e^{-i(m-n)\pi}}{i(m-n)} = 0.$$

We conclude that the set

$$\{f_n \in L^2([-\pi, \pi]) | \, n \in \mathbb{Z}\}$$

is an orthonormal set of the Hilbert space $L^2([-\pi, \pi])$.

Example 6.5.2   We studied the Hermite polynomials $H_n(x)$,

$$H_n(x) = (-1)^n e^{x^2} \frac{d^n}{dx^n} \left(e^{-x^2} \right)$$

and we proved

$$\int_{-\infty}^{\infty} H_n(x) H_m(x) e^{-x^2} dx = \begin{cases} 0, \ m \neq n \\ 2^n \sqrt{\pi} \, n!, \ m = n \end{cases}$$

If we consider the functions $\varphi_n(x) = e^{-x^2/2} H_n(x)$, $n \in \mathbb{N}$, it is possible to observe that they are part of an orthogonal system in $L^2(\mathbb{R})$ because the inner product

$$\langle \varphi_n | \varphi_m \rangle = \int_{-\infty}^{\infty} H_n(x) H_m(x) e^{-x^2} dx$$

has the result expressed above. It becomes clear that, if we choose

$$\Psi_n(x) = \frac{1}{\sqrt{2^n n! \sqrt{\pi}}} e^{-x^2/2} H_n(x), \ n \in \mathbb{N},$$

we highlighted an *orthonormal system* of $L^2(\mathbb{R})$.

Summary of Lecture 30. In this lecture, we identify the full mathematical meaning of the formulas we are developing. The Hermite polynomials are involved in the functions $\varphi_n(x) = e^{-x^2/2} H_n(x), \ n \in \mathbb{N}$ which are part of an orthogonal system in $L^2(\mathbb{R})$ with respect to the usual inner product because

$$\int_{-\infty}^{\infty} H_n(x) H_m(x) e^{-x^2} dx = \begin{cases} 0, \ m \neq n \\ 2^n \sqrt{\pi} \, n!, \ m = n \end{cases}$$

If we choose

$$\Psi_n(x) = \frac{1}{\sqrt{2^n n! \sqrt{\pi}}} e^{-x^2/2} H_n(x), \ n \in \mathbb{N},$$

we highlighted an orthonormal system of $L^2(\mathbb{R})$ which is involved in the solution of quantum harmonic oscillator.

6.6   Lecture 31: Linear Operators, Eigenvalues, Eigenvectors for the Schrödinger Equation

Let us consider again the free particle problem for the time independent Schrödinger equation

$$H\Psi = -\frac{\hbar^2}{2m} \frac{d^2\Psi}{dx^2} + V(x)\Psi$$

for a free particle with $V(x) = 0$. It becomes

$$\frac{d^2\Psi}{dx^2} = -\frac{2m}{\hbar^2} H\Psi,$$

which can be written with respect to the wavenumber k as

$$\frac{d^2\Psi}{dx^2} = -k^2\Psi.$$

Here we used $H = T = \dfrac{p^2}{2m}$. Suppose that the particle moves along a line segment of length L. A general solution is

$$\Psi(x) = A_1 e^{ikx} + A_2 e^{-ikx}$$

and it appears as a linear combination of two wave-like solutions. If we call them pure states, the general solution is a superposition of the two pure states involved. We will understand later what they represent. The momentum is

$$p = \frac{nh}{L}$$

therefore, if we replace in the energy formula, we obtain

$$H = H_n = \frac{h^2}{2mL^2} n^2, \ n = 1, 2, 3, \ldots.$$

The energy of the free particle is quantized and if the particle moves between two energy levels photons are highlighted because

$$\Delta H_{nm} = H_n - H_m = h\nu,$$

where $\nu = \dfrac{h}{2mL^2}(n^2 - m^2)$. What kind of mathematical structure do we have here?

If we write the equation

$$\frac{d^2 \Psi}{dx^2} = -k^2 \Psi$$

in the form

$$\frac{d^2}{dx^2} \Psi = -k^2 \Psi \, ,$$

we have, in left-hand side, the operator $\mathbb{H} := \dfrac{d^2}{dx^2}$, which is linear. It acts on an element of the Hilbert space, here denoted by Ψ. In the right side, a real number is multiplied by the element Ψ. The real value has the behavior of an *eigenvalue* and Ψ becomes an *eigenvector*. Now let us write the basic facts of the theory we need.

A *linear operator* on a Hilbert space $\mathbb{V}$ to $\mathbb{V}$ is a transformation that satisfies the requirement

$$L|\alpha_1 u + \alpha_2 v\rangle = L|\alpha_1 u\rangle + L|\alpha_2 v\rangle, \ \text{for all} \ \alpha_1, \alpha_2 \in \mathbb{C}, \ |u\rangle, |v\rangle \in \mathbb{V}.$$

Since $L : \mathbb{V} \to \mathbb{V}$, we can denote $L|u\rangle \in \mathbb{V}$ by $|Lu\rangle \in \mathbb{V}$.

A vector $|u\rangle$ is called an eigenvector of the operator L if $L|u\rangle = \lambda|u\rangle$ for some $\lambda \in \mathbb{C}$.

Such a λ is called an eigenvalue corresponding to the eigenvector $|u\rangle$.

The operator $L^\dagger$ is called the *adjoint* to L if, for all $|u\rangle, |v\rangle \in \mathbb{V}$, it is

$$\langle v|L|u\rangle = \langle u|L^\dagger|v\rangle^*.$$

We have to observe that this is only a notation, what we really have is the inner product acting for some elements, that is

$$\langle v|Lu\rangle = \langle u|L^\dagger v\rangle^*,$$

where * means the complex conjugate. Therefore the last expression can be written as

$$\langle v|Lu\rangle = \langle L^\dagger v|u\rangle.$$

By definition, an operator L is called *Hermitian or self-adjoint operator* if $L = L^\dagger$, that is

$$\langle v|Lu\rangle = \langle Lv|u\rangle.$$

Exercise 6.6.1 *All eigenvalues of a Hermitian operator are real.*
Hint. Let us start from the relation involving the Hermitian operator, the eigenvector and the eigenvalue, $L|u\rangle = \lambda|u\rangle$). Then

$$\langle Lu|Lu\rangle = \langle Lu|\lambda u\rangle = \lambda\langle Lu|u\rangle = \lambda\langle u|Lu\rangle = \lambda^2\langle u|u\rangle,$$

that is

$$\lambda^2 = \frac{\langle Lu|Lu\rangle}{\langle u|u\rangle} \in \mathbb{R}_+,$$

therefore λ belongs to $\mathbb{R}$.

Another result is very interesting when we are talking about eigenvectors and eigenvalues.

Exercise 6.6.2 *Consider a Hermitian operator L and two distinct real eigenvalues $\lambda_u,\ \lambda_v$ corresponding to two eigenvectors $|u\rangle$ and $|v\rangle$. Show that the corresponding eigenvectors are orthogonal, that is $\langle u|v\rangle = 0$.*
Hint. From $L|u\rangle = \lambda_u|u\rangle$ and $L|v\rangle = \lambda_v|v\rangle$, we get

$$\langle v|Lu\rangle = \lambda_u\langle v|u\rangle \text{ and } \langle u|Lv\rangle = \lambda_v\langle u|v\rangle.$$

Since $\lambda_v \in \mathbb{R}$, after complex-conjugating the last equality, we have

$$\langle u|Lv\rangle^* = \lambda_v^*\langle u|v\rangle^* = \lambda_v\langle u|v\rangle^* = \lambda_v\langle v|u\rangle.$$

Now, using the fact that L is Hermitian, we have $\langle u|Lv\rangle^* = \langle v|L^\dagger u\rangle = \langle v|Lu\rangle$ *that is*

$$\langle v|Lu\rangle = \lambda_v\langle v|u\rangle.$$

It results

$$0 = (\lambda_u - \lambda_v)\langle v|u\rangle \text{ which means } \langle v|u\rangle = 0.$$

Therefore the Schrödinger equation, in this case, is an operator equation. It highlights orthogonal eigenvectors corresponding to distinct real eigenvalues. In the particular case when $A_2 = 0$, the eigenvectors, that is the pure states we discussed earlier, are $\Psi_n(x) = A_1 e^{inx}$ which can be written as $\Psi_n(x) = A_1 e^{in\frac{2\pi}{L}x}$. In this way, an orthonormal eigenvectors system, corresponding to the eigenvalues $-n^2\frac{4\pi^2}{L^2}$, appears.

An important remark is necessary. Studying differential equations in the Hilbert space L^2 is not very easy. The equations makes sense in a subspace $H^2([0, L])$ of the Hilbert space $L^2([0, L])$ where the functions can be derived twice and the inner product is

$$\langle u|v\rangle_{H^2} = \langle u|v\rangle_{L^2} + \langle Du|Dv\rangle_{L^2} + \langle D^2u|D^2v\rangle_{L^2}.$$

The derivatives appear in the weak sense, that is $D^m u$ is the function v which makes true the equality

$$\int_a^b u(x)D^m\varphi(x)dx = (-1)^m \int_a^b v(x)\varphi(x)dx, \ m = 1, 2.$$

In the case when the functions are continuous differentiable twice the weak derivative $D^1 u$ is $\frac{du}{dx}$ and $D^2 u$ is $\frac{d^2u}{dx^2}$. $H^2([0, L])$ is a Sobolev space and it can be proved that it is a Hilbert space. Sobolev spaces appeared when mathematicians observed that the functions of class $\mathcal{C}^1$ or $\mathcal{C}^2$ are not the right spaces to study differential equations. The weak solutions appear when there are no strong solutions for differential equations. In our case, the solution is a strong one, that is we deal with standard derivatives. Let us observe that if $u, v \in \mathcal{C}^2$ are wave functions as Ψ_n is, it makes sense according to "the convenient border condition" to obtain

$$\left\langle v \left| \frac{d^2}{dx^2}u \right.\right\rangle_{L^2([0,L])} = \int_0^L v^*(x)\frac{d^2u}{dx^2}(x)dx = \int_0^L \frac{d^2v^*}{dx^2}(x)u(x)dx = \left\langle \frac{d^2}{dx^2}v \left| u \right.\right\rangle_{L^2([0,L])},$$

the equality between the two integrals being a consequence of integration by parts applied twice and to the fact that $v(0) = u(0) = v(L) = u(L) = 0$. Therefore the operator $\mathbb{H} := \frac{d^2}{dx^2}$ is Hermitian for such functions.

The technical part, which allows to extend it to a Hermitian operator in L^2, is because the previous computations make sense in $\mathcal{C}^2([0, L])$. The main idea can be explained as follows. It can be proved that, for any $u, v \in L^2([0, L])$, there exist the sets u_n, $v_n \in \mathcal{C}^\infty([0, L])$ such that $u_n \to u$ and $v_n \to v$ and $v_k(0) = u_k(0) = v_k(L) = u_k(L) = 0$, $k \in \mathbb{N}$.

According to the previous part, for all $n \in \mathbb{N}$ and for all u_n, $v_n \in \mathcal{C}^\infty([0, L])$ such that $v_n(0) = u_n(0) = v_n(L) = u_n(L) = 0$, $n \in \mathbb{N}$, it is

$$\left\langle v_n \mid \frac{d^2}{dx^2} u_n \right\rangle_{L^2([0,L])} = \int_0^L v_n^*(x) \frac{d^2 u_n}{dx^2}(x) dx = \int_0^L \frac{d^2 v_n^*}{dx^2}(x) u_n(x) dx = \left\langle \frac{d^2}{dx^2} v_n \mid u_n \right\rangle_{L^2([0,L])}.$$

Since the inner product is continuous in both variables, we have

$$\left\langle v_n \mid \frac{d^2}{dx^2} u_n \right\rangle_{L^2([0,L])} \to \left\langle v \mid \frac{d^2}{dx^2} u \right\rangle_{L^2([0,L])}$$

and

$$\left\langle \frac{d^2}{dx^2} v_n \mid u_n \right\rangle_{L^2([0,L])} \to \left\langle \frac{d^2}{dx^2} v \mid u \right\rangle_{L^2([0,L])}.$$

It results

$$\left\langle v \mid \frac{d^2}{dx^2} u \right\rangle_{L^2([0,L])} = \left\langle \frac{d^2}{dx^2} v \mid u \right\rangle_{L^2([0,L])},$$

that is the Hamilton operator is a Hermitian operator in $L^2([0, L])$.

Now we have a complete view over the study of free particle in Quantum Mechanics. For the others problems we studied, we can identify linear operators which are Hermitian in the sense explained above. The exploration of the related mathematical models can be achieved following the ideas of this lecture.

Exercise 6.6.3 *1. Check if the matrix*

$$L = \begin{pmatrix} 1 & 1 \\ 0 & 1 \end{pmatrix}$$

is Hermitian with respect the Hilbert structure of the two dimensional Euclidean space.
2. Check if the eigenvalues of L are real.
Conclude that if the eigenvalues are real, the operator is not necessary a Hermitian one.
Hint. 1. Compute $\langle Lu|v \rangle$ *and* $\langle u|Lv \rangle$ *if*

$$L = \begin{pmatrix} 1 & 1 \\ 0 & 1 \end{pmatrix}.$$

You have to compare $(u_1 + u_2)v_1 + u_2 v_2$ with $(v_1 + v_2)u_1 + v_2 u_2$.
2. You have to solve the equation $\det(L - \lambda I_2) = 0$, *that is* $(\lambda - 1)^2 - 1 = 0$.

Summary of Lecture 31 We considered the Schrödinger equation in the case of the free particle problem for $V(x) = 0$. It can be considered as an operator equation

$$\frac{d^2}{dx^2}\Psi = -k^2\Psi$$

written with respect to the wavenumber k. The operator is $\mathbb{H} := \dfrac{d^2}{dx^2}$, the real eigenvalues are $-k^2$, that is $-n^2\dfrac{4\pi^2}{L^2}$, and the pure states are the wave-like eigenvectors $\Psi_n(x) = A_1 e^{in\frac{2\pi}{L}x}$. We can look at the above Schrödinger equation as at the operator equation

$$\mathbb{H}\Psi = -k^2\Psi.$$

The Hilbert space structure involved is $L^2([0, L])$ and the solutions belong to $\mathcal{C}^\infty([0, L])$.

During the lecture, we discussed about Hermitian operators in Hilbert spaces, and we proved that their eigenvalues are real. For two different real eigenvalues, the eigenvectors are orthogonal with respect to the inner product.

An operator L is called Hermitian or self-adjoint operator if $L = L^\dagger$, that is

$$\langle v|Lu \rangle = \langle Lv|u \rangle.$$

The eigenvalues of a Hermitian operator are real because

$$\lambda^2 = \frac{\langle Lu|Lu \rangle}{\langle u|u \rangle} \in \mathbb{R}_+.$$

In the case of the free particle in Quantum Mechanics, let us observe that if $u, v \in \mathcal{C}^2[0, L]$ respecting convenient (null) boundary conditions, it makes sense

$$\left\langle v \left| \frac{d^2}{dx^2}u \right. \right\rangle_{L^2([0,L])} = \int_0^L v^*(x)\frac{d^2u}{dx^2}(x)dx = \int_0^L \frac{d^2v^*}{dx^2}(x)u(x)dx = \left\langle \frac{d^2}{dx^2}v \left| u \right. \right\rangle_{L^2([0,L])}$$

the equality between the two integrals being a consequence of integration by parts applied twice. Therefore, the operator $\mathbb{H} := \dfrac{d^2}{dx^2}$ is Hermitian for such functions. There is a very technical part which allows to extend it at a Hermitian operator in $L^2[0, L]$ because the previous computations make sense in $\mathbb{C}^2([0, L])$. It can be proved that for any $u, v \in L^2([0, L])$ there exist the sets $u_n, v_n \in \mathbb{C}^\infty([0, L])$ verifying $v_n(0) = u_n(0) = v_n(L) = u_n(L) = 0, \ n \in \mathbb{N}$ such that $u_n \to u, \ v_n \to v$ and

$$\left\langle v \mid \frac{d^2}{dx^2} u \right\rangle_{L^2([0,L])} = \left\langle \frac{d^2}{dx^2} v \mid u \right\rangle_{L^2([0,L])}.$$

These results constitute a complete view over the study of free particle in Quantum Mechanics. For the others problems studied, we can identify linear operators which are Hermitian in the sense explained above. Real eigenvalues and eigenvectors can be derived for the previous examples.

Chapter 7
From Monochromatic Plane Waves to Wave Packets

God runs electromagnetics by wave theory on Monday,
Wednesday, and Friday,
and the Devil runs them by quantum theory on Tuesday,
Thursday, and Saturday.

William Bragg

7.1 Lecture 32: Again on the de Broglie Hypothesis. Wave-Particle Duality and Wave Packets

The above formal considerations can be applied to better understand the wave-particle duality and then the concept of matter wave. Let us start with the de Broglie hypothesis.

The photoelectric effect proved that the electromagnetic radiation is made of particles called photons having an energy E described by the formula

$$E = h\nu,$$

where ν is the frequency of the wave and the momentum p is described by

$$p = \hbar k,$$

where k is the wave number. In both formulas, we have the Planck constant h or $\hbar = \dfrac{h}{2\pi}$.

To summarize the previous results, we have the following formulas

$$\nu = \frac{1}{T}; \; k = \frac{2\pi}{\lambda}; \; \omega = 2\pi\nu,$$

where T is the period of the wave, λ is the wavelength and ω is the angular velocity.

© The Author(s), under exclusive license to Springer Nature Switzerland AG 2021
S. Capozziello and W.-G. Boskoff, *A Mathematical Journey to Quantum Mechanics*,
UNITEXT for Physics, https://doi.org/10.1007/978-3-030-86098-1_7

Therefore we have also

$$E = h\nu = \frac{h}{2\pi}2\pi\nu = \hbar\omega.$$

Interference appeared in the two-split experiment proved the wave character of electromagnetic radiation. Therefore the electromagnetic radiation has a dual nature, the so called wave-particle duality. This property was transferred by de Broglie to all particles. They have to act in fact under the wave-particle duality and we saw this looking at electrons. All elementary particles, that is all matter, must obey wave-particle duality and we underlined this in the previous sections. In the case of electron, de Broglie stated the relation

$$\lambda = \frac{h}{p} = \frac{h}{mv}.$$

A very important issue appears. A particle, say an electron, cannot be simply described by a wave having the form

$$\Psi(t, x) = Ae^{i(kx - wt)}.$$

Why? Let us look at the speed of the wave,

$$v = \frac{x}{t} = \frac{\omega}{k} = \frac{\hbar\omega}{\hbar k} = \frac{E}{p}.$$

In the case of photons, it works because $c = \dfrac{E}{p}$ is valid involving both the photon energy and the photon momentum as we previously discussed. In the case of other particles as electron, it is,

$$v = \frac{E}{p} = \frac{mc^2}{mv} = \frac{c^2}{v} > c.$$

This cannot be true, therefore we have to think at another way to express the "carrier" of the electron.

Before presenting "other possible carriers", let us look at de Broglie attempts to offer a road through Special Relativity to explain this situation. Consider

$$E = mc^2 = \frac{m_0 c^2}{\sqrt{1 - v^2/c^2}}$$

and

$$\mathbf{p} = \frac{m_0 \mathbf{v}}{\sqrt{1 - v^2/c^2}},$$

whose one dimensional version is

$$p = \frac{m_0 v}{\sqrt{1 - v^2/c^2}}.$$

It results

$$\frac{E}{p} = \frac{mc^2}{mv} = \frac{\dfrac{m_0 c^2}{\sqrt{1 - v^2/c^2}}}{\dfrac{m_0 v}{\sqrt{1 - v^2/c^2}}} = \frac{c^2}{v} = v_p,$$

the last equality being a definition of an abstract entity called v_p suggested from the above definition of v. According to de Broglie, it is possible to define a "group velocity" as

$$v_g = \frac{\partial \omega}{\partial k} = \frac{\partial(\hbar \omega)}{\partial(\hbar k)} = \frac{\partial E}{\partial p}.$$

Then he attributed a relativistic feature to v_g by computing

$$v_g = \frac{\partial E}{\partial p} = \frac{\partial}{\partial p}\sqrt{p^2 c^2 + m_0^2 c^4} = \frac{pc^2}{\sqrt{p^2 c^2 + m_0^2 c^4}} = \frac{pc^2}{E}.$$

Therefore he concluded

$$v_g = \frac{\partial E}{\partial p} = \frac{pc^2}{E} = \frac{c^2}{\dfrac{E}{p}} = \frac{c^2}{v_p} = v,$$

that is this "relativistic" group velocity is the one which equals v. v_p remained only a mathematical step to offer a physical meaning to v_g. He completed his theory with the relativistic versions of the formulas presented at the beginning of this lecture: The relation

$$\lambda = \frac{h}{p}$$

now becomes

$$\lambda = \frac{h}{p} = \frac{h}{m_0 v}\sqrt{1 - \frac{v^2}{c^2}}$$

and the relation

$$E = h\nu$$

is now seen as

$$\nu = \frac{E}{h} = \frac{m_0 c^2}{h} \frac{1}{\sqrt{1 - \dfrac{v^2}{c^2}}}.$$

However, a problem still remained. What does the group velocity refer to?

Let us define instead of a wave having the wavenumber k_0, an infinity of waves each one having the wavenumber in an interval $\left(k_0 - \dfrac{\Delta k}{2}, k_0 + \dfrac{\Delta k}{2} \right)$. The carrier will be

$$\Psi(t, x) = A \int_{k_0 - \Delta k/2}^{k_0 + \Delta k/2} e^{i(kx - \omega t)} dk = A \int_{k_0 - \Delta k/2}^{k_0 + \Delta k/2} e^{i(px - Et)/\hbar} dk,$$

called a *wave-packet* corresponding to the wavenumber interval $\left(k_0 - \dfrac{\Delta k}{2}, k_0 + \dfrac{\Delta k}{2} \right)$. In the case of the wave studied before, k was a constant, that is p was a constant and the speed was obtained by what happened at a crest of the wave, therefore from the condition $kx - \omega t = 0$ appeared. In this case, the *speed of the wave-packet* can be expressed from the speed formula

$$x = v_g t$$

in its differential form at the center of the packet (for $k = k_0$ or equivalently, for $p = p_0$) from the condition

$$\frac{d(px - Et)}{dp} \bigg|_{p=p_0} = 0.$$

Therefore

$$v_g = \frac{x}{t} = \frac{dE}{dp} = \frac{d\omega}{dk}.$$

It results

$$v_g = \frac{dE}{dp} = \frac{\frac{1}{2} d(mv^2)}{d(mv)} = v,$$

that is the speed of the packet equals the speed of the particle. The mathematical description of the wave packet can be improved as we will see later. The wave packet will be related to the Fourier transforms and the Heisenberg uncertainty principle.

Summary of Lecture 32. A photon can be described by a wave in the form

$$\Psi(t, x) = A e^{i(kx - \omega t)}$$

because its speed is

$$v = \frac{x}{t} = \frac{\omega}{k} = \frac{\hbar\omega}{\hbar k} = \frac{E}{p} = c.$$

However, an electron cannot be described by a standard monochromatic planar wave because the speed of such a wave should be

$$v = \frac{x}{t} = \frac{\omega}{k} = \frac{\hbar\omega}{\hbar k} = \frac{E}{p} = \frac{mc^2}{mv} = \frac{c^2}{v} > c.$$

Therefore we consider the wave replaced by something new,

$$\Psi(t, x) := A \int_{k_0-\Delta k/2}^{k_0+\Delta k/2} e^{i(kx-\omega t)} dk = A \int_{k_0-\Delta k/2}^{k_0+\Delta k/2} e^{i(px-Et)/\hbar} dk,$$

called a wave packet corresponding to the wavenumber interval $\left(k_0 - \frac{\Delta k}{2}, k_0 + \frac{\Delta k}{2}\right)$.

The speed of the packet is derived from the condition

$$\left.\frac{d(px - Et)}{dp}\right|_{p=p_0} = 0.$$

Therefore

$$v_g = \frac{x}{t} = \frac{dE}{dp} = \frac{d\omega}{dk}$$

that is

$$v_g = \frac{dE}{dp} = \frac{\frac{1}{2}d(mv^2)}{d(mv)} = v.$$

The speed of the packet now equals the speed of the particle. The mathematical description of the wave packet can be improved as we will see later replacing A from outside of the integral by $A(k)$ inside the integral.

During the lecture, we presented the de Broglie attempts to offer an explanation to the fact that a wave in the form

$$\Psi(t, x) = A e^{i(kx-\omega t)}$$

can however be a mathematical representation for a particle. He used a relativistic explanation about how this could be possible.

7.2   Lecture 33: More About Electron in an Atom

Let us think again at the electron as a particle confined to its circular trajectory. Denote by L the length of the circle. Ψ has the property

$$\Psi(x) = \Psi(x + L), \text{ for all } x \in [0, L].$$

Let us suppose that $V(x) = 0$, that is $H = E$ and the corresponding time-independent Schrödinger equation is

$$-\frac{\hbar^2}{2m}\frac{d^2\Psi}{dx^2}(x) = H\Psi(x),$$

We expect to have $H \geq 0$. Indeed

$$-\frac{\hbar^2}{2m}\int_0^L \Psi^*(x)\frac{d^2\Psi}{dx^2}(x)dx = H\int_0^L \Psi^*(x)\Psi(x)dx = H.$$

and

$$H = -\frac{\hbar^2}{2m}\int_0^L \Psi^*(x)\frac{d^2\Psi}{dx^2}(x)dx = -\frac{\hbar^2}{2m}\int_0^L \left(\frac{d}{dx}\left(\Psi^*(x)\frac{d\Psi}{dx}(x)\right) - \frac{d\Psi^*}{dx}(x)\frac{d\Psi}{dx}(x)\right)dx =$$

$$= \Psi^*(L)\frac{d\Psi}{dx}(L) - \Psi^*(0)\frac{d\Psi}{dx}(0) + \frac{\hbar^2}{2m}\int_0^L \left|\frac{d\Psi}{dx}\right|^2 dx = \frac{\hbar^2}{2m}\int_0^L \left|\frac{d\Psi}{dx}\right|^2 dx \geq 0.$$

The Schrödinger equation becomes

$$\frac{d^2\Psi}{dx^2}(x) = -\frac{2mH}{\hbar^2}\Psi(x),$$

where $\dfrac{2mH}{\hbar^2}$ is a positive constant denoted by α^2. The solution depends on a constant amplitude A,

$$\Psi(x) = Ae^{i\alpha x}$$

and has to fulfill

$$e^{i\alpha(x+L)} = e^{i\alpha x} \text{ that is } \alpha L = 2\pi n, \ n \in \mathbb{Z}.$$

Therefore α is a quantized wavenumber.
For each

$$\alpha_n = \frac{2\pi n}{L} \text{ the solution is } \Psi_n(x) = Ae^{i\alpha_n x}.$$

Since

$$\int_0^L \Psi_n^*(x)\Psi_n(x)dx = A^2\int_0^L dx = 1,$$

it results

$$A = \frac{1}{\sqrt{L}}.$$

Therefore the eigenvectors are

$$\Psi_n(x) = \frac{1}{\sqrt{L}} e^{2n\pi i x/L},$$

and the associated energies, the eigenvalues, are

$$H_n = \frac{2\hbar^2 \pi^2 n^2}{mL^2}.$$

There are a lot of similitudes between this situation of a particle moving on a circle and the results we obtained in the case of the free particle which moves on a line segment in Lecture 23. Here, there are infinite energy levels and we can still make comments on the solution. We observe that

$$\int_0^L \Psi_n^*(x)\Psi_l(x)dx = \frac{1}{L}\int_0^L e^{2(m-n)\pi i x/L}dx = \delta_{mn}.$$

Therefore, the eigenvectors are orthogonal and the general wave function, which is periodic, is the superposition

$$\psi(x) = \sum_{n \in \mathbb{Z}} a_n \Psi_n(x)$$

with the coefficients

$$a_n = \int_0^L \Psi_n^*(x)\psi(x)dx$$

which will be related to the Fourier series. It is worth noticing that any linear combination of functions, in mathematical terms, is a superposition in Physics language. The term "superposition" will be defined and discussed in detail in Lecture 41.

Summary of Lecture 33. Consider a particle with a circular trajectory of length L in absence of a potential $V(x)$. The corresponding time-independent Schrödinger equation is

$$\frac{d^2\Psi}{dx^2}(x) = -\frac{2mH}{\hbar^2}\Psi(x),$$

where $\dfrac{2mH}{\hbar^2}$ is a positive constant denoted by α^2. The solution depends on a constant amplitude A,

$$\Psi(x) = Ae^{i\alpha x}$$

and has to fulfill the condition

$$e^{i\alpha(x+L)} = e^{i\alpha x} \quad \text{that is } \alpha L = 2\pi n, \ n \in \mathbb{Z}.$$

Therefore the eigenvectors are

$$\Psi_n(x) = \frac{1}{\sqrt{L}} e^{2n\pi i x/L}$$

and the associated energies, the eigenvalues, are

$$H_n = \frac{2\hbar^2 \pi^2 n^2}{m L^2}.$$

We observe that

$$\int_0^L \Psi_n^*(x) \Psi_l(x) dx = \frac{1}{L} \int_0^L e^{2(m-n)\pi i x/L} dx = \delta_{mn},$$

that is the eigenvectors are orthogonal and the general wave function, which is periodic, is the superposition

$$\psi(x) = \sum_{n \in \mathbb{Z}} a_n \Psi_n(x)$$

having the coefficients

$$a_n = \int_0^L \Psi_n^*(x) \psi(x) dx.$$

Chapter 8
The Heisenberg Uncertainty Principle and the Mathematics Behind

It is the theory that decides what we can observe.

Albert Einstein

8.1 Lecture 34: Wave Packets and the Schrödinger Equation

Starting from the present considerations, we intend to arrive to Lecture 36 when the concept of Gaussian wave packet will allow us to establish the Heisenberg Uncertainty Principle for position and momentum

$$\Delta x \, \Delta p = \hbar,$$

i.e. the position and the momentum of a particle cannot be determined with the same degree of precision. There exists a limit in the accuracy of the prediction for both variables involved in the previous equality and it is easy to understand why. It means that if more precisely the position of the particle is determined, then less precisely the momentum is, and vice versa. According to the previous considerations, we know that a wave packet of the form

$$\Psi(t, x) = A \int_{k_0 - \Delta k/2}^{k_0 + \Delta k/2} e^{i(kx - \omega t)} dk$$

describes better a wave because the group velocity represents the velocity of the particle. We can even think at a model of wave packet where A is not a constant, but depends on the wave number k, that is $A = A(k)$. Therefore the corresponding $A(k)$ wave packet is

$$\Psi(t, x) = \int_{k_0 - \Delta k/2}^{k_0 + \Delta k/2} A(k) e^{i(kx - \omega t)} dk$$

© The Author(s), under exclusive license to Springer Nature Switzerland AG 2021
S. Capozziello and W.-G. Boskoff, *A Mathematical Journey to Quantum Mechanics*,
UNITEXT for Physics, https://doi.org/10.1007/978-3-030-86098-1_8

or we can also consider models where the particle is described on the entire real line $\mathbb{R}$. We can also take into account the form

$$\Psi(t, x) = \int_{-\infty}^{\infty} \phi(p) e^{i(px - Ht)/\hbar} dp \,,$$

depending on the momentum. In such a case there are two issues. The integral must exist, that is $\phi(p)$ has to be well chosen. In the previous examples $\Psi(0, x)$ must verify the Schrödinger time independent equation offering us information about the energy of the free particle. This is also related to

$$\Psi(0, x)\Psi^*(0, x) = 1 \,,$$

a condition we interpret as the fact that "the particle has to be somewhere on the real line".

In order to go forward, let us suppose that $\phi(p) := e^{-(p-p_0)/2(\Delta p)^2}$. Here p_0 is a given momentum and Δp is a momentum chosen in a small interval centered at p_0. A constant can appear in the calculations. Let us discard it for the moment because we can put it in front of the integral after we complete the computations having in mind the condition which must be satisfied, that is

$$\Psi(0, x)\Psi^*(0, x) = 1.$$

In Lecture 36, we will show that

$$\int_{-\infty}^{\infty} e^{-au^2 + bu} du = \sqrt{\frac{\pi}{a}} \, e^{b^2/4a} \,.$$

Therefore

$$\Psi(0, x) = \int_{-\infty}^{\infty} e^{-(p-p_0)^2/2(\Delta p)^2} e^{i(px)/\hbar} dp$$

can be computed for $u := p - p_0$, $a := \dfrac{1}{2(\Delta p)^2}$, $b := \dfrac{ix}{\hbar}$ and the result is

$$\Psi(0, x) = \sqrt{2\pi} \, \Delta p \, e^{i(p_0 x)/\hbar} e^{-(\Delta p)^2 x^2/2\hbar^2} \,.$$

Furthermore, it is

$$\Psi(0, x)\Psi^*(0, x) = 2\pi(\Delta p)^2 e^{-(\Delta p)^2 x^2/\hbar^2} \,,$$

that is the constant K_0 has to fulfill the condition

$$K_0 K_0^* 2\pi(\Delta p)^2 e^{-(\Delta p)^2 x^2/\hbar^2} = 1.$$

We will clarify in details this result. It remains to discuss the Schrödinger equation verified by $\Psi(0, x)$. Some basic computations lead to

$$\frac{d\Psi}{dx}(0, x) = \left(\frac{ip_0}{\hbar} - \frac{(\Delta p)^2}{\hbar^2} x \right) \Psi(0, x)$$

and

$$\frac{d^2\Psi}{dx^2} = -\frac{(\Delta p)^2}{\hbar^2} \Psi(0, x) + \left(\frac{ip_0}{\hbar} - \frac{(\Delta p)^2}{\hbar^2} x \right)^2 \Psi(0, x)$$

that is

$$-\frac{\hbar^2}{2m} \frac{d^2\Psi}{dx^2} + \frac{1}{2m} \left(ip_0 - \frac{(\Delta p)^2}{\hbar} x \right)^2 \Psi(0, x) = \frac{(\Delta p)^2}{2m} \Psi(0, x).$$

Choosing a complex potential energy of the form

$$V := \frac{1}{2m} \left(ip_0 - \frac{(\Delta p)^2}{\hbar} x \right)^2$$

the Schrödinger equation is fulfilled. As we will discuss later, such a situation is possible if the operator equation

$$\left(-\frac{\hbar^2}{2m} \frac{d^2}{dx^2} + V \right) \Psi = \frac{(\Delta p)^2}{2m} \Psi$$

has real eigenvalues.

Summary of Lecture 34. Wave packets have to be adopted because they give the correct speed of a particle. In fact, the speed of a particle equals the speed of the related wave packet. Considering a wave packet in the form

$$\Psi(t, x) = \int_{-\infty}^{\infty} \phi(p) e^{i(px - Ht)/\hbar} dp$$

for $\phi(p) := e^{-(p - p_0)/2(\Delta p)^2}$, we can show that

$$\Psi(0, x) = \sqrt{2\pi}\, \Delta p \; e^{i(p_0 x)/\hbar} e^{-(\Delta p)^2 x^2/2\hbar^2}.$$

A constant K_0 must be introduced and determined by the condition

$$K_0 \Psi(0, x) K_0^* \Psi^*(0, x) = K_0 K_0^* 2\pi (\Delta p)^2 e^{-(\Delta p)^2 x^2/\hbar^2} = 1.$$

We can show that $\Psi(0, x)$ verifies the Schrödinger equation

$$-\frac{\hbar^2}{2m}\frac{d^2\Psi}{dx^2}(0, x) + \frac{1}{2m}\left(ip_0 - \frac{(\Delta p)^2}{\hbar}x\right)^2 \Psi(0, x) = \frac{(\Delta p)^2}{2m}\Psi(0, x).$$

for a complex potential energy of the form

$$V := \frac{1}{2m}\left(ip_0 - \frac{(\Delta p)^2}{\hbar}x\right)^2.$$

As we will discuss below, such a situation is possible if the operator equation

$$\left(-\frac{\hbar^2}{2m}\frac{d^2}{dx^2} + V\right)\Psi = \frac{(\Delta p)^2}{2m}\Psi$$

has real eigenvalues. By these preliminaries, we can make some steps towards understanding the Heisenberg Uncertainty Principle.

8.2   Lecture 35: The Wave Function Ψ Solution of the Schrödinger Equation

The above discussion on the plane wave function of x allows us to generalize the result to waves depending on t and all space coordinates x, y, z.

Let us denote by $\mathbf{p} = (p_x, p_y, p_z)$ the given momentum vector. Its magnitude is $p := |\mathbf{p}| = \sqrt{\mathbf{p}_x^2 + \mathbf{p}_y^2 + \mathbf{p}_z^2}$. Now $\mathbf{k} := (k_x, k_y, k_z)$ is the wave propagation vector or simply, the wave vector; $k := |\mathbf{k}| = \sqrt{\mathbf{k}_x^2 + \mathbf{k}_y^2 + \mathbf{k}_z^2}$ is its length; its connection to the momentum vector is given by

$$\mathbf{p} := \hbar\mathbf{k}.$$

If the space coordinates are related to a position vector $\mathbf{r} := (x, y, z)$, by definition, the wave function is

$$\Psi(t, \mathbf{r}) = Ae^{i(\mathbf{k}\cdot\mathbf{r}-\omega t)}.$$

Of course, this definition generalizes the one dimensional case seen previously to vector components. Therefore

$$k = |k| = \frac{|p|}{\hbar} = \frac{2\pi}{\lambda},$$

where λ is the wavelength.

We know that the energy corresponding to the wave is the total energy H. Exactly as in the plane waves description, it is $\omega = \dfrac{H}{\hbar}$. Let us observe that the wave function can also be described by

$$\Psi(t, \mathbf{r}) = A e^{i(\mathbf{p}\cdot\mathbf{r} - Ht)/\hbar}.$$

Now we can use a similar argument to "derive" the Schrödinger equation in the case of the wave function

$$\Psi(t, \mathbf{r}) = A e^{i(\mathbf{k}\cdot\mathbf{r} - \omega t)}$$

or in the case of the three-dimensional wave packet. The wave packet, in this case, is

$$\Psi(t, \mathbf{r}) = \frac{1}{\sqrt{(2\pi\hbar)^3}} \iiint_{\mathbb{R}^3} e^{i(\mathbf{p}\cdot\mathbf{r} - Ht)/\hbar} \phi(\mathbf{p}) d\mathbf{p},$$

where $d\mathbf{p} = dp_x dp_y dp_z$. The function ϕ has to be chosen such that the triple integral exists. Furthermore, in order to express a quantum particle, the function Ψ has to verify the Schrödinger equation.

In fact, the difference between this section and the previous one is related to the fact that we can compute $\dfrac{\partial\Psi^2}{\partial x^2}$; $\dfrac{\partial\Psi^2}{\partial y^2}$, $\dfrac{\partial\Psi^2}{\partial z^2}$, therefore the Laplace operator ∇^2 appears. For the time-dependent equation, we see that both Ψ and the potential V depends on $(t, \mathbf{r})$, therefore the postulated Schrödinger equation is

$$i\hbar \frac{\partial}{\partial t} \Psi(t, \mathbf{r}) = \left(-\frac{\hbar^2}{2m} \nabla^2 + V(t, \mathbf{r}) \right) \Psi(t, \mathbf{r}).$$

The time-independent Schrödinger equation has now the form

$$H\Psi(\mathbf{r}) = \left(-\frac{\hbar^2}{2m} \nabla^2 + V(\mathbf{r}) \right) \Psi(\mathbf{r}).$$

We can consider the physical meaning of Ψ. To step into this direction, we need to define the Gauss wave packets and, according these, we will observe the probabilistic meaning of Ψ.

Summary of Lecture 35. Let us consider a three-dimensional space with coordinates (x, y, z). The position x in the one dimensional case is replaced by the position vector $\mathbf{r} = (x, y, z)$. The momentum p and the wavenumber k are replaced by the vectors $\mathbf{p}$ and $\mathbf{k}$ connected by the relation $\mathbf{p} = \hbar\mathbf{k}$. The Laplace operator ∇^2 replaces, in the three-dimensional case, the derivative

$\dfrac{d^2}{dx^2}$. In this case, the time-independent Schrödinger equation is

$$H\Psi(\mathbf{r}) = \left(-\frac{\hbar^2}{2m}\nabla^2 + V(\mathbf{r})\right)\Psi(\mathbf{r}).$$

A solution is the wave function

$$\Psi(t, \mathbf{r}) = Ae^{i(\mathbf{k}\cdot\mathbf{r} - \omega t)}.$$

In the case of three-dimensional wave packet, the situation is more complicated. The wave packet is

$$\Psi(t, \mathbf{r}) = \frac{1}{\sqrt{(2\pi\hbar)^3}} \iiint_{\mathbb{R}^3} e^{i(\mathbf{p}\cdot\mathbf{r} - Ht)/\hbar}\phi(\mathbf{p})d\mathbf{p},$$

where $d\mathbf{p} = dp_x dp_y dp_z$ and the function ϕ is chosen in such a way that the triple integral exist. H is the energy. The function Ψ has to be the solution of the Schrödinger equation and we will see below its probabilistic meaning.

8.3   Lecture 36: The Gauss Wave Packet and the Heisenberg Uncertainty Principle

Let us take into account a specific wave packet describing a particle belonging to a region of $x-$axis where no potential acts on it, that is $H = E = E(p) = \dfrac{p^2}{2m}$. Consequently, $E = \hbar\omega$ and $p = \hbar k$. The formula to be considered is

$$\Psi(t, x) = \frac{1}{\sqrt{2\pi\hbar}} \int_{-\infty}^{\infty} e^{i(px - E(p)t)/\hbar}\phi(p)dp.$$

It is very similar to the wave-packet, solution of the Schrödinger equation, when we used

$$\Psi(t, x) = \int_{k_0-\Delta k/2}^{k_0+\Delta k/2} A(k)e^{i(kx - \omega t)}dk$$

for a convenient $A(k) = A\left(\dfrac{p}{\hbar}\right)$ and $dk = \dfrac{1}{\hbar}dp$. The difference is the interval where the integral acts.

Now, let us suppose a positive function $\phi(p)$ having one crest only and two asymptotic branches in the $p-$axis. Furthermore, let us suppose that the maximum of the crest is at p_0, therefore the maximum value is $\phi(p_0)$ and there exists a symmetry of $\phi(p)$ with respect to the line $p = p_0$. Now, let us consider a very small interval $(p_0 - \Delta p, p_0 + \Delta p)$ where the values of $\phi(p)$ are "almost $\phi(p_0)$". We have described "a sort of Gauss function" having its crest at $(p_0, \phi(p_0))$.

First of all, we have to observe that $|\Psi(t, x)|$ has its maximum when $p_x x - E(p_x)t$ is almost constant in the interval $(p_0 - \Delta p, p_0 + \Delta p)$. This happens because, in general, $|\int f| \leq \int |f|$, with equality when f is a constant. The immediate consequence is that the condition

$$\frac{d(px - E(p)t)}{dp}\bigg|_{p=p_0} = 0$$

determines the speed of the wave packet at its center. It is

$$0 = \frac{d(px - E(p)t)}{dp}\bigg|_{p=p_0} = \left(x - \frac{dE(p)}{dp}t\right)\bigg|_{p=p_0} = \left(x - \frac{d\left(\frac{p^2}{2m}\right)}{dp}t\right)\bigg|_{p=p_0} = x - \frac{p_0}{m}t.$$

Therefore the speed of the wave packet is $v_g = \dfrac{p_0}{m}$. It is worth noticing that $E(p_0) = \dfrac{p_0^2}{2m}$. Taking into account the polynomial form of the energy with respect to p, we can write

$$E(p) = E(p_0) + \frac{1}{1!}\frac{dE(p)}{dp}\bigg|_{p=p_0}(p - p_0) + \frac{1}{2!}\frac{d^2E(p)}{dp^2}\bigg|_{p=p_0}(p - p_0)^2,$$

that is

$$E(p) = \frac{p_0^2}{2m} + \frac{1}{1!}v_g(p - p_0) + \frac{1}{2!}\frac{1}{m}(p - p_0)^2.$$

We may consider p close to p_0 such that $(p - p_0)^2$ can be neglected. In this case

$$E(p) = \frac{p_0^2}{2m} + v_g(p - p_0)$$

and

$$e^{i(px-E(p)t)/\hbar} = e^{i(px-[E(p_0)+v_g(p-p_0)]t)/\hbar}.$$

If we look at the exponent only, we can observe that

$$
\begin{aligned}
px - [E(p_0) + v_g(p - p_0)]t &= px - E(p_0)t - v_g tp + v_g tp_0 \\
&= (p - p_0)(x - v_g t) + p_0 x - E(p_0)t
\end{aligned}
$$

therefore the new form of the wave packet is

$$\Psi(t, x) = \frac{1}{\sqrt{2\pi\hbar}} \int_{-\infty}^{\infty} e^{i(px - E(p)t)/\hbar} \phi(p) dp$$

$$= \frac{1}{\sqrt{2\pi\hbar}} e^{i(p_0 x - E(p_0)t)/\hbar} \int_{-\infty}^{\infty} e^{i(p - p_0)(x - v_g t)/\hbar} \phi(p) dp$$

and this happens because we work with respect to a small interval centered at p_0, exactly as we considered the form of the wave packet with respect to k in a small neighborhood of k_0.

Exercise 8.3.1 *Show that*

$$\int_{-\infty}^{\infty} e^{-u^2} du = \sqrt{\pi}.$$

Let us first observe that

$$\left(\int_{-\infty}^{\infty} e^{-u^2} du\right)^2 = \left(\int_{-\infty}^{\infty} e^{-x^2} dx\right)\left(\int_{-\infty}^{\infty} e^{-y^2} dy\right) = \iint_{\mathbb{R}^2} e^{-(x^2+y^2)} dx dy.$$

To compute the last integral, we write, for $(x, y) \in (-\infty, +\infty) \times (-\infty, +\infty)$, that

$$\left(\int_{-\infty}^{\infty} e^{-u^2} du\right)^2 = \iint_{\mathbb{R}^2} e^{-(x^2+y^2)} dx dy = \int_0^{\infty} \int_0^{2\pi} e^{-r^2} r dr d\theta = 2\pi \int_0^{\infty} e^{-r^2} r dr = \pi,$$

where the last double integral is obtained after we use the change of coordinates $x = r\cos\theta, \ y = r\sin\theta, \ dx dy = r dr d\theta, \ r \in (0, +\infty), \ \theta \in (0, 2\pi).$

Another useful exercise is

Exercise 8.3.2 *Show that*

$$\int_{-\infty}^{\infty} e^{-au^2 + bu} du = \sqrt{\frac{\pi}{a}} \, e^{b^2/4a}.$$

It is easy to see that if Re $a \neq 0$

$$\int_{-\infty}^{\infty} e^{-au^2 + bu} du = e^{b^2/4a} \int_{-\infty}^{\infty} e^{-a(u - b/2a)^2} d(u - b/2a) = \sqrt{\frac{1}{a}} \, e^{b^2/4a} \int_{-\infty}^{\infty} e^{-v^2} dv,$$

the last integral being calculated in the previous exercise.

Now, we continue our study in the case when ϕ is a Gauss function of the form

$$\phi(p) := C e^{-(p - p_0)^2/2(\Delta p)^2},$$

where Δp is a p of the interval centered at p_0 and C is determined from the condition

$$\int_{-\infty}^{\infty} |\phi(p)|^2 dp = 1.$$

Let us first write the form of Ψ,

$$\Psi(t, x) = \frac{1}{\sqrt{2\pi\hbar}} e^{i(p_0 x - E(p_0)t)/\hbar} C \int_{-\infty}^{\infty} e^{i(p-p_0)(x-v_g t)/\hbar} \cdot e^{-(p-p_0)^2/2(\Delta p)^2} dp.$$

Since

$$|\phi(p)|^2 = C^2 e^{-(p-p_0)^2/(\Delta p)^2},$$

after we denote by $u := \dfrac{p - p_0}{\Delta p}$, $du = \dfrac{1}{\Delta p} dp$, it remains to compute

$$1 = \int_{-\infty}^{\infty} |\phi(p)|^2 dp = C^2 \Delta p \int_{-\infty}^{\infty} e^{-u^2} du = C^2 \Delta p \sqrt{\pi}.$$

The positive value of C is

$$C = \frac{1}{\sqrt{\Delta p} \sqrt[4]{\pi}}.$$

So, finally we have

$$\Psi(t, x) = \frac{1}{\sqrt{\Delta p} \sqrt[4]{\pi}} \frac{1}{\sqrt{2\pi\hbar}} e^{i(p_0 x - E(p_0)t)/\hbar} \int_{-\infty}^{\infty} e^{-(p-p_0)^2/2(\Delta p)^2} \cdot e^{i(p-p_0)(x-v_g t)/\hbar} dp.$$

Let us consider the time $t = 0$ to look at the Ψ properties:

$$\Psi(0, x) = \frac{1}{\sqrt{\Delta p} \sqrt[4]{\pi}} \frac{1}{\sqrt{2\pi\hbar}} \cdot e^{i p_0 x/\hbar} \int_{-\infty}^{\infty} e^{-(p-p_0)^2/2(\Delta p)^2} \cdot e^{i(p-p_0)x/\hbar} d(p - p_0).$$

We have

$$\Psi(0, x) = \frac{1}{\sqrt{\Delta p} \sqrt[4]{\pi}} \frac{1}{\sqrt{2\pi\hbar}} e^{i p_0 x/\hbar} \int_{-\infty}^{\infty} e^{-u^2/2(\Delta p)^2} e^{iux/\hbar} du,$$

that is

$$\Psi(0, x) = \frac{1}{\sqrt{\Delta p} \sqrt[4]{\pi}} \frac{1}{\sqrt{2\pi\hbar}} e^{i p_0 x/\hbar} \int_{-\infty}^{\infty} e^{-au^2 + bu} du.$$

We can use the second above exercise to compute the integral considering

$$a = \frac{1}{2(\Delta p)^2}, \quad b = \frac{ix}{\hbar}.$$

Denoting by K_0 the constant in front of the integral, we obtain

$$\Psi(0, x) = K_0 e^{ip_0 x/\hbar} e^{-x^2(\Delta p)^2/2\hbar^2}.$$

If $\Psi^*(0, x)$ is the complex conjugate of the complex number $\Psi(0, x)$, we have

$$|\Psi(0, x)|^2 = \Psi(0, x) \cdot \Psi^*(0, x) = K_0^2 e^{-x^2(\Delta p)^2/\hbar^2}.$$

Let us denote by Δx the value of x such that the exponential function, from the right member, reaches the value $1/e$. Therefore, in the case of the Gauss wave packet, we obtain the equality

$$\Delta x \Delta p = \hbar.$$

This equality looks like a condition imposed both to the position and to the momentum of the particle described by the Gauss wave packet. Let us remember again how we obtained this equality: Manipulating the Gauss wave packet, we first choose a Δp in the neighborhood of p_0 and then, for this given Δp, we find a corresponding Δx looking at the time when the exponential function $e^{-x^2(\Delta p)^2/\hbar^2}$ reaches exactly the value $1/e$, that is after the exponential function decreases from its maximum value to the value $1/e$.

The condition obtained is $\Delta x \Delta p = \hbar$. So, let us see how important is this equality. Suppose we choose $\Delta p_1 > \Delta p$. The corresponding Δx_1 has to satisfy $\Delta x_1 < \Delta x$ because

$$\Delta p_1 \Delta x_1 = \Delta p \Delta x = \hbar.$$

That is, if we decrease Δp, that is if we are interested to determine more accurately the momentum of a particle, the position of the particle belongs to a greater interval, therefore we loose the possibility to know accurately the position of the particle and vice versa, if we determine more precisely the position of the particle, we loose the possibility to determine with the same accuracy its momentum.

This is the Heisenberg Uncertainty Principle: it is a limit in our possibility to determine, at the same time, the position and the momentum of a particle. A general discussion about the Heisenberg Uncertainty Principle will be presented later when we discuss the Dirac formalism.

In other words, we lose the classical determinism in the world of particles. The wave $\Psi(t, x)$, describing a particle, is called the particle wave function and plays the role of a probability amplitude, while

$$|\Psi(t, x)|^2 = \Psi(t, x) \cdot \Psi^*(t, x)$$

is the probability to find the particle at a given point. Accepting this interpretation, we step into the Quantum World where the determinism of the Classical Mechanics is replaced by the probability to locate a particle related to its corresponding wave.

Summary of Lecture 36. In this lecture, we studied the Gauss wave packet

$$\Psi(t, x) = \frac{1}{\sqrt{2\pi\hbar}} \int_{-\infty}^{\infty} e^{i(px - E(p)t)/\hbar} \phi(p) dp,$$

in the case where the function ϕ is a Gauss function of the form

$$\phi(p) = Ce^{-(p-p_0)^2/2(\Delta p)^2},$$

where Δp is a p of an interval centered at p_0 and C is determined by the condition

$$\int_{-\infty}^{\infty} |\phi(p)|^2 dp = 1.$$

After some physical considerations, we found, at $t = 0$, that

$$|\Psi(0, x)|^2 = \Psi(0, x) \cdot \Psi^*(0, x) = K_0^2 e^{-x^2(\Delta p)^2/\hbar^2}.$$

Denoting by Δx the value of x such that the exponential function from the right member reaches the value $1/e$, in the case of the Gauss wave packet, we obtain the equality

$$\Delta x \Delta p = \hbar.$$

This is the Heisenberg Uncertainty Principle obtained by using the Gauss wave packet: it is a limit in our possibility to determine, at the same time, the position and the momentum of a particle.

The wave function is related by the probability amplitude which provides the connection between the state of a system and the results of observations. The squared modulus

$$|\Psi(t, x)|^2 = \Psi(t, x) \cdot \Psi^*(t, x)$$

is the probability density. It is postulated to be the probability to find the particle in a given "volume" $d\mathbf{r} = dx$ centered at x.

8.4   Lecture 37: The Mathematics Behind the Wave Packets. The Fourier Series and the Fourier Transforms

The motivations for introducing the Fourier transforms are related to the Fourier series which allow to describe complicated periodic functions with respect to waves represented by superpositions of $\sin kx$ and $\cos kx$. In order to have a representation of the concept, let us draw first a linear segment $[0, 2\pi]$ and then the function $\sin x$

corresponding to this interval. After, let us draw again the same interval and the function $\sin 2x$. Continue the stack with the same interval and the function $\sin 3x$ and so on. Now, at each step, we consider instead of $\sin x$, $\sin 2x$, $\sin 3x$, .., relative weights for each sinusoid, that is $b_1 \sin x$, $b_2 \sin 2x$, $b_3 \sin 3x$, etc. For each point $x_0 \in [0, 2\pi]$ consider the sum

$$b_1 \sin x + b_2 \sin 2x + b_3 \sin 3x + \dots.$$

It makes sense if

$$\sum_{k=1}^{\infty} |b_k| < \infty.$$

Let us do the same in the case of $\cos x$ function, this time with coefficients a_n, that is the new sum will be

$$a_0 + a_1 \cos x + a_2 \cos 2x + a_3 \cos 3x + \dots$$

which makes sense under the similar condition

$$\sum_{k=0}^{\infty} |a_k| < \infty.$$

At the end, we have constructed the function

$$g(x) := \sum_{k=0}^{\infty} a_k \cos kx + \sum_{k=1}^{\infty} b_k \sin kx.$$

$g(x)$ can be thought as a periodic function corresponding to $\mathbb{R}$ if we stick the image on $[0, 2\pi]$ along the x axis. Now, we may ask if a given periodic function $f : [0, 2\pi] \to \mathbb{R}$ can be approximated or even represented by

$$\sum_{k=0}^{\infty} a_k \cos kx + \sum_{k=1}^{\infty} b_k \sin kx.$$

Let us describe the problem in a more general frame because the functions f, we are using in physics, can have different periods with respect to 2π.

A periodic function $f : \mathbb{R} \to \mathbb{R}$, with the fundamental period T, satisfies for all $x \in \mathbb{R}$ the equality $f(x + T) = f(x)$.

A Fourier series with period T is the function

$$g(x) := \sum_{n \in \mathbb{N}} a_n \cos \left(\frac{2\pi n x}{T} \right) + \sum_{n \in \mathbb{N}, \, n \neq 0} b_n \sin \left(\frac{2\pi n x}{T} \right).$$

It is better to write it in the form

$$g(x) := \sum_{n=0}^{\infty} a_n \cos\left(\frac{2\pi nx}{T}\right) + \sum_{n=1}^{\infty} b_n \sin\left(\frac{2\pi nx}{T}\right)$$

and to have in mind that the coefficients are chosen to satisfy the relations

$$\sum_{k=0}^{\infty} |b_k| < \infty; \quad \sum_{k=1}^{\infty} |a_k| < \infty.$$

The numbers a_n and b_n are called *Fourier coefficients*. If a given periodic function f is given, can it be approximate by a Fourier series? The answer is affirmative if $f \in \mathcal{C}^{\infty}(\mathbb{R})$, that is if f is smooth. The coefficients determined by the function f are

$$a_0 = \frac{1}{T} \int_0^T f(x) dx;$$

$$a_n = \frac{2}{T} \int_0^T f(x) \cos\left(\frac{2\pi nx}{T}\right) dx;$$

$$b_n = \frac{2}{T} \int_0^T f(x) \sin\left(\frac{2\pi nx}{T}\right) dx,$$

consequently, in the first case presented here, they are

$$a_0 = \frac{1}{2\pi} \int_0^{2\pi} f(x) dx;$$

$$a_n = \frac{1}{\pi} \int_0^{2\pi} f(x) \cos nx\, dx;$$

$$b_n = \frac{1}{\pi} \int_0^{2\pi} f(x) \sin nx\, dx.$$

The explanation of this result is the following: If

$$f(x) = \sum_{k=0}^{\infty} a_k \cos kx + \sum_{k=1}^{\infty} b_k \sin kx,$$

it is

$$\int_0^{2\pi} f(x) \cos nx\, dx = \sum_{k=0}^{\infty} a_k \int_0^{2\pi} \cos kx \cos nx\, dx + \sum_{k=1}^{\infty} b_k \int_0^{2\pi} \sin kx \cos nx\, dx =$$

$$= a_n \int_0^{2\pi} \cos^2 nx \, dx = a_n \int_0^{2\pi} \frac{1 + \cos 2nx}{2} dx = \pi a_n.$$

We have used the two trigonometric formulas $2 \cos kx \cos nx = \cos(k - n)x + \cos(k + n)x$ and $2 \sin kx \cos nx = \sin(k - n)x + \sin(k + n)x$. If we integrate $f(x)$ written as a Fourier series, it results

$$\int_0^{2\pi} f(x)dx = \int_0^{2\pi} a_0 dx + \sum_{k=1}^{\infty} a_k \int_0^{2\pi} \cos kx \, dx + \sum_{k=1}^{\infty} b_k \int_0^{2\pi} \sin kx \, dx = 2\pi a_0,$$

therefore

$$a_0 = \frac{1}{2\pi} \int_0^{2\pi} f(x)dx.$$

There is a *Dirichlet theorem* which asserts that, if f is smooth and periodic, then f can be identified with its Fourier series, that is

$$f(x) = \sum_{k=0}^{\infty} a_k \cos kx + \sum_{k=1}^{\infty} b_k \sin kx$$

with the coefficients described by previous formulas. The result remains valid even if the function f is smooth on each interval $[0, x_1), [x_1, x_2), ..., [x_k, 2\pi]$ where k is finite and there exist finite values for the left limits in $x_m, \ m = 1, 2, ..., k$ denoted by $f(x_m - 0)$.

The road to the Fourier transforms passes through the complex form of the Fourier series. A remark is necessary at this point. Sometimes in Physics, it is better to work with complex numbers because many concepts have a more elegant form if expressed in such way. Furthermore, the Mathematics of complex numbers includes several applications to algebra, geometry and calculus which can be directly applied to Physics.

With these considerations in mind, let us take into account the same smooth and periodic function $f : \mathbb{R} \to \mathbb{R}$. Consider the set of integers $\mathbb{Z}$ and the function

$$\sum_{n \in \mathbb{Z}} f_n e^{i \frac{2\pi n x}{T}}.$$

Let us now suppose that the complex Fourier series converges to f, that is we have

$$f(x) := \sum_{n \in \mathbb{Z}} f_n e^{i \frac{2\pi n x}{T}}.$$

To determine the coefficients f_n, we follow the idea previously considered in the case of real Fourier series. This time we use the multiplication by $e^{-i\frac{2\pi mx}{T}}$ and integrate after:

$$\int_0^T f(x)e^{-i\frac{2\pi mx}{T}}\, dx = \int_0^T \left(\sum_{n\in\mathbb{Z}} f_n e^{i\frac{2\pi nx}{T}} e^{-i\frac{2\pi mx}{T}} \right) dx$$

$$= \sum_{n\in\mathbb{Z}} \int_0^T f_n e^{i\frac{2\pi(n-m)x}{T}}\, dx =$$

$$= \sum_{n\in\mathbb{Z}} f_n \int_0^T e^{i\frac{2\pi(n-m)x}{T}}\, dx.$$

It is

$$\int_0^T e^{i\frac{2\pi(n-m)x}{T}}\, dx = \begin{cases} 0, & m \neq n \\ T, & m = n \end{cases}$$

which implies

$$f_m = \frac{1}{T}\int_0^T f(x)e^{-i\frac{2\pi mx}{T}}\, dx.$$

In some problems, the interval $[0, T]$ is replaced by an interval $[a, b]$ such that $b - a = T$. In this case, if f is the complex Fourier series

$$f(x) = \frac{1}{b-a}\sum_{n\in\mathbb{Z}} f_n e^{2\pi inx/(b-a)},$$

then

$$f_m = \int_a^b f(x)e^{-2\pi imx/(b-a)}\, dx.$$

An important comment is now in order. The coefficients f_n are complex numbers and f_{-m} is the complex conjugate of f_m because

$$f_{-m} = \frac{1}{T}\int_0^T f(x)e^{i\frac{2\pi mx}{T}}\, dx = f_m^*.$$

Therefore

$$A_m = f_{-m}e^{-i\frac{2\pi mx}{T}} + f_m e^{i\frac{2\pi mx}{T}} = f_m^* e^{-i\frac{2\pi mx}{T}} + f_m e^{i\frac{2\pi mx}{T}}, \quad m \in \mathbb{N},\ m \geq 1$$

has the property $A_m^* = A_m$, that is a real number. If we take into account the formulas

$$\cos y = \frac{e^{iy} + e^{-iy}}{2}$$

and

$$\sin y = \frac{e^{iy} - e^{-iy}}{2i},$$

we can arrange A_m as

$$b_m \cos \frac{2\pi mx}{T} + a_m \sin \frac{2\pi mx}{T},$$

therefore, we can switch from the Fourier series with complex coefficients to the Fourier series with real coefficients.

Let us connect this subject to the Hilbert spaces. Consider an orthogonal system of elements in the Hilbert space $\mathbb{H}$, that is

$$(e_n)_{n \in \mathbb{N}} \subset \mathbb{H}, \ \langle e_n, e_m \rangle = 0, \text{ if } m \neq n.$$

If the values of the inner product are real, for $x \in \mathbb{H}$, we can define the real numbers

$$c_n = \frac{1}{||e_n||^2} \langle x, e_n \rangle$$

which are the Fourier coefficients for the Fourier series

$$\sum_{n \in \mathbb{N}} c_n e_n$$

corresponding to x with respect to the orthogonal system $(e_n)_{n \in \mathbb{N}}$.

Suppose that the given orthogonal system has the following property:
If $\langle y, e_n \rangle = 0$ for all $n \in \mathbb{N}$, then $y = 0$. If the orthogonal system $(e_n)_{n \in \mathbb{N}}$ has the previous property, it can be proven that any $x \in \mathbb{H}$ can be written as its associated Fourier series, that is

$$x = \sum_{n \in \mathbb{N}} \frac{\langle x, e_n \rangle}{||e_n||^2} e_n.$$

More specifically, under the same conditions, the *Plancharel-Parseval equality* holds. That is

$$||x||^2 = \sum_{n \in \mathbb{N}} c_n^2 ||e_n||^2.$$

Considering now the set

$$\{F_n \in L^2([-\pi, \pi]) | \ n \in \mathbb{Z}\} \quad \text{where} \quad F_n(x) = e^{inx},$$

it is easy to see that it is an orthogonal system. For $m = n$, we have

$$\langle F_n | F_n \rangle = \int_{-\pi}^{\pi} F_n^*(x) F_n(x) dx = \int_{-\pi}^{\pi} e^{-inx} e^{inx} dx = 2\pi,$$

and for $m \neq n$,

$$\langle F_n | F_m \rangle = \frac{1}{2\pi} \int_{-\pi}^{\pi} F_n^*(x) F_m(x) dx = \frac{1}{2\pi} \int_{-\pi}^{\pi} e^{-inx} e^{imx} dx = \frac{1}{2\pi} \int_{-\pi}^{\pi} e^{i(m-n)x} dx =$$

$$= \frac{1}{2\pi} \frac{e^{i(m-n)\pi} - e^{-i(m-n)\pi}}{i(m-n)} = 0.$$

Therefore, the formulas for the Fourier coefficients, attached to a function $f : \mathbb{R} \to \mathbb{R}$, are

$$c_n = \frac{1}{2\pi} \int_{-\pi}^{\pi} f(x) e^{inx} dx, \ n \in \mathbb{Z}$$

and the *Plancharel-Parseval equality* becomes

$$\sum_{n \in \mathbb{Z}} c_n^2 = \frac{1}{2\pi} \int_{-\pi}^{\pi} f^2(x) dx.$$

We are now ready to define the Fourier transforms. The idea comes from the Fourier series. A Fourier series decomposes a periodic function into a sum of sinusoidal functions. A Fourier transform decomposes any function in a sum of sinusoidal functions, therefore it is the extension of the Fourier series to non-periodic functions. If you look at the formulas

$$f(x) := \sum_{n \in \mathbb{Z}} f_n e^{i \frac{2\pi n x}{T}}$$

and

$$f_m = \frac{1}{T} \int_0^T f(x) e^{-i \frac{2\pi m x}{T}} dx,$$

the following definitions are the consequences.

The *Fourier transform* of a function f of variable x is the function $\hat{f}$ of variable k defined by

$$\mathcal{F}(f(x)) = \hat{f}(k) := \int_{-\infty}^{\infty} f(x) e^{-ikx} dx.$$

Let us observe that we offered two notations for the Fourier transform of $f(x)$. Both are important because the first one is related to the meaning and the second one can be used in complicate formulas, or even in the formula known as *Fourier inverse theorem*, where the function $f(x)$ is reconstructed starting the original Fourier transform $\hat{f}(k)$:

$$\mathcal{F}^{-1}(\hat{f}(k)) = f(x) = \frac{1}{2\pi} \int_{-\infty}^{\infty} \hat{f}(k)e^{ikx}dk.$$

It is important to say that the Fourier transforms act on functions $f : \mathbb{R} \to \mathbb{C}$ from $L^1(\mathbb{R})$.

It appears as the *Fourier pair*

$$f \longleftrightarrow \hat{f},$$

or, adopting the other notation, as

$$\mathcal{F}^{-1}(\hat{f}(k)) \longleftrightarrow \mathcal{F}(f(x)).$$

The sufficient conditions for the existence of the Fourier transform of f are
1. f has a finite number of discontinuities;
2. $f \in L^1(\mathbb{R})$.

With the above considerations in mind, let us define the *Dirac delta function* as a generalized function which is 0 everywhere in $\mathbb{R}$ except for a single point, say 0, where its "value" is $+\infty$, having the supplementary property

$$\int_{-\infty}^{\infty} \delta(x)dx = 1.$$

Obviously, it is not a "function" in the proper term, but this object can be thought as the possibility to "charge" a single point from $\mathbb{R}$. An equivalent definition is

$$\delta_a(x) = \lim_{a \to 0} \frac{1}{|a|\sqrt{\pi}} e^{-x^2/a^2}.$$

It is worth noticing that Fourier introduced it as the expression

$$\delta(x - x') = \frac{1}{2\pi} \int_{-\infty}^{\infty} \cos(kx - kx')dk,$$

while Cauchy expressed it in the exponential form

$$f(x) = \frac{1}{2\pi} \int_{-\infty}^{\infty} e^{ikx} \left(\int_{-\infty}^{\infty} f(x')e^{-ikx'}dx' \right) dk$$

that is

$$\delta(x - x') = \int_{-\infty}^{\infty} e^{ik(x-x')}dk.$$

This object is a distribution and there is a mathematical domain developed by Laurent Schwartz where the properties of such generalized functions are studied. Let us look at the details of Cauchy exponential form for Dirac's function. From the definitions of direct and inverse Fourier transforms, it results

$$f(x) = \frac{1}{2\pi} \int_{-\infty}^{\infty} \hat{f}(k)e^{ikx}dk = \frac{1}{2\pi} \int_{-\infty}^{\infty} \left(e^{ikx} \int_{-\infty}^{\infty} f(x')e^{-ikx'}dx' \right) dk =$$

$$= \frac{1}{2\pi} \int_{-\infty}^{\infty} \left(f(x') \int_{-\infty}^{\infty} e^{ik(x-x')}dk \right) dx' = \frac{1}{2\pi} \int_{-\infty}^{\infty} f(x')\delta(x-x')dx'.$$

We have

$$\int_{-\infty}^{\infty} |f(x)|^2 dx = \int_{-\infty}^{\infty} f^*(x)f(x)dx =$$

$$= \int_{-\infty}^{\infty} \left(\frac{1}{2\pi} \int_{-\infty}^{\infty} \hat{f}^*(k)e^{-ikx}dk \right) \left(\frac{1}{2\pi} \int_{-\infty}^{\infty} \hat{f}(k')e^{ik'x}dk' \right) dx =$$

$$= \frac{1}{4\pi^2} \int_{-\infty}^{\infty} \left(\left(\int_{-\infty}^{\infty} \hat{f}^*(k)\hat{f}(k') \int_{-\infty}^{\infty} e^{i(k'-k)x}dx \right) dk' \right) dk =$$

$$= \frac{1}{2\pi} \int_{-\infty}^{\infty} \hat{f}^*(k) \left(\frac{1}{2\pi} \int_{-\infty}^{\infty} \hat{f}(k')\delta(k'-k)dk' \right) dk =$$

$$= \frac{1}{2\pi} \int_{-\infty}^{\infty} \hat{f}^*(k)\hat{f}(k)dk = \frac{1}{2\pi} \int_{-\infty}^{\infty} |\hat{f}(k)|^2 dk,$$

This is the central theorem of this theory and it is called *Plancherel-Parseval's formula*. The Plancherel-Parseval formula

$$\int_{-\infty}^{\infty} |f(x)|^2 dx = \frac{1}{2\pi} \int_{-\infty}^{\infty} |\hat{f}(k)|^2 dk$$

is used to prove Pinsky's theorem which is presented below.

Let us consider first the Gauss wave packet

$$\Psi(t, x) = \frac{1}{\sqrt{2\pi\hbar}} \int_{-\infty}^{\infty} e^{i(px-E(p)t)/\hbar}\phi(p)dp.$$

We start with $t = 0$ and

$$\Psi(0, x) = \frac{1}{\sqrt{2\pi\hbar}} \int_{-\infty}^{\infty} e^{ipx/\hbar}\phi(p)dp.$$

$\Psi(0, x)$ is the inverse Fourier transform of $\phi(p)$ modulo a constant. In the case when the function ϕ is a Gauss function of the form

$$\phi(p) = Ce^{-(p-p_0)^2/2(\Delta p)^2},$$

where Δp is a p in the interval centered at p_0 and C is determined from the condition

$$\int_{-\infty}^{\infty} |\phi(p)|^2 dp = 1,$$

the function

$$\Psi(0, x) = \frac{C}{\sqrt{2\pi\hbar}} \int_{-\infty}^{\infty} e^{ipx/\hbar} e^{-(p-p_0)^2/2(\Delta p)^2} dp = \frac{Ce^{p_0x/h}}{\sqrt{2\pi\hbar}} \int_{-\infty}^{\infty} e^{i(p-p_0)x/\hbar} e^{-(p-p_0)^2/2(\Delta p)^2} dp$$

is the inverse Fourier transform of the Gauss function

$$\phi(p) = Ce^{-(p-p_0)^2/2(\Delta p)^2},$$

again modulo a constant. The inverse Fourier transform can be computed, after convenient substitutions, taking into account the integral

$$\int_{-\infty}^{\infty} e^{-au^2+bu} du = \sqrt{\frac{\pi}{a}} e^{-b^2/2a^2}.$$

According to these results and deginitions, If

$$\int_{-\infty}^{\infty} |\phi(p)|^2 dp = 1,$$

then from the Plancherel-Parseval theorem, it follows that

$$\int_{-\infty}^{\infty} |\Psi(0, x)|^2 dx = 2\pi.$$

The spread around $p = 0$ is measured by the dispersion about zero, that is

$$D_0(\phi) = \int_{-\infty}^{\infty} p^2 |\phi(p)|^2 dp.$$

The functions $p\phi(p)$ and $\phi'(p)$ are square integrable and the equality

$$D_0(\phi) D_0(\Psi(0, x)) = \frac{k_0\pi}{2},$$

can be proven, where k_0 is a constant. This is the Heisenberg Uncertainty Principle expressed with respect to the Fourier transforms.

This result is a particular case of the *Pinsky Theorem* which asserts that if the Fourier pair f and $\hat{f}$ are normalized functions and f fulfills the above properties for ϕ, then

$$D_0(f)D_0(\hat{f}) \geq \frac{\pi}{2},$$

the equality is attained in the case when f is a normalized Gauss function.

Let us sketch a proof in the special case when we suppose, for $f : \mathbb{R} \to \mathbb{C}$, that

$$\int_{-\infty}^{\infty} f^2(x)dx = 1$$

and the supplementary condition $\lim_{|x| \to +\infty} xf^2(x) = 0$.

We want, under these conditions, to prove that

$$\int_{-\infty}^{\infty} x^2 f^2(x)dx \int_{-\infty}^{\infty} k^2 |\hat{f}(k)|^2 dk \geq \frac{\pi}{2}.$$

Let us denote by f' the derivative of f. Consider the integral $I(\alpha)$ which has the property

$$I(\alpha) := \int_{-\infty}^{\infty} (\alpha x f(x) + f'(x))^2 dx \geq 0 \text{ for all } \alpha \in \mathbb{R},$$

that is

$$\left(\int_{-\infty}^{\infty} x f(x) f'(x)dx \right)^2 - \int_{-\infty}^{\infty} x^2 f^2(x)dx \int_{-\infty}^{\infty} (f'(x))^2 dx \leq 0.$$

From

$$\int_{-\infty}^{\infty} x f(x) f'(x)dx = x f^2(x)|_{-\infty}^{+\infty} - \int_{-\infty}^{\infty} (x f(x) f'(x) + f^2(x))dx$$

$$= -1 - \int_{-\infty}^{\infty} x f(x) f'(x)dx,$$

it results

$$\int_{-\infty}^{\infty} x f(x) f'(x)dx = -\frac{1}{2},$$

i.e.

$$\frac{1}{4} \leq \int_{-\infty}^{\infty} x^2 f^2(x)dx \int_{-\infty}^{\infty} (f'(x))^2 dx = \left(\int_{-\infty}^{\infty} x^2 f^2(x)dx \right) \left(\frac{1}{2\pi} \int_{-\infty}^{\infty} |-ik\hat{f}(k)|^2 dk \right),$$

the last integral being the consequence of Plancherel-Parseval's formula applied to
the derivative and the trivial Fourier pair $f'(x) \longleftrightarrow -ik\hat{f}(k)$. It is

$$\frac{\pi}{2} \le \left(\int_{-\infty}^{\infty} x^2 f^2(x)dx\right)\left(\int_{-\infty}^{\infty} k^2|\hat{f}(k)|^2 dk\right).$$

Therefore, at least at the level of basic knowledge, it is clear that the Heisenberg
Uncertainty Principle has its roots in the Mathematics of Fourier transforms. Another
proof of Uncertainty Principle will be presented in Lecture 43.

Summary of Lecture 37. The Fourier series allow to describe complicated
periodic functions with respect to simple waves represented by $\sin kx$ and
$\cos kx$ while the Fourier transforms decompose any function, not only the
periodic ones, with respect to sinusoidal functions.

Consider a given periodic function $f : [0, 2\pi] \to \mathbb{R}$ which can be represented as the real Fourier series

$$\sum_{k=0}^{\infty} a_k \cos kx + \sum_{k=1}^{\infty} b_k \sin kx.$$

It can be proven that the Fourier coefficients of such a representation are

$$a_0 = \frac{1}{2\pi}\int_0^{2\pi} f(x)dx;$$

$$a_n = \frac{1}{\pi}\int_0^{2\pi} f(x)\cos nx dx;$$

$$b_n = \frac{1}{\pi}\int_0^{2\pi} f(x)\sin nx dx.$$

The theory can be presented in a more general form. A periodic function
$f : \mathbb{R} \to \mathbb{R}$, with the fundamental period T, satisfies, for all $x \in \mathbb{R}$, the equality
$f(x + T) = f(x)$.

A Fourier series with the period T is the function

$$g(x) := \sum_{n=0}^{\infty} a_n \cos\left(\frac{2\pi nx}{T}\right) + \sum_{n=1}^{\infty} b_n \sin\left(\frac{2\pi nx}{T}\right)$$

and the coefficients are chosen to satisfy the relations

$$\sum_{k=0}^{\infty} |b_k| < \infty; \quad \sum_{k=1}^{\infty} |a_k| < \infty.$$

If $f \in \mathcal{C}^{\infty}(\mathbb{R})$, that is f is smooth, f can be represented as a real Fourier series with the coefficients

$$a_0 = \frac{1}{T} \int_0^T f(x)dx;$$

$$a_n = \frac{2}{T} \int_0^T f(x) \cos\left(\frac{2\pi n x}{T}\right) dx;$$

$$b_n = \frac{2}{T} \int_0^T f(x) \sin\left(\frac{2\pi n x}{T}\right) dx.$$

The road to Fourier transforms passes through the complex form of the Fourier series. Let us consider the same smooth and periodic function $f : \mathbb{R} \to \mathbb{R}$. Consider the set of integers $\mathbb{Z}$ and the function

$$\sum_{n \in \mathbb{Z}} f_n e^{i\frac{2\pi n x}{T}}.$$

Suppose that the complex Fourier series converges to f, that is

$$f(x) := \sum_{n \in \mathbb{Z}} f_n e^{i\frac{2\pi n x}{T}}.$$

The coefficients are proven to be

$$f_m = \frac{1}{T} \int_0^T f(x) e^{-i\frac{2\pi m x}{T}} dx.$$

The coefficients f_n are complex numbers and f_{-m} is the complex conjugate of f_m because

$$f_{-m} = \frac{1}{T} \int_0^T f(x) e^{i\frac{2\pi m x}{T}} dx = f_m^*.$$

Therefore

$$A_m = f_{-m} e^{-i\frac{2\pi m x}{T}} + f_m e^{i\frac{2\pi m x}{T}} = f_m^* e^{-i\frac{2\pi m x}{T}} + f_m e^{i\frac{2\pi m x}{T}}, \quad m \in \mathbb{N}, \ m \geq 1$$

has the property $A_m^* = A_m$, that is a real number. If we take into account the formulas

$$\cos y = \frac{e^{iy} + e^{-iy}}{2}$$

and

$$\sin y = \frac{e^{iy} - e^{-iy}}{2i},$$

we can arrange A_m as

$$b_m \cos \frac{2\pi m x}{T} + a_m \sin \frac{2\pi m x}{T},$$

therefore we can switch from the Fourier series with complex coefficients to the Fourier series with real coefficients.

The theory can be formulated in the Hilbert spaces. For an orthogonal system $(e_n)_{n\in\mathbb{N}}$ for $x \in \mathbb{H}$, the associated Fourier series can be written, that is

$$\sum_{n\in\mathbb{N}} \frac{\langle x, e_n \rangle}{||e_n||^2} e_n.$$

If the Hilbert space has the property $\langle y, e_n \rangle = 0$ it implies $y = 0$, then, for any x, it is

$$x = \sum_{n\in\mathbb{N}} \frac{\langle x, e_n \rangle}{||e_n||^2} e_n$$

and the Plancharel-Parseval equality

$$||x||^2 = \sum_{n\in\mathbb{N}} c_n^2 ||e_n||^2$$

holds. In the Hilbert space $L^2([-\pi, \pi])$, we can consider the set

$$\{F_n \in L^2([-\pi, \pi]) | \, n \in \mathbb{Z}\} \text{ where } F_n(x) = e^{inx}.$$

It is easy to see that it is an orthogonal system. Therefore, the Fourier coefficients attached to a function $f : \mathbb{R} \to \mathbb{R}$ are

$$c_n = \frac{1}{2\pi} \int_{-\pi}^{\pi} f(x)e^{inx}dx, \, n \in \mathbb{Z}$$

and the Plancharel-Parseval equality becomes

$$\sum_{n \in \mathbb{Z}} c_n^2 = \frac{1}{2\pi} \int_{-\pi}^{\pi} f^2(x) dx.$$

The Fourier transform of a function f of variable x is the function $\hat{f}$ of variable k defined by

$$\mathcal{F}(f(x)) = \hat{f}(k) := \int_{-\infty}^{\infty} f(x) e^{-ikx} dx.$$

The function $f(x)$ is reconstructed starting from the original Fourier transform $\hat{f}(k)$:

$$\mathcal{F}^{-1}(\hat{f}(k)) = f(x) = \frac{1}{2\pi} \int_{-\infty}^{\infty} \hat{f}(k) e^{ikx} dk.$$

It is important to say that the Fourier transforms act on functions $f : \mathbb{R} \to \mathbb{C}$ from $L^1(\mathbb{R})$. The result is a Fourier pair

$$f \longleftrightarrow \hat{f},$$

or, using the other notation,

$$\mathcal{F}^{-1}(\hat{f}(k)) \longleftrightarrow \mathcal{F}(f(x)).$$

The sufficient conditions for the existence of Fourier transform of f are
1. f has a finite number of discontinuities;
2. $f \in L^1(\mathbb{R})$.
The central theorem of this theory is the Plancherel-Parseval formula

$$\int_{-\infty}^{\infty} |f(x)|^2 dx = \frac{1}{2\pi} \int_{-\infty}^{\infty} |\hat{f}(k)|^2 dk.$$

Consider the Gauss wave packet

$$\Psi(t, x) = \frac{1}{\sqrt{2\pi\hbar}} \int_{-\infty}^{\infty} e^{i(px - E(p)t)/\hbar} \phi(p) dp.$$

in the case $t = 0$, i.e.

$$\Psi(0, x) = \frac{1}{\sqrt{2\pi\hbar}} \int_{-\infty}^{\infty} e^{ipx/\hbar} \phi(p) dp.$$

$\Psi(0, x)$ is the inverse Fourier transform of $\phi(p)$ modulo the constant considered in the above formula.

In the case the function ϕ is a Gauss function of the form

$$\phi(p) = Ce^{-(p-p_0)^2/2(\Delta p)^2},$$

where Δp is a p on an interval centered at p_0 and C is determined from the condition

$$\int_{-\infty}^{\infty} |\phi(p)|^2 dp = 1,$$

the function

$$\Psi(0, x) = \frac{C}{\sqrt{2\pi\hbar}} \int_{-\infty}^{\infty} e^{ipx/\hbar} \cdot e^{-(p-p_0)^2/2(\Delta p)^2} dp =$$

$$= \frac{Ce^{p_0 x/\hbar}}{\sqrt{2\pi\hbar}} \int_{-\infty}^{\infty} e^{i(p-p_0)x/\hbar} \cdot e^{-(p-p_0)^2/2(\Delta p)^2} dp$$

is the inverse Fourier transform of the Gauss function

$$\phi(p) = Ce^{-(p-p_0)^2/2(\Delta p)^2},$$

again modulo a constant. The inverse Fourier transform can be computed after convenient substitutions taking into account the integral

$$\int_{-\infty}^{\infty} e^{-au^2+bu} du = \sqrt{\frac{\pi}{a}} e^{-b^2/2a^2}.$$

Since

$$\int_{-\infty}^{\infty} |\phi(p)|^2 dp = 1,$$

from the Plancherel-Parseval theorem, it follows that

$$\int_{-\infty}^{\infty} |\Psi(0, x)|^2 dx = 2\pi.$$

The spread around $p = 0$ is measured by the dispersion about zero, that is

$$D_0(\phi) = \int_{-\infty}^{\infty} p^2 |\phi(p)|^2 dp.$$

It can be proven that the functions $p\phi(p)$ and $\phi'(p)$ are square integrable and

$$D_0(\phi)D_0(\Psi(0, x)) = \frac{k_0\pi}{2},$$

where k_0 is a constant. In fact this is the Heisenberg Uncertainty Principle expressed with respect to the Fourier transforms.

To represent this concept, the first integral can be imagined as a number a corresponding to a length and the second integral is a number b corresponding to the width of a rectangle. The numbers are linked by the direct and inverse Fourier transforms. These Fourier transforms act in such a way that if a becomes a', then, it is mandatory that b becomes b' with $ab = a'b'$, that is the area of the rectangle is preserved.

This result is a particular case of the Pinsky theorem which asserts that if the Fourier pair f and $\hat{f}$ are normalized functions and f fulfills the above properties of ϕ, then

$$D_0(f)D_0(\hat{f}) \geq \frac{\pi}{2},$$

the equality is attained in the case when f is a normalized Gauss function.

In conclusion, we can state that the Heisenberg Uncertainty Principle can be based on the properties of Fourier transforms.

Chapter 9
The Principles of Quantum Mechanics

When the solution is simple, God is answering.

Albert Einstein

9.1 Lecture 38: Operators in Quantum Mechanics

In Lecture 31, we discussed how the Schrödinger equation can be transformed in a problem of eigenvalues and eigenvectors and we highlighted a linear operator derived from the equation. Let us reload the idea because we want to formalize the role of operators related to Quantum Mechanics. To advance in this subject, we start from the wave function

$$\Psi(t, x) = e^{i(kx - \omega t)}$$

where ω and k are determined by the formulas $p = \hbar k$ and $H = E = \hbar \omega = \dfrac{p^2}{2m}$.

Since

$$\frac{\partial}{\partial x} \Psi(t, x) = ik \Psi(t, x),$$

it results

$$\frac{\hbar}{i} \frac{\partial}{\partial x} \Psi(t, x) = p \Psi(t, x).$$

We can define a *momentum operator*, denoted $\hat{p}$, which extracts the information p from the wave function Ψ, that is

$$\hat{p}\Psi = p\Psi.$$

S. Capozziello and W.-G. Boskoff, *A Mathematical Journey to Quantum Mechanics*, UNITEXT for Physics, https://doi.org/10.1007/978-3-030-86098-1_9

This operator is defined as

$$\hat{p} := \frac{\hbar}{i} \frac{\partial}{\partial x}.$$

We can also write the previous formula as

$$\hat{p} := -i\hbar \frac{d}{dx}$$

for a one-dimensional problem. Therefore the wave function Ψ becomes an eigenvector and p the corresponding eigenvalue for the momentum operator $\hat{p}$. Considering the interpretation of Ψ as a "state function" (see Lecture 41 below), we can call Ψ, appearing in the relation $\hat{p}\Psi = p\Psi$, an eigenstate of $\hat{p}$ corresponding to the eigenvalue p. In the same way, we can extract the information "energy" from Ψ. It is

$$\frac{\partial}{\partial t}\Psi(t, x) = -i\omega\Psi(t, x),$$

i.e.

$$i\hbar\frac{\partial}{\partial t}\Psi(t, x) = E\Psi(t, x).$$

Since the particle energy is related to its momentum, we can write

$$E\Psi(t, x) = \frac{p^2}{2m}\Psi(t, x) = \frac{p}{2m}(p\Psi(t, x)) = \frac{p}{2m}\left(\frac{\hbar}{i}\frac{\partial}{\partial x}\Psi(t, x)\right) = \frac{1}{2m}\frac{\hbar}{i}\frac{\partial}{\partial x}\left(\frac{\hbar}{i}\frac{\partial}{\partial x}\Psi(t, x)\right) =$$

$$= \frac{1}{2m}\left(\frac{\hbar}{i}\frac{\partial}{\partial x}\left(\frac{\hbar}{i}\frac{\partial}{\partial x}\right)\right)\Psi(t, x) = \frac{1}{2m}\left(\hat{p}\cdot\hat{p}\right)\Psi(t, x) = \frac{\hat{p}^2}{2m}\Psi(t, x),$$

which implies the definition of the operator energy $\hat{E}$ as

$$\hat{E} := \frac{\hat{p}^2}{2m} = -\frac{\hbar^2}{2m}\frac{\partial^2}{\partial x^2} \quad \text{or} \quad \hat{E} := -\frac{\hbar^2}{2m}\frac{d^2}{dx^2},$$

the last equality holds for one-dimensional problems. The result is a relation implying the *energy operator* $\hat{E}$, whose eigenstate Ψ corresponds to the "energy eigenvalue" E,

$$\hat{E}\Psi(t, x) = E\Psi(t, x)$$

together with its corresponding form in derivatives

$$i\hbar\frac{\partial}{\partial t}\Psi(t, x) = -\frac{\hbar^2}{2m}\frac{\partial^2}{\partial x^2}\Psi(t, x).$$

Therefore the Schrödinger equation for a particle is described by the energy operator $\hat{E}$ in the form

$$i\hbar\frac{\partial}{\partial t}\Psi(t, x) = \hat{E}\Psi(t, x).$$

These results can be stated as follow: the operators $\hat{p}$ and $\hat{E}$ represent the spatial and time translations, respectively, for the wave function $\Psi(t, x)$.

Let us reload the discussion from Lecture 31 now in the new language we developed. The Schrödinger equation is linear and allows more general solutions than the only de Broglie wave function $e^{i(kx-\omega t)}$. Any superposition

$$\Psi(t, x) = e^{i(k_1 x - \omega_1 t)} + e^{i(k_2 x - \omega_2 t)}$$

of de Broglie waves is a solution, but even if each term corresponds to a given momentum, the sum does not correspond to a definite momentum or a definite energy, too.

To see this fact, let us consider

$$\hat{p}\Psi(t, x) = \frac{\hbar}{i}\frac{\partial}{\partial x}\Psi(t, x) = \hbar k_1 e^{i(k_1 x - \omega_1 t)} + \hbar k_2 e^{i(k_2 x - \omega_2 t)} \neq \hbar k(e^{i(k_1 x - \omega_1 t)} + e^{i(k_2 x - \omega_2 t)}).$$

The wave packet

$$\Psi(t, x) = \int_{-\infty}^{\infty} e^{i(kx - \omega(k)t)}\phi(k)dk$$

is the general solution of the Schrödinger equation for any function $\phi(k)$. In fact, if we know an initial wave function $\Psi(0, x)$, we can construct its Fourier transform

$$\phi(k) = \int_{-\infty}^{\infty} e^{-ikx}\Psi(0, x)dx.$$

Now, with this given $\phi(k)$, we can construct the evolution in time of $\Psi(0, x)$, that is $\Psi(t, x)$.

We continue considering the case when the particle is moving under the action of an external potential $V(t, x)$. In the previous case, the total energy H was only the kinetic energy $E = \dfrac{p^2}{2m}$. In this case, it is $H = E + V = \dfrac{p^2}{2m} + V(t, x)$.

The postulated Schrödinger time independent equation, that we presented in Lecture 23, is

$$H\Psi = -\frac{\hbar^2}{2m}\frac{d^2\Psi}{dx^2} + V(x)\Psi.$$

By definition, the energy operator is the Hamiltonian operator, denoted by $\hat{H}$, that is

$$\hat{H} := -\frac{\hbar^2}{2m}\frac{\partial^2}{\partial x^2} + \hat{V}(x).$$

Let us continue with the *position operator* $\hat{x}$ which acts on $f(x)$ by the rule

$$\hat{x} f(x) := x f(x).$$

It is easy to observe that

$$\hat{x}^2 f(x) = \hat{x}(x f(x)) := x^2 f(x),$$

so, in general,

$$\hat{x}^k f(x) := x^k f(x).$$

For two linear operators $\bar{X}$ and $\bar{Y}$, we can define the *commutator operator*

$$[\bar{X}, \bar{Y}] := \bar{X}\bar{Y} - \bar{Y}\bar{X}.$$

It is easy to observe that the commutator of two linear operators is again a linear operator and $[\bar{X}, \bar{Y}] = -[\bar{Y}, \bar{X}]$. Of course, if the two operators commute, then $[\bar{X}, \bar{Y}] = \hat{0}$.

Exercise 9.1.1 If $\hat{p}$ is the momentum operator and $\hat{x}$ is the position operator, we have

$$[\hat{p}, \hat{x}] = -i\hbar.$$

Solution: Let us consider the one-dimensional case, so it is up to us if we use the formula with ∂ or the one using d.

$$[\hat{p}, \hat{x}] f(x) = (\hat{p}\hat{x} - \hat{x}\hat{p}) f(x) = \hat{p}\hat{x} f(x) - \hat{x}\hat{p} f(x) = \hat{p}(\hat{x} f(x)) - \hat{x}(\hat{p} f(x)) =$$

$$\left(\frac{\hbar}{i}\frac{\partial}{\partial x}\right)\hat{x} f(x) - \hat{x}\left(\frac{\hbar}{i}\frac{\partial}{\partial x}\right) f(x) = \frac{\hbar}{i} f(x) + \frac{\hbar}{i} x \left(\frac{\partial f}{\partial x}\right) - \frac{\hbar}{i} x \left(\frac{\partial f}{\partial x}\right) = -i\hbar f(x).$$

Looking at the next exercise we understand why the operators position and momentum, used in Quantum Mechanics cannot be represented as matrices with elements in $\mathbb{C}$. Matrices are linear operators and we can define eigenvalues and eigenvectors for matrices. It makes sense to define the commutator of two matrices and, if the matrices commute, the result is the 0 matrix. However an identity, as the one obtained in the previous exercise, is impossible to be obtained. Let us underline this idea: there is no matrix representation for the two previous operators, that is for the momentum and position operators.

Exercise 9.1.2 Show that it is not possible to find two matrices $X, Y \in M_n(\mathbb{C})$ such that

$$XY - YX = I_n.$$

Hint: It is easy to see that the trace of a matrix has the property

$$Tr(XY - YX) = 0.$$

Therefore, if the equality

$$XY - YX = I_n$$

is satisfied for a pair of matrices X and Y then the statement

$$0 = Tr I_n = n,$$

is false.

Let us continue with the wave function seen in Lecture 34, i.e.

$$\Psi(t, \mathbf{r}) = A e^{i(\mathbf{k} \cdot \mathbf{r} - \omega t)},$$

for $A = 1$, that is

$$\Psi(t, \mathbf{r}) = e^{i(k_x x + k_y y + k_z z - \omega t)}.$$

This is the wave function associated to the momentum $\mathbf{p} = \hbar \mathbf{k}$ and $H = E = \hbar \omega = \dfrac{\mathbf{p}^2}{2m} = \dfrac{1}{2m} \mathbf{p} \cdot \mathbf{p}$. It is easy to observe

$$\frac{\hbar}{i} \nabla \Psi = \frac{\hbar}{i} (i k_x, i k_y, i k_z) \Psi = \mathbf{p} \Psi.$$

Therefore the information "$\mathbf{p}$" is extracted from the wave function Ψ by the operator $\hat{\mathbf{p}}$ defined as

$$\hat{\mathbf{p}} := \frac{\hbar}{i} \nabla.$$

In the following, we need to change the classical notations for the coordinates and momentum by $(x_1, x_2, x_3) := (x, y, z)$ and $(p_1, p_2, p_3) := (p_x, p_y, p_z)$ to define

$$\hat{p}_k = \frac{\hbar}{i} \frac{\partial}{\partial x_k}, \ k = 1, 2, 3.$$

The position operator $\hat{\mathbf{x}}$ acts on $f(x)$ under the rule

$$\hat{\mathbf{x}} f(x) := (\hat{x}_1, \hat{x}_2, \hat{x}_3) f(x) = (x_1, x_2, x_3) f(x),$$

that is

$$\hat{x}_k f(x) = x_k f(x), \ k = 1, 2, 3.$$

It is important to observe that both the position operator $\hat{x}$ or the momentum operator $\hat{p}$ do not depend on time. The above relation

$$[\hat{p}, \hat{x}] = -i\hbar,$$

leads to

$$[\hat{p}_j, \hat{x}_k] = -i\hbar\delta_{jk}, \; j, k = 1, 2, 3,$$

where δ_{jk} is the Kroneker delta. To extract, also in this case, the information "energy", we have

$$\frac{\partial}{\partial t}\Psi = -i\omega\Psi,$$

i.e.

$$i\hbar\frac{\partial}{\partial t}\Psi = E\Psi = \frac{\mathbf{p}^2}{2m}\Psi,$$

therefore

$$E\Psi = \frac{1}{2m}\mathbf{p}\cdot\mathbf{p}\Psi = \frac{1}{2m}\mathbf{p}\cdot\frac{\hbar}{i}\nabla\Psi = \frac{1}{2m}\frac{\hbar}{i}\nabla\cdot\mathbf{p}\Psi =$$

$$= \frac{1}{2m}\frac{\hbar}{i}\nabla\cdot\frac{\hbar}{i}\nabla\Psi = \frac{1}{2m}\hat{\mathbf{p}}\cdot\hat{\mathbf{p}}\Psi = \frac{\hat{\mathbf{p}}^2}{2m}\Psi,$$

which implies the definition of the operator energy $\hat{E}$ in terms of Laplacian ∇^2

$$\hat{E} := \frac{\hat{\mathbf{p}}^2}{2m} = \frac{1}{2m}\frac{\hbar}{i}\nabla\cdot\frac{\hbar}{i}\nabla = -\frac{\hbar^2}{2m}\nabla^2.$$

Summary of Lecture 38. Starting from the wave function

$$\Psi(t, x) = e^{i(kx-\omega t)}$$

where ω and k are determined by the formulas $p = \hbar k$ and $H = E = \hbar\omega = \frac{p^2}{2m}$, we can define a momentum operator, denoted $\hat{p}$, which extracts the information p from the wave function Ψ,

$$\hat{p}\Psi = p\Psi.$$

This operator is defined as

$$\hat{p} := \frac{\hbar}{i}\frac{\partial}{\partial x} = -i\hbar\frac{d}{dx},$$

where the last equality is written for a one-dimensional problem. The same notation can be considered for the energy operator in one-dimensional case.

Therefore, the definition of the energy operator $\hat{E}$ as

$$\hat{E} := \frac{\hat{p}^2}{2m} = -\frac{\hbar^2}{2m}\frac{\partial^2}{\partial x^2} = -\frac{\hbar^2}{2m}\frac{d^2}{dx^2}$$

leads to its eigenstate Ψ corresponding to the "energy eigenvalue" E, that is

$$\hat{E}\Psi(t,x) = E\Psi(t,x).$$

It can be defined the position operator $\hat{x}$ which acts on $f(x)$ under the rule

$$\bar{x}f(x) := xf(x).$$

The most important relation implying $\hat{p}$ and $\hat{x}$ is

$$[\hat{p},\hat{x}] = -i\hbar, \quad \text{that is} \quad [\hat{x},\hat{p}] = i\hbar$$

which allows to eliminate a description of Quantum Mechanics by matrices of $M_n(\mathbb{C})$. If the wave function is defined in three-dimensional space, it is,

$$\Psi(t,\mathbf{r}) = e^{i(k_x x + k_y y + k_z z - \omega t)},$$

where $\mathbf{p} = \hbar\mathbf{k}$ and $H = E = \hbar\omega = \dfrac{\mathbf{p}^2}{2m} = \dfrac{1}{2m}\mathbf{p}\cdot\mathbf{p}$. It is easy to observe that

$$\frac{\hbar}{i}\nabla\Psi = \frac{\hbar}{i}(ik_x, ik_y, ik_z)\Psi = \mathbf{p}\Psi.$$

Therefore the information "$\mathbf{p}$" is extracted from the wave function Ψ by the operator $\hat{\mathbf{p}}$ defined as

$$\hat{\mathbf{p}} := \frac{\hbar}{i}\nabla.$$

Each component satisfies the relation

$$\hat{p}_k = \frac{\hbar}{i}\frac{\partial}{\partial x_k}, \quad k = 1,2,3.$$

The position operator $\hat{\mathbf{x}}$ acts on $f(x)$ under the rule

$$\hat{\mathbf{x}}f(x) := (\hat{x}_1, \hat{x}_2, \hat{x}_3)f(x) = (x_1, x_2, x_3)f(x),$$

that is $\hat{x}_k f(x) = x_k f(x)$, $k = 1,2,3$. In this case, the following relations hold

$$[\hat{p}_j, \hat{x}_k] = -i\hbar\delta_{jk}, \; j, k = 1, 2, 3,$$

where δ_{jk} is the Kroneker delta symbol.

The definition of the operator energy $\hat{E}$ in terms of the Laplace operator ∇^2 is

$$\hat{E} := \frac{\hat{\mathbf{p}}^2}{2m} = \frac{1}{2m}\frac{\hbar}{i}\nabla \cdot \frac{\hbar}{i}\nabla = -\frac{\hbar^2}{2m}\nabla^2$$

being related to Schrödinger equation when the potential is $V = 0$,

$$i\hbar\frac{\partial}{\partial t}\Psi = \hat{E}\Psi = -\frac{\hbar^2}{2m}\nabla^2\Psi.$$

9.2 Lecture 39: The Relation $\phi^* \nabla^2 \Psi - \Psi \nabla^2 \phi^* = \text{div}(\phi^* \nabla \Psi - \Psi \nabla \phi^*)$ and Its Consequences

Let us start from the time independent Schrödinger equation for a particle in presence of a potential $V(\mathbf{r})$, as in Lecture 35:

$$H\Psi(\mathbf{r}) = \left(-\frac{\hbar^2}{2m}\nabla^2 + V(\mathbf{r})\right)\Psi(\mathbf{r}).$$

Here, it is $\mathbf{r} = (x, y, z)$. The momentum vector $\mathbf{p}$ and wavenumber vector $\mathbf{k}$ are connected by the relation $\mathbf{p} = \hbar\mathbf{k}$ and a solution is the the wave function

$$\Psi(t, \mathbf{r}) = Ae^{i(\mathbf{k}\cdot\mathbf{r} - \omega t)}.$$

As we did it in the previous lecture, we can consider the Hamilton operator $\hat{H}$, here

$$\hat{H} := -\frac{\hbar^2}{2m}\nabla^2 + V(\mathbf{r})$$

and the time independent Schrödinger equation appears as an operator problem related to eigenvectors and eigenvalues, that is

$$\hat{H}\Psi = H\Psi.$$

Obviously, $\hat{H}$ is a linear operator.

Exercise 9.2.1 For two wave functions Ψ and ϕ the following equality holds

$$\phi^* \nabla^2 \Psi - \Psi \nabla^2 \phi^* = \mathrm{div}(\phi^* \nabla \Psi - \Psi \nabla \phi^*).$$

To show this, we start from a function $f : \mathbb{R}^3 \to C$, whose gradient is the vector $\nabla f \in \mathbb{C}^3$ established by the formula

$$\nabla f = \left(\frac{\partial f}{\partial x}, \frac{\partial f}{\partial y}, \frac{\partial f}{\partial z} \right).$$

Then, for a vector $A(x, y, z) = (A_x(x, y, z), A_y(x, y, z), A_z(x, y, z)) \in \mathbb{C}^3$ its divergence is expressed by the formula

$$\mathrm{div}\, A = \frac{\partial A_x}{\partial x} + \frac{\partial A_y}{\partial y} + \frac{\partial A_z}{\partial z}.$$

First, it is easy to see that the divergence applied to the gradient operator leads to the Laplace operator,

$$\mathrm{div}\, \nabla f = \mathrm{div} \left(\frac{\partial f}{\partial x}, \frac{\partial f}{\partial y}, \frac{\partial f}{\partial z} \right) = \frac{\partial}{\partial x} \left(\frac{\partial f}{\partial x} \right) + \frac{\partial}{\partial x} \left(\frac{\partial f}{\partial y} \right) + \frac{\partial}{\partial x} \left(\frac{\partial f}{\partial z} \right)$$

$$= \frac{\partial^2 f}{\partial x^2} + \frac{\partial^2 f}{\partial y^2} + \frac{\partial^2 f}{\partial z^2} = \nabla^2 f.$$

Then, if we take into account that

$$\mathrm{div}(\phi^* \nabla \Psi) = \mathrm{div} \left(\phi^* \frac{\partial \Psi}{\partial x}, \phi^* \frac{\partial \Psi}{\partial y}, \phi^* \frac{\partial \Psi}{\partial z} \right) = \frac{\partial}{\partial x} \left(\phi^* \frac{\partial \Psi}{\partial x} \right) + \frac{\partial}{\partial y} \left(\phi^* \frac{\partial \Psi}{\partial y} \right)$$

$$+ \frac{\partial}{\partial z} \left(\phi^* \frac{\partial \Psi}{\partial z} \right) = \frac{\partial \phi^*}{\partial x} \frac{\partial \Psi}{\partial x} + \frac{\partial \phi^*}{\partial y} \frac{\partial \Psi}{\partial y} + \frac{\partial \phi^*}{\partial z} \frac{\partial \Psi}{\partial z} + \phi^* \nabla^2 \Psi$$

and

$$\mathrm{div}(\Psi \nabla \phi^*) = \frac{\partial \Psi}{\partial x} \frac{\partial \phi^*}{\partial x} + \frac{\partial \Psi}{\partial y} \frac{\partial \phi^*}{\partial y} + \frac{\partial \Psi}{\partial z} \frac{\partial \phi^*}{\partial z} + \Psi \nabla^2 \phi^*,$$

it results

$$\mathrm{div}(\phi^* \nabla \Psi - \Psi \nabla \phi^*) = \phi^* \nabla^2 \Psi - \Psi \nabla^2 \phi^*. \ \square$$

In Lecture 31 we showed that the Hamilton operator, in the case of a free particle, denoted there by $\mathbb{H} := \dfrac{d^2}{dx^2}$, is Hermitian on $L^2([0, L])$. We started from a computation as

$$\left\langle v_n \,\bigg|\, \frac{d^2}{dx^2} u_n \right\rangle_{L^2([0,L])} = \int_0^L v_n^*(x) \frac{d^2 u_n}{dx^2}(x) dx = \int_0^L \frac{d^2 v_n^*}{dx^2}(x) u_n(x) dx = \left\langle \frac{d^2}{dx^2} v_n \,\bigg|\, u_n \right\rangle_{L^2([0,L])}$$

made for two sets $u_n, v_n \in \mathcal{C}^\infty([0, L])$ which converge to two functions u and v of $L^2([0, L])$. It is important to say that, in Lecture 31, all functions we used had "proper boundary conditions" that is they vanish in $x = 0$ and $x = L$. Finally we succeeded to show that

$$\left\langle v \mid \frac{d^2}{dx^2}u \right\rangle_{L^2([0,L])} = \left\langle \frac{d^2}{dx^2}v \mid u \right\rangle_{L^2([0,L])},$$

that is the Hamiltonian $\mathbb{H} := \dfrac{d^2}{dx^2}$ is a Hermitian operator in $L^2([0, L])$. We intend to show that $\hat{H}$ has the same property, now in $L^2(\mathbb{R}^3)$.

Proposition 9.2.2 *The Hamilton linear operator* $\hat{H} = -\dfrac{\hbar^2}{2m} \nabla^2 + V(t, \mathbf{r})$ *is a Hermitian operator in* $L^2(\mathbb{R}^3)$.

Proof Since $V(\mathbf{r}) \in \mathbb{R}$, it results, for all $\phi, \Psi \in L^2(\mathbb{R}^3)$ that

$$\langle \phi, V\Psi \rangle = \iiint_{\mathbb{R}^3} \phi^*(\mathbf{r})(V\Psi)(\mathbf{r})d\mathbf{r} = \iiint_{\mathbb{R}^3} (V\phi)^*(\mathbf{r})\Psi(\mathbf{r})d\mathbf{r} = \langle V\phi, \Psi \rangle.$$

To conclude, it is enough to prove that the Laplace operator is a Hermitian operator, that is, for all $\phi, \Psi \in L^2(\mathbb{R}^3)$, we have

$$\langle \phi, \nabla^2\Psi \rangle = \iiint_{\mathbb{R}^3} \phi^*(\mathbf{r}) \nabla^2 \Psi(\mathbf{r})d\mathbf{r} = \iiint_{\mathbb{R}^3} \nabla^2\phi^*(\mathbf{r})\Psi(\mathbf{r})d\mathbf{r} = \langle \nabla^2\phi, \Psi \rangle.$$

The idea is the same as in Lecture 31. We have to use an important result adapted now for $L^2(\mathbb{R}^3)$. It is a density property for the compact support $\mathcal{C}^\infty$ functions, denoted by $\mathcal{C}_0^\infty(\mathbb{R}^3)$ in $L^2(\mathbb{R}^3)$. By definition, the support of a function $f : \mathbb{R}^3 \to \mathbb{C}$ is the topological closure of the set of all $x \in \mathbb{R}^3$ such that $f(x) \neq 0$. It is enough to prove the property for two sets of functions ϕ_n and Ψ_n from $\mathcal{C}_0^\infty(\mathbb{R}^3)$ which converges to the two functions ϕ and Ψ of $L^2(\mathbb{R}^3)$. Therefore we have to prove

$$\langle \phi_n, \nabla^2\Psi_n \rangle = \iiint_{\mathbb{R}^3} \phi_n^*(\mathbf{r}) \nabla^2 \Psi_n(\mathbf{r})d\mathbf{r} = \iiint_{\mathbb{R}^3} \nabla^2\phi_n^*(\mathbf{r})\Psi_n(\mathbf{r})d\mathbf{r} = \langle \nabla^2\phi_n, \Psi_n \rangle.$$

The integrals are in fact calculated on a compact subset D_n in $\mathbb{R}^3$ because both functions have compact supports. We can even consider D_n as a sphere such that both functions are 0 on its surface, denoted by S_n, and outside it. We have

$$\iiint_{D_n} \left[\phi_n^* \nabla^2 \Psi_n - \Psi_n \nabla^2 \phi_n^* \right] d\mathbf{r} = \iiint_{D_n} \text{div}(\phi_n^* \nabla \Psi_n - \Psi_n \nabla \phi_n^*)d\mathbf{r}.$$

The Gauss divergence theorem states

$$\iiint_{D_n} \text{div}(\phi_n^* \triangledown \Psi_n - \Psi_n \triangledown \phi_n^*) d\mathbf{r} = \iint_{S_n} (\phi_n^* \triangledown \Psi_n - \Psi_n \triangledown \phi_n^*) \cdot \vec{n} dS.$$

We want to show that the last integral is 0. Using the Mean Value Theorem for multiple integrals applied twice, it results that u_0 and l_0 exist in S_n such that

$$\iiint_{D_n} \left[\phi_n^* \triangledown^2 \Psi_n - \Psi_n \triangledown^2 \phi_n^* \right] d\mathbf{r} = \phi_n^*(u_0) \iint_{S_n} \triangledown \Psi_n \cdot \vec{n} dS - \Psi_n(l_0) \iint_{S_n} \triangledown \phi_n^* \cdot \vec{n} dS.$$

The right member becomes 0. Since the difference between the two integrals is 0 and converges at

$$\langle \phi, \triangledown^2 \Psi \rangle - \langle \triangledown^2 \phi, \Psi \rangle$$

we obtain the desired result

$$\langle \phi, \triangledown^2 \Psi \rangle = \langle \triangledown^2 \phi, \Psi \rangle.$$

$\square$

Consider now the time-dependent Schrödinger equation

$$i\hbar \frac{\partial}{\partial t} \Psi(t, \mathbf{r}) = \left(-\frac{\hbar^2}{2m} \triangledown^2 + V(t, \mathbf{r}) \right) \Psi(t, \mathbf{r}),$$

as considered in Lecture 38, together with its complex conjugate

$$-i\hbar \frac{\partial}{\partial t} \Psi^*(t, \mathbf{r}) = \left(-\frac{\hbar^2}{2m} \triangledown^2 + V(t, \mathbf{r}) \right) \Psi^*(t, \mathbf{r}).$$

Multiply the first relation by Ψ^* and the second one by Ψ. It results

$$\left(\Psi^* \frac{\partial}{\partial t} \Psi + \Psi \frac{\partial}{\partial t} \Psi^* \right) = -\frac{\hbar}{2im} \left(\Psi^* \triangledown^2 \Psi - \Psi \triangledown^2 \Psi^* \right) = -\frac{\hbar}{2im} \text{div}(\Psi^* \triangledown \Psi - \Psi \triangledown \Psi^*),$$

that is

$$\frac{\partial}{\partial t} |\Psi|^2 = -\frac{\hbar}{2im} \text{div}(\Psi^* \triangledown \Psi - \Psi \triangledown \Psi^*).$$

If we denote by

$$\vec{J} := \frac{\hbar}{2im} (\Psi^* \triangledown \Psi - \Psi \triangledown \Psi^*),$$

the last equation becomes

$$\frac{\partial}{\partial t} |\Psi|^2 + \text{div} \, \vec{J} = 0,$$

which is a conservation law. We proceed as in Proposition 9.2.2 assuming the conservation law, written with respect to the set of functions Ψ_n from $\mathcal{C}_0^\infty(\mathbb{R}^3)$ which converges to the function Ψ in $L^2(\mathbb{R}^3)$, i.e. we write it as

$$\frac{\partial}{\partial t}|\Psi_n|^2 = -\frac{\hbar}{2im}\operatorname{div}(\Psi_n^* \nabla \Psi_n - \Psi_n \nabla \Psi_n^*) = -\operatorname{div} J_n.$$

Using the set $D_n \subset \mathbb{R}^3$ described in Proposition 9.2.2, it follows

$$\iiint_{D_n} \frac{\partial}{\partial t}|\Psi_n|^2 d\mathbf{r} + \iiint_{D_n} \operatorname{div} \vec{J}_n d\mathbf{r} = 0$$

and then

$$\frac{\partial}{\partial t}\iiint_{D_n}|\Psi_n|^2 d\mathbf{r} + \iint_{S_n} \vec{J}_n \cdot \vec{n} dS = 0.$$

The same argument used in Proposition 9.2.2 makes the last integral to be 0. When $n \to \infty$, it results

$$\frac{\partial}{\partial t}\iiint_{\mathbb{R}^3}|\Psi|^2 d\mathbf{r} = 0.$$

Therefore

$$\iiint_{\mathbb{R}^3}|\Psi|^2 d\mathbf{r} = \text{constant} = A.$$

It is clear the possibility to act on Ψ such that the constant becomes 1, i.e. this result expresses the idea that the particle must be somewhere in the entire space. Choosing

$$\Psi_1 := \frac{1}{A}\Psi,$$

we obtain

$$\iiint_{\mathbb{R}^3}|\Psi_1|^2 d\mathbf{r} = 1.$$

In this context, it is possible to offer another interpretation for the functions Ψ. We will refer to it as a state function which will be discussed in Lecture 41. To anticipate, $|\Psi|^2$ will be related to the probability to find the particle in a given volume and $\vec{J}$ is the probability current density. If we look back, we understand how important is the relation $\phi^* \nabla^2 \Psi - \Psi \nabla^2 \phi^* = \operatorname{div}(\phi^* \nabla \Psi - \Psi \nabla \phi^*)$ to prove that $\hat{H}$ is a Hermitian operator and the conservation law $\frac{\partial}{\partial t}\iiint_{\mathbb{R}^3}|\Psi|^2 d\mathbf{r} = 0$. For both results, the key concept is the density of $\mathcal{C}_0^\infty(\mathbb{R}^3)$ in $L^2(\mathbb{R}^3)$.

Summary of the Lecture 39. After we prove that, for two wave functions Ψ and ϕ, the following equality holds

$$\phi^* \nabla^2 \Psi - \Psi \nabla^2 \phi^* = \text{div}(\phi^* \nabla \Psi - \Psi \nabla \phi^*),$$

we demonstrated that $\hat{H} = -\dfrac{\hbar^2}{2m} \nabla^2 + V(t, \mathbf{r})$ is a Hermitian operator in $L^2(\mathbb{R}^3)$.

Since $V(\mathbf{r}) \in \mathbb{R}$, it results

$$\langle \phi, V\Psi \rangle = \iiint_{\mathbb{R}^3} \phi^*(\mathbf{r})(V\Psi)(\mathbf{r})d\mathbf{r} = \iiint_{\mathbb{R}^3} (V\phi)^*(\mathbf{r})\Psi(\mathbf{r})d\mathbf{r} = \langle V\phi, \Psi \rangle.$$

It remains to prove that the Laplace operator is Hermitian satisfying the relation

$$\langle \phi, \nabla^2 \Psi \rangle = \iiint_{\mathbb{R}^3} \phi^*(\mathbf{r}) \nabla^2 \Psi(\mathbf{r})d\mathbf{r} = \iiint_{\mathbb{R}^3} \nabla^2 \phi^*(\mathbf{r})\Psi(\mathbf{r})d\mathbf{r} = \langle \nabla^2 \phi, \Psi \rangle.$$

The last equality is a consequence of the result presented at the beginning of this summary. Using the time-dependent Schrödinger equation, we can derive

$$\frac{\partial}{\partial t}|\Psi|^2 = -\frac{\hbar}{2im} \text{div}(\Psi^* \nabla \Psi - \Psi \nabla \Psi^*),$$

which leads to the conservation law

$$\frac{\partial}{\partial t} \iiint_{\mathbb{R}^3} |\Psi|^2 d\mathbf{r} = 0,$$

which implies to interpret Ψ as a state function as we will see in Lecture 41.

9.3 Lecture 40. Similarities with Hamiltonian Formalism of Classical Mechanics

At the beginning of this book, in the first six lectures, we discussed the Lagrangian and the Hamiltonian formalisms of Classical Mechanics. Let us now connect them to Quantum Mechanics. The connection can be realized through the *Hamilton–Jacobi equation*. We have to consider the Hamilton principal function $S(q_k, t)$ and its total differential

$$dS = \sum_k \frac{\partial S}{\partial q_k} dq_k + \frac{\partial S}{\partial t} dt.$$

The time derivative of Hamilton principal function is

$$\frac{dS}{dt} = \sum_k \frac{\partial S}{\partial q_k} \dot{q}_k + \frac{\partial S}{\partial t}.$$

Considering now $p_k = \dfrac{\partial S}{\partial q_k}$ and $\dfrac{\partial S}{\partial t} = -H$, where H is the classical Hamiltonian, we have

$$\frac{dS}{dt} = \sum_k \frac{\partial S}{\partial q_k} \dot{q}_k + \frac{\partial S}{\partial t} = \sum_k p_k \dot{q}_k - H.$$

Alternatively, the right member, according to the Legendre transform, is the Lagrangian L. We obtain

$$\frac{dS}{dt} = \sum_k p_k \dot{q}_k - H = L$$

and

$$S = \int L \, dt,$$

that is the Hamilton principal function S of a given action. From

$$p_k = \frac{\partial S}{\partial q_k}, \quad k = 1, 2, 3$$

after replacing $x := q_1$, $y = q_2$, $z = q_3$, $p_x := p_1$, $p_y := p_2$, $p_z := p_3$, we have the Hamiltonian written in the form

$$H = \frac{1}{2m}(p_x^2 + p_y^2 + p_z^2) + V(t, x, y, z) = -\frac{\partial S}{\partial t}$$

transformed into the Hamilton–Jacobi equation of Classical Mechanics,

$$-\frac{\partial S}{\partial t} = \frac{1}{2m}\left[\left(\frac{\partial S}{\partial x}\right)^2 + \left(\frac{\partial S}{\partial y}\right)^2 + \left(\frac{\partial S}{\partial z}\right)^2\right] + V t, x, y, z).$$

For the Schrödinger equation, we can proceed in the following way. If

$$\Psi(t, \mathbf{r}) := e^{iS(t,\mathbf{r})/\hbar}$$

where S is a complex function depending on t and $\mathbf{r} = (x, y, z)$, we have

$$\frac{\partial \Psi}{\partial t} = \frac{i}{\hbar}\frac{\partial S}{\partial t}\Psi; \quad \frac{\partial \Psi}{\partial x} = \frac{i}{\hbar}\frac{\partial S}{\partial x}\Psi; \quad \frac{\partial^2 \Psi}{\partial x^2} = \frac{i}{\hbar}\frac{\partial^2 S}{\partial x^2}\Psi - \frac{1}{\hbar^2}\left(\frac{\partial S}{\partial x}\right)^2\Psi$$

and the same holds for the variables y and z. Replacing in the time-dependent Schrödinger equation

$$i\hbar\frac{\partial}{\partial t}\Psi(t, \mathbf{r}) = \left(-\frac{\hbar^2}{2m}\nabla^2 + V(t, \mathbf{r})\right)\Psi(t, \mathbf{r})$$

the following equality appears:

$$-\frac{\partial S}{\partial t} = \frac{1}{2m}\left[\left(\frac{\partial S}{\partial x}\right)^2 + \left(\frac{\partial S}{\partial y}\right)^2 + \left(\frac{\partial S}{\partial z}\right)^2\right] - \frac{i\hbar}{2m}\nabla^2 S + V(t, x, y, z).$$

This is the Hamilton–Jacobi equation for Quantum Mechanics. On the other hand, the classical Hamilton–Jacobi equation is recovered as soon as the term in $\hbar$ is negligible. According to this result, we realize that the Hamilton–Jacobi formalism works in Quantum Mechanics and it is directly connected to the same formalism in Classical Mechanics.

Summary of Lecture 40. The Hamilton–Jacobi equation of Classical Mechanics

$$-\frac{\partial S}{\partial t} = \frac{1}{2m}\left[\left(\frac{\partial S}{\partial x}\right)^2 + \left(\frac{\partial S}{\partial y}\right)^2 + \left(\frac{\partial S}{\partial z}\right)^2\right] + V(t, x, y, z)$$

depends on the Hamilton principal function S. This is an action and depends on coordinates and time. It can be obtained starting from the classical Hamiltonian and replacements suggested by Legendre's transform.

The equivalent of Hamilton–Jacobi equation in Quantum Mechanics is

$$-\frac{\partial S}{\partial t} = \frac{1}{2m}\left[\left(\frac{\partial S}{\partial x}\right)^2 + \left(\frac{\partial S}{\partial y}\right)^2 + \left(\frac{\partial S}{\partial z}\right)^2\right] - \frac{i\hbar}{2m}\nabla^2 S + V(t, x, y, z).$$

It is obtained by replacing, in the time-dependent Schrödinger equation,

$$i\hbar\frac{\partial}{\partial t}\Psi(t, \mathbf{r}) = \left(-\frac{\hbar^2}{2m}\nabla^2 + V(t, \mathbf{r})\right)\Psi(t, \mathbf{r})$$

the wave function

$$\Psi(t, \mathbf{r}) := e^{iS(t,\mathbf{r})/\hbar}$$

where S is a complex function depending on t and $\mathbf{r} = (x, y, z)$. The difference between classical and quantum formulation is due to the term containing $\hbar$ which is neglected in Classical Mechanics.

9.4   Lecture 41: From the Wave Function to the Quantum State. The Postulates of Quantum Mechanics

The Uncertainty Principle, formulated by Heisenberg in 1927, and the conservation law for Ψ (see Lecture 39) offer the possibility to set up a paradigm on how to interpret the concepts of Quantum Mechanics. The so-called *Copenhagen Interpretation* of Quantum Mechanics [26], proposed by Max Born, allows to understand the meaning of state function or quantum state. Essentially, the main statements are:

1. Each wave function Ψ is a pure state function and, from a mathematical point of view, it represents an element of a Hilbert finite/infinite/real/complex space.
2. Each superposition of pure state functions is also a state function of the system. They are elements of a Hilbert space and the coefficients of pure states, involved in the superposition, are real/complex arbitrary numbers.
3. Mixed states are state functions. They correspond to a probabilistic mixture of pure states. In fact, mixed states represent the probabilistic degree of knowledge of a statistical ensemble of independent systems. From a mathematical point of view, the mixed states Ψ can be expressed with respect to pure states Ψ_n whose coefficients a_n are real numbers called probability amplitudes. The square of such probability amplitude, $p_n := |a_n|^2 = a_n a_n^*$, means the probability to find the system Ψ in the pure state Ψ_n.

These statements have to be understood with respect to experimental results. Suppose we study a large number of independent identical systems, each system consisting of one particle moving under the same force acting for all systems. We call this situation as identically prepared systems. It is clear that a wave function Ψ describes all this set of systems, that is Ψ can describe the state of the set of identically prepared systems. This justifies why Ψ is often called the quantum state of the system or a quantum vector state.

We can consider particles in a one-dimensional space or in a three-dimensional space. The wave function Ψ depends then on (t, x) or on (t, x, y, z).

Born's approach to the problem is: The state function Ψ points out the probability of finding the particle, at a given time, in a given space region. Let us suppose that the Hilbert space we use for modeling the experiment is $L^2(\mathbb{R})$. Taking into account this approach, it is mandatory to have

$$\int_{-\infty}^{\infty} |\Psi(t, x)|^2 dx = \int_{-\infty}^{\infty} \Psi^*(t, x)\Psi(t, x)dx = 1,$$

that is the particle is somewhere on the real line. If measurements are made for all systems, the probability to find at the time t the particle, represented by Ψ, inside the interval centered at x_0 and having the dimension $2a$ denoted by U is

$$P(t, U) = \int_{x_0-a}^{x_0+a} |\Psi(t, x)|^2 dx = \int_{x_0-a}^{x_0+a} \Psi^*(t, x)\Psi(t, x)dx.$$

Therefore, the state function of the system is the state function of a set of identically prepared systems and the probability to find the particle in an interval can be computed by the above formula.

Let us now look what happens when we consider the Hilbert space $L^2(\mathbb{R}^3)$. Since the probability to find the particle in the entire space is 1, we have, according to both the conservation law derived in Lecture 39 and the inner product, that

$$\iiint_{\mathbb{R}^3} |\Psi(t,\mathbf{r})|^2 d\mathbf{r} = \iiint_{\mathbb{R}^3} \Psi^*(t,\mathbf{r})\Psi(t,\mathbf{r})d\mathbf{r} = 1,$$

where $d\mathbf{r} = dxdydz$. Exactly as in the one-dimensional case, these functions are square-integrable functions, that is

$$\iiint_{\mathbb{R}^3} |f(\mathbf{r})|^2 d\mathbf{r} < +\infty$$

can be replaced by $\Psi = Af$, where $A \in \mathbb{C}$, such that

$$AA^* \iiint_{\mathbb{R}^3} |f(\mathbf{r})|^2 d\mathbf{r} = 1.$$

We say that f is normed at unity by the complex number A and so the integral condition related to the probability meaning makes sense.

Furthermore, we can observe that, if Ψ is a square-integrable function, then $e^{i\beta}\Psi$ is a square-integrable function and it can be normed at unity by the same complex number A.

It becomes clear that Quantum Mechanics is related to the $L^2(\mathbb{R}^3)$ space which can be organized as a Hilbert space in the same way as we did for $L^2(\mathbb{R})$. Here the probability to have the particle at time t in a region centered at (x_0, y_0, z_0) with dimensions $2a, 2b, 2c$ denoted by V, is

$$P(t,V) - \iiint_V |\Psi(t,r)|^2 d\mathbf{r} = \iiint_V \Psi^*(t,\mathbf{r})\Psi(t,\mathbf{r})d\mathbf{r}.$$

Therefore we can observe that the Born postulate regarding the probability is mathematically justified and strengthened by the conservation law discussed in Lecture 39.

Let us see if the previous description of the state function works in the case of double-split experiment describing the interference. Here we consider the Hilbert space $\mathbb{C}$. If $\Psi_1(t,x)$ is a possible state of the system and another state is $\Psi_2(t,x)$, then any other linear combination

$$\Psi(t,x) = c_1\Psi_1(t,x) + c_2\Psi_2(t,x)$$

where c_1 and c_2 are complex numbers, can describe any other possible state of the system.

By definition, this state of the system is called a superposition of the two given states $\Psi_1(t, x)$ and $\Psi_2(t, x)$. The quantity $|\Psi|^2$ is the probability to find the particle in a volume associated to a point as we previously discussed. Here it is given by the complex number Ψ,

$$
\begin{aligned}
|\Psi|^2 &= [c_1^* \Psi_1^* + c_2^* \Psi_2^*][c_1 \Psi_1 + c_2 \Psi_2] \\
&= |c_1 \Psi_1|^2 + |c_2 \Psi_2|^2 + 2Re[c_1 c_2^* \Psi_1 \Psi_2^*] \neq |c_1 \Psi_1|^2 + |c_2 \Psi_2|^2
\end{aligned}
$$

as we expected.

The interference can be derived if we consider the state functions in the forms,

$$
\Psi_1 = |\Psi_1(t, x)| e^{i\alpha_1} \quad \text{and} \quad \Psi_2 = |\Psi_2(t, x)| e^{i\alpha_2},
$$

that is

$$
\Psi(t, x) = c_1 |\Psi_1(t, x)| e^{i\alpha_1} + c_2 |\Psi_2(t, x)| e^{i\alpha_2}.
$$

In this case

$$
|\Psi|^2 = |c_1 \Psi_1|^2 + |c_2 \Psi_2|^2 + 2Re[\, c_1 c_2^* \, |\Psi_1| |\Psi_2| e^{i(\alpha_1 - \alpha_2)} \,]
$$

where

$$
2Re[\, c_1 c_2^* \, |\Psi_1| |\Psi_2| e^{i(\alpha_1 - \alpha_2)} \,]
$$

is the interference term obtained by the superposition of the given wave states, and it is easier to understand it if one consider the case $c_1, c_2 \in \mathbb{R}$.

Now if we look at the Schrödinger equation, we observe that if Ψ_1 is a solution and Ψ_2 is another solution, any linear combination

$$
\Psi(t, \mathbf{r}) = c_1 \Psi_1(t, \mathbf{r}) + c_2 \Psi_2(t, \mathbf{r})
$$

where c_1 and c_2 are complex numbers is a solution of the Schrödinger equation

$$
i\hbar \frac{\partial}{\partial t} \Psi(t, \mathbf{r}) = \left(-\frac{\hbar^2}{2m} \nabla^2 + V(t, \mathbf{r}) \right) \Psi(t, \mathbf{r}).
$$

But only in few situations the combination $c_1 \Psi_1(t, \mathbf{r}) + c_2 \Psi_2(t, \mathbf{r})$ can be written as $A e^{i(\mathbf{k} \cdot \mathbf{r} - \omega t)}$. All other solutions, imagined in the above form, do not describe classical waves. This means that Quantum Mechanics begins when these mixed solutions describe realistic systems, and this means that Ψ has to be seen as a state of the system. In summary we can only predict the probability to find a particle in a given volume.

In the Copenhagen Interpretation of Quantum Mechanics, as formulated by Max Born, *"the quantum state is a mathematical entity which provides the probability distribution for each possible measurement on the system together with the evolution of the system by the Schrödinger equation".*

Let us recall some examples to underline other state functions we discussed along our lectures. In Lecture 24, Lecture 25, Lecture 3, and Lecture 33, we gave information on pure states or superpositions of pure states which are solutions of the Schrödinger equation. We identified even the related Hilbert spaces. The eigenvalues are associated to possible values of the energy levels of particles.

In Lecture 24, where we studied a particle in a box, we have a time independent Schrödinger equation

$$\frac{d^2\Psi}{dx^2} = \frac{2m}{\hbar^2}(V - H)\Psi.$$

$V - H$ was a positive quantity because we supposed the particle in the box. The previous equation was written as

$$\frac{d^2\Psi}{dx^2} = K^2\Psi,$$

where

$$K^2 = \frac{2m}{\hbar^2}(V - H)$$

and the solution was

$$\Psi(x) = A_1 e^{Kx} + A_2 e^{-Kx}.$$

Describing a particle in the box it resulted $A_1 = 0$. If not, as $x \to +\infty$, the solution Ψ approaches to $+\infty$. Therefore, inside the box, the solution was

$$\Psi(x) = A_2 e^{-Kx},$$

that is a pure state.

In the case $V(x) = 0$, for the particle inside the box, we have

$$\frac{d^2\Psi}{dx^2} = -k^2\Psi, \quad k^2 = \frac{2m}{\hbar^2}H.$$

The solution is similar to the case of the free particle in Lecture 23, but some conditions have to be imposed to preserve the particle inside the box. This happens when the box is infinitely deep. We write

$$\Psi(x) = A_1 e^{ikx} + A_2 e^{-ikx} = C_1 \cos kx + C_2 \sin kx, \quad C_1, C_2 \in \mathbb{C}.$$

The constants are identified imposing some physical conditions and we found

$$\Psi(x) = C_2 \sin \frac{n\pi}{L}x.$$

The energy of the particle,

$$H = T = \frac{p^2}{2m} = \frac{h^2\pi^2}{2mL^2}n^2,$$

is quantized. Introducing $P(x)$ as the probability to have the particle in the interval $x - dx/2, x + dx/2$ since the particle has always to be inside the box, the constant C_2 was determined from the condition

$$\int_0^L |\Psi(x)|^2 dx = 1,$$

the final result being

$$\Psi(x) = \sqrt{\frac{2}{L}} \sin \frac{n\pi}{L}x.$$

The description of this case was related to the Hilbert space $L^2([0, L])$ and the solution, initially a superposition, was transformed by the initial conditions into a pure state. Therefore Lecture 24 is a preliminary case of the axiomatic frame described in this Lecture.

The same situation works for the quantum harmonic oscillator described in Lectures 25 and 31. We started from

$$\Psi(t, x) = e^{-iEt/\hbar}\Psi(x)$$

as the pure state function describing an electron in the harmonic oscillator. Since the solution Ψ_n, corresponding to the n-energy level, is related to the Hermite polynomials which are usually denoted by H_n, in Lecture 25 we have used the letter E instead of H to describe the total energy. Replacing the pure state into the Schrödinger time independent equation

$$i\hbar\frac{d\Psi}{dt}(t, x) = -\frac{\hbar^2}{2m}\frac{d^2\Psi}{dx^2}(t, x) + V(x)\Psi(t, x)$$

corresponding to this case when $V(t, x) = V(x) = \frac{1}{2}m\omega^2 x^2$, we obtained, after denoting by

$$u := \sqrt{\frac{m\omega}{\hbar}}x; \quad \varepsilon := \frac{2E}{\hbar\omega},$$

the equation to solve

$$\frac{d^2\Psi}{du^2}(u) + (\varepsilon - u^2)\Psi(u) = 0.$$

This equation can be solved if $\varepsilon = \varepsilon_n = 2n + 1$. Therefore the n-level quantized energy was

$$E_n = \left(n + \frac{1}{2} \right) \hbar \omega$$

for

$$\Psi_n(x) = \frac{1}{\sqrt{2^n n!}} \left(\frac{m\omega}{\pi\hbar} \right)^{1/4} H_n \left(\sqrt{\frac{m\omega}{\hbar}} x \right) e^{-m\omega x^2/2\hbar}$$

and

$$\Psi_n(t,x) = e^{-iE_n t/\hbar} \frac{1}{\sqrt{2^n n!}} \left(\frac{m\omega}{\pi\hbar} \right)^{1/4} H_n \left(\sqrt{\frac{m\omega}{\hbar}} x \right) e^{-m\omega x^2/2\hbar}.$$

Since the Schrödinger equation is linear, the general solution was described by a superposition of pure states, that is a sum of all modes of n-oscillations, that is

$$\Psi(t,x) = \sum_n c_n e^{-iE_n t/\hbar} \Psi_n(x).$$

The Hermite polynomials are involved in the functions

$$\varphi_n(u) = e^{-u^2/2} H_n(u), \ n \in \mathbb{N}$$

which are part of an orthogonal system in $L^2(\mathbb{R})$ with respect to the usual inner product because we have proved that

$$\int_{-\infty}^{\infty} H_n(u) H_l(u) e^{-u^2} du = \begin{cases} 0, \ n \neq l \\ 2^n \sqrt{\pi} \, n!, \ l = n \end{cases}$$

After a convenient coordinate change in the previous formulas

$$x := \sqrt{\frac{\hbar}{mw}} u,$$

we found

$$\Psi_n(x) = \frac{1}{\sqrt{2^n n!}} \left(\frac{m\omega}{\pi\hbar} \right)^{1/4} H_n \left(\sqrt{\frac{m\omega}{\hbar}} x \right) e^{-m\omega x^2/2\hbar}$$

highlighting an orthonormal system of $L^2(\mathbb{R})$. As a consequence, the description of the quantum harmonic oscillator solutions is related to the probability condition required by Born's postulate. With these examples in mind, we can write the

Postulates of the Copenhagen Interpretation of Quantum Mechanics:

1. **The state of a quantum system is completely described by the state function Ψ which represents an element of a Hilbert space. The state function Ψ is normed at unity, that is|, $|\Psi|^2 = \langle \Psi | \Psi \rangle = 1$.**
   The above examples confirm this first postulate.

2. **The state function Ψ points out the probability of finding the particle, at a given time, in a given space region.**
   It depends on the Hilbert space inner product how the probability formula looks like. As we saw, in the particular case of the Hilbert space $L^2(\mathbb{R}^3)$, the probability is

$$P(t, V) = \iiint_V |\Psi(t, \mathbf{r})|^2 d\mathbf{r} = \iiint_V \Psi^*(t, \mathbf{r})\Psi(t, \mathbf{r}) d\mathbf{r}.$$

If we work in the Hilbert space $L^2(\mathbb{R})$, the probability to find the particle described by the state function Ψ, now in the interval V, is

$$P(t, V) = \int_V |\Psi(t, x)|^2 dx = \int_V \Psi^*(t, x)\Psi(t, x) dx.$$

Since the probability to find the particle somewhere on the line is 1, the state function which describes the particle has to fulfill the property

$$P(t, \mathbb{R}) = \int_{-\infty}^{\infty} |\Psi(t, x)|^2 dx = \int_{-\infty}^{\infty} \Psi^*(t, x)\Psi(t, x) dx = 1.$$

The same must happen when we study a quantum particle in the three dimensional space. Since the probability to find the particle in the entire space is 1, if we suppose that the entire space is $\mathbb{R}^3$ and the states are elements in the Hilbert space $L^2(\mathbb{R}^3)$, it must be

$$\iiint_{\mathbb{R}^3} |\Psi(t, \mathbf{r})|^2 d\mathbf{r} = \iiint_{\mathbb{R}^3} \Psi^*(t, \mathbf{r})\Psi(t, \mathbf{r}) d\mathbf{r} = 1.$$

According to the first axiom, the state function Ψ has to fulfill this property in order to express correctly the above probability. Such a state function is called *normalized state function*.

Finally, let us suppose to work in a two-dimensional real Hilbert space and the state function Ψ is described by a column vector having two complex components a and b. Therefore, according to the previous description, it is mandatory to have $aa^* + bb^* = 1$. If the components are real numbers the previous condition becomes $a^2 + b^2 = 1$.

3. **The state function Ψ evolves in time according to the time-dependent Schrödinger equation**

$$i\hbar \frac{\partial}{\partial t}\Psi(t, \mathbf{r}) = \left(-\frac{\hbar^2}{2m}\nabla^2 + V(t, \mathbf{r})\right)\Psi(t, \mathbf{r}).$$

The state function has to be related to the evolution of the quantum system and such an evolution has to be in agreement with experiments. We measure possible values of a dynamical variables as position, momentum, energy and so on. Therefore, the result of an experiment is something we measure by a device, for example on a screen where the measure appears as a real number or as some other information which can be converted in real numbers. If we measure the momentum, from

$$\frac{\hbar}{i} \nabla \Psi = \frac{\hbar}{i}(ik_x, ik_y, ik_z)\Psi = \mathbf{p}\Psi$$

the information "$\mathbf{p}$" is extracted from the wave function Ψ on which acts the operator $\hat{\mathbf{p}}$ defined by

$$\hat{\mathbf{p}} := \frac{\hbar}{i} \nabla .$$

Each component satisfies the equation

$$\hat{p}_k = \frac{\hbar}{i}\frac{\partial}{\partial x_k}, \; k = 1, 2, 3.$$

Working with standard coordinates means to replace x_1 by x, etc., that is there exists a momentum operator $\hat{p}_x$ corresponding to the real value "momentum" p_x in the direction x. In the case of a single particle, we have proved that this is $-i\hbar\frac{\partial}{\partial x}$. Therefore each component leads to a linear operator, called observable, as $\hat{p}_x$ is, and an extracted real eigenvalue p_x, i.e. $\hat{p}_x\Psi = p_x\Psi$.

4. **The average value $\langle\hat{A}\rangle$ corresponding to the operator $\hat{A}$ and to the normalized state function Ψ is**

$$\langle\hat{A}\rangle := \langle\Psi|\hat{A}|\Psi\rangle.$$

$\langle\hat{A}\rangle$ **is also called the expectation value, or simply, the mean value of the operator $\hat{A}$. This expectation value results after a measurement.**

It is important to stress that this notion is related to the normalized state functions, therefore each time we have expectation values, they are related to normalized state functions. The concept represents the "average" of all possible outcomes values of measurements related to a given experiment. If we make measurements of the position of an electron, the expectation value is related to the average position of it. It will be denoted as $\langle\hat{x}\rangle$ and, according to the previous formula, for the state Ψ, its value is

$$\langle\hat{x}\rangle := \langle\Psi|\hat{x}|\Psi\rangle.$$

If we remember that the position operator $\hat{x}$ acts on $\Psi(x)$ according to the rule

$$\hat{x}\Psi(x) := x\Psi(x),$$

in $L^2(\mathbb{R})$, it can be computed using the formula

$$\langle \hat{x} \rangle := \langle \Psi | \hat{x} | \Psi \rangle = \int_{-\infty}^{\infty} \Psi^*(x) x \Psi(x) dx = \int_{-\infty}^{\infty} x \Psi^*(x) \Psi(x) dx = \int_{-\infty}^{\infty} x P(x) dx$$

which represents the average position of the electron after a large number of measurements made on identically prepared systems.

If we consider the momentum operator corresponding to the same case, since the momentum operator is defined by

$$\hat{p} := \frac{\hbar}{i} \frac{\partial}{\partial x} ,$$

its expectation value, corresponding to the normalized state function Ψ, is

$$\langle \hat{p} \rangle = -i\hbar \int_{-\infty}^{\infty} \Psi^*(x) \frac{\partial \Psi(x)}{\partial x} dx.$$

The expectation value strengthens the probabilistic interpretation of "objects" in Quantum Mechanics. At the same time, together with the Ehrenfest theorem presented in the next lecture, it is the way to see the basic formulas of Classical Mechanics

$$m \frac{dx}{dt} = p \text{ and } \frac{dp}{dt} = -\frac{dV}{dx} = F$$

expressed in quantum mechanical form

$$m \frac{d}{dt} \langle \hat{x} \rangle = \langle \hat{p} \rangle \text{ and } \frac{d}{dt} \langle \hat{p} \rangle = \left\langle -\frac{\partial \hat{V}}{\partial x}(\hat{x}) \right\rangle = \langle \hat{F} \rangle.$$

Therefore the rate of change in time of the position operator expectation value is the ratio between the momentum operator expectation value and the particle mass. Analogously, the rate of change in time of the momentum operator expectation value is the expectation value of the force which acts on the particle.
It is worth stressing that not all operators has a real expectation value but only Hermitian operators. Indeed,

$$\langle \hat{A} \rangle^* = \langle \Psi | \hat{A} | \Psi \rangle^* = \langle \Psi | \hat{A} \Psi \rangle^* = \langle \hat{A} \Psi | \Psi \rangle = \langle \Psi | \hat{A} \Psi \rangle = \langle \Psi | \hat{A} | \Psi \rangle = \langle \hat{A} \rangle.$$

An operator which has a real expectation value is called an "observable" and it is described in the next postulates.

5. **Every observable is given by a Hermitian operator acting on a Hilbert space whose elements are the states Ψ.**

Let us remember that we have proved that the Hamiltonian operator is a Hermitian one if the Hilbert space of states is $L^2(\mathbb{R}^3)$. We know that its eigenvalues are real numbers. So, the next postulate results natural:

6. **In any measurement of the observable $\hat{A}$, the only observed values are real eigenvalues a such that**

$$\hat{A}\Psi = a\Psi.$$

7. **If the states Ψ_n describe a quantum system, then the eigenvalues a_n satisfy the relation**

$$\hat{A}\Psi_n = a_n\Psi_n$$

   and an arbitrary state of the system is

$$\Psi = \sum_n c_n\Psi_n,$$

   where c_n are related to a_n in the following way: the probability of observing the eigenvalue a_n is $c_n^* c_n$. Therefore, Hermitian operators are the mathematical way to describe measurements in Quantum Mechanics.

8. **In Quantum Mechanics two observables which cannot be simultaneously measured are expressed by the operators $\hat{A}$, $\hat{B}$ such that**

$$[\hat{A}, \hat{B}] = \hat{A}\hat{B} - \hat{B}\hat{A} \neq \hat{0}.$$

The commutator is related to the *order of measurements* in Quantum Mechanics. If the order does not matter, the commutator is $\hat{0}$. Let us explain this statement.

Consider the observables $\hat{A}$ and $\hat{B}$ and suppose we would like to make a measurement with respect to a state of the system, here denoted by Ψ. This means that each one of them offers an information in this state, that is $\hat{A}\Psi = a\Psi$ and $\hat{B}\Psi = b\Psi$. Now consider an experiment which consists in finding simultaneously information on the state Ψ. The same operators extract information in the order: first $\hat{B}$ then $\hat{A}$. Therefore

$$(\hat{A}\hat{B})\Psi = \hat{A}(\hat{B}\Psi) = \hat{A}(b\Psi) = b\hat{A}\Psi = ba\Psi.$$

Now, consider the order: first $\hat{A}$ then $\hat{B}$.

$$(\hat{B}\hat{A})\Psi = \hat{B}(\hat{A}\Psi) = \hat{B}(a\Psi) = a\hat{B}\Psi = ab\Psi.$$

If we extract the second relation from the first relation we obtain

$$(\hat{A}\hat{B} - \hat{B}\hat{A})\Psi = (ba - ab)\Psi = 0 \cdot \Psi = 0,$$

relation which implies

$$[\hat{A}, \hat{B}] = \hat{0}.$$

In fact we see that the order does not matter if there is the same state which gives information about both observables.

We know that there are observables whose commutator is not $\hat{0}$. An example of such observables are the position operator $\hat{x}$ and the momentum operator $\hat{p}$ because we have proved

$$[\hat{x}, \hat{p}] = i\hbar.$$

The order matter so we cannot extract simultaneously information about position and momentum.

Let us provide, in the one-dimensional case, an example with two observables which commute. In Lecture 38, we have considered the energy operator in the absence of a potential field $V(x)$ as

$$\hat{E} := -\frac{\hbar^2}{2m}\frac{d^2}{dx^2}.$$

Let us prove that

$$[\hat{p}, \hat{E}] = \hat{0}.$$

Indeed,

$$[\hat{p}, \hat{E}]\Psi = (\hat{p}\hat{E} - \hat{E}\hat{p})\Psi = -i\hbar\frac{d}{dx}\left(-\frac{\hbar^2}{2m}\frac{d^2}{dx^2}\Psi\right) + \frac{\hbar^2}{2m}\frac{d^2}{dx^2}\left(-i\hbar\frac{d}{dx}\Psi\right) =$$

$$= \frac{i\hbar^3}{2m}\left[\frac{d}{dx}\left(\frac{d^2}{dx^2}\Psi\right) - \frac{d^2}{dx^2}\left(\frac{d}{dx}\Psi\right)\right] = 0$$

that is, the momentum operator and the energy operator commute. Thus, we can find the same eigenfunctions states with definite real values eigenvalues for the two observables, while in the previous case, with the position and momentum operators, it was not possible.

A further comment is necessary. If the particle moves in a potential $V(x)$, the total energy is now the Hamiltonian H. The attached observable is

$$\hat{H} = \hat{E} + \hat{V},$$

i.e.

$$[\hat{p}, \hat{H}]\Psi = [\hat{p}, \hat{E} + \hat{V}]\Psi = [\hat{p}, \hat{E}]\Psi + [\hat{p}, \hat{V}] = [\hat{p}, \hat{V}] = -i\hbar\Psi\frac{d\hat{V}}{dx} \neq 0.$$

This means that the total energy and momentum of a particle ca not be determined precisely and simultaneously if the particle moves in a potential $V(x)$.

Summary of Lecture 41. Quantum Mechanics can be formulated, in a probabilistic interpretation, according to the following statements:

1. The state of a quantum system is completely described by the state function Ψ which represents an element of a Hilbert space. The state function Ψ is normed at unity, that is $|\Psi|^2 = \langle\Psi|\Psi\rangle = 1$.
2. The state function Ψ points out the probability of finding the particle, at a given time, in a given space region.
3. The state function Ψ evolves in time according to the time-dependent Schrödinger equation

$$i\hbar\frac{\partial}{\partial t}\Psi(t, \mathbf{r}) = \left(-\frac{\hbar^2}{2m}\nabla^2 + V(t, \mathbf{r})\right)\Psi(t, \mathbf{r}).$$

4. The expectation value $\langle\hat{A}\rangle$, corresponding to the operator $\hat{A}$ and to the normalized state Ψ, is

$$\langle\hat{A}\rangle := \langle\Psi|\hat{A}|\Psi\rangle.$$

5. Every observable is a Hermitian operator on a Hilbert space whose elements are the states Ψ.
6. In any measurement of the observable $\hat{A}$, the only values observed are the real eigenvalues a such that

$$\hat{A}\Psi = a\Psi.$$

7. If the states Ψ_n describe the system, then the eigenvalues a_n satisfy

$$\hat{A}\Psi_n = a_n\Psi_n$$

and an arbitrary state of the system is

$$\Psi = \sum_n c_n\Psi_n,$$

where c_n are related to a_n in the following way: the probability to observe
the eigenvalue a_n is $c_n^* c_n$.

8. In Quantum Mechanics, two observables which cannot be simultaneously
measured are expressed by the operators $\hat{A}$ and $\hat{B}$ such that

$$[\hat{A}, \hat{B}] = \hat{A}\hat{B} - \hat{B}\hat{A} \neq \hat{0}.$$

Chapter 10
Consequences of Quantum Mechanics Principles

Anything one man can imagine,
other men can make real.

Jules Verne

10.1 Lecture 42: The Ehrenfest Theorem

In Lecture 41, we discussed about the expectation value of a given physical quantity and how this important concept relates the formalism of Classical Mechanics to Quantum Mechanics. To give an image, the evolution of a particle is understood through the evolution of its expectation value. If you read the George Gamow book *"The wonderful world of Mr. Tompkins"*, you can image the character trying to kill the quantum fly using a quantum splash and hitting the dense part of the shadow cloud which represents the quantum fly description. The foundation of this description is based on the following

Theorem 10.1.1 (Ehrenfest's theorem) *Consider an observable $\hat{A}$ in a quantum system whose Hamilton operator is $\hat{H}$. Then, the time-dependent Schrödinger equation leads to*

$$\frac{d}{dt}\left\langle \hat{A} \right\rangle = \frac{1}{i\hbar}\left\langle [\hat{A}, \hat{H}] \right\rangle + \left\langle \frac{\partial \hat{A}}{\partial t} \right\rangle.$$

Proof

$$\frac{d}{dt}\left\langle \hat{A} \right\rangle = \frac{d}{dt}\left\langle \Psi \left| \hat{A} \right| \Psi \right\rangle = \left\langle \frac{\partial}{\partial t}\Psi \left| \hat{A} \right| \Psi \right\rangle + \left\langle \Psi \left| \frac{\partial}{\partial t}\hat{A} \right| \Psi \right\rangle + \left\langle \Psi \left| \hat{A} \right| \frac{\partial}{\partial t}\Psi \right\rangle =$$

$$= \left\langle \frac{\partial}{\partial t}\Psi \left| \hat{A} \right| \Psi \right\rangle + \left\langle \frac{\partial \hat{A}}{\partial t} \right\rangle + \left\langle \Psi \left| \hat{A} \right| \frac{\partial}{\partial t}\Psi \right\rangle.$$

© The Author(s), under exclusive license to Springer Nature Switzerland AG 2021
S. Capozziello and W.-G. Boskoff, *A Mathematical Journey to Quantum Mechanics*,
UNITEXT for Physics, https://doi.org/10.1007/978-3-030-86098-1_10

Let us remember that the Hamiltonian operator is Hermitian. From time-dependent Schrödinger equation written in the form

$$\frac{\partial \Psi}{\partial t} = \frac{1}{i\hbar} \hat{H} \Psi$$

it results

$$\frac{d}{dt} \left\langle \hat{A} \right\rangle = \left\langle \frac{1}{i\hbar} \hat{H} \Psi \left| \hat{A} \right| \Psi \right\rangle + \left\langle \frac{\partial \hat{A}}{\partial t} \right\rangle + \left\langle \Psi \left| \hat{A} \right| \frac{1}{i\hbar} \hat{H} \Psi \right\rangle =$$

$$= \left\langle \Psi \left| -\frac{1}{i\hbar} \hat{H} \hat{A} \right| \Psi \right\rangle + \left\langle \frac{\partial \hat{A}}{\partial t} \right\rangle + \left\langle \Psi \left| \frac{1}{i\hbar} \hat{A} \hat{H} \right| \Psi \right\rangle = \frac{1}{i\hbar} \left\langle [\hat{A}, \hat{H}] \right\rangle + \left\langle \frac{\partial \hat{A}}{\partial t} \right\rangle.$$

$\square$

Corollary 10.1.2 *If the operator $\hat{A}$ does not depend on t, the statement of the Ehrenfest theorem becomes*

$$\frac{d}{dt} \left\langle \hat{A} \right\rangle = \frac{1}{i\hbar} \left\langle [\hat{A}, \hat{H}] \right\rangle.$$

The last equation states that the rate of change in time of the expectation value of the operator $\hat{A}$ can be described with respect to the expectation value of a bracket involving the initial operator and the Hamiltonian operator of the system. It recalls us a relation like

$$\frac{dF}{dt} = [F, H],$$

discussed in Lecture 6. In the one-dimensional case studied in Lecture 38, since both $\hat{x}$ and $\hat{p}$ do not depend on t, we already obtained two important formulas,

$$\frac{d}{dt} \left\langle \hat{x} \right\rangle = \frac{1}{i\hbar} \left\langle [\hat{x}, \hat{H}] \right\rangle \quad \text{and} \quad \frac{d}{dt} \left\langle \hat{p} \right\rangle = \frac{1}{i\hbar} \left\langle [\hat{p}, \hat{H}] \right\rangle.$$

Furthermore, let us consider the case of a single particle whose Hamiltonian operator is

$$\hat{H} = \frac{\hat{p}^2}{2m} + \hat{V}(\hat{x}).$$

Theorem 10.1.3

$$m\frac{d}{dt} \left\langle \hat{x} \right\rangle = \left\langle \hat{p} \right\rangle.$$

Proof First of all, let us observe that $[\hat{x}, \hat{V}(\hat{x})] = 0$. Therefore

$$\frac{d}{dt}\langle\hat{x}\rangle = \frac{1}{i\hbar}\left\langle\left[\hat{x}, \hat{H}\right]\right\rangle = \frac{1}{i\hbar}\left\langle\left[\hat{x}, \frac{\hat{p}^2}{2m} + \hat{V}(\hat{x})\right]\right\rangle = \frac{1}{i\hbar}\left\langle\left[\hat{x}, \frac{\hat{p}^2}{2m}\right] + \left[\hat{x}, \hat{V}(\hat{x})\right]\right\rangle = \frac{1}{2mi\hbar}\left\langle\left[\hat{x}, \hat{p}^2\right]\right\rangle.$$

Then

$$\left[\hat{x}, \hat{p}^2\right] = \hat{x}\hat{p}\hat{p} - \hat{p}\hat{p}\hat{x} = \hat{x}\hat{p}\hat{p} - \hat{p}\hat{x}\hat{p} + \hat{p}\hat{x}\hat{p} - \hat{p}\hat{p}\hat{x} = \left[\hat{x}, \hat{p}\right]\hat{p} + \hat{p}\left[\hat{x}, \hat{p}\right] = 2i\hbar\hat{p},$$

that is

$$\frac{d}{dt}\langle\hat{x}\rangle = \frac{1}{m}\langle\hat{p}\rangle$$

which ends the proof. $\qquad\qquad\square$

According to the statement of this theorem, the rate of change in time of the expectation value of the position of a particle, times its mass, equals the expectation value of the momentum operator, that is

$$m\frac{d}{dt}\langle\hat{x}\rangle = \langle\hat{p}\rangle .$$

This is a quantum view of the momentum definition from Classical Mechanics, i.e,

$$m\frac{dx}{dt} = p.$$

In fact it is the same formula, but in Quantum Mechanics, the "objects" are expectations values of quantum operators. Let us obtain now the Quantum Mechanics equivalent of Newton's second principle

$$\frac{dp}{dt} = F.$$

Theorem 10.1.4

$$\frac{d}{dt}\langle\hat{p}\rangle = \left\langle-\frac{\partial\hat{V}}{\partial x}(\hat{x})\right\rangle = \langle\hat{F}\rangle.$$

Proof From

$$\frac{d}{dt}\langle\hat{p}\rangle = \frac{1}{i\hbar}\left\langle\left[\hat{p}, \hat{H}\right]\right\rangle$$

it results

$$\frac{d}{dt}\langle\hat{p}\rangle = \frac{1}{i\hbar}\left\langle\left[\hat{p}, \frac{\hat{p}^2}{2m} + \hat{V}(\hat{x})\right]\right\rangle = \frac{1}{i\hbar}\left\langle\left[\hat{p}, \hat{V}(\hat{x})\right]\right\rangle.$$

It remains to compute $\left\langle\left[\hat{p},\hat{V}(\hat{x})\right]\right\rangle$. Let us take into account that $\hat{p}$ operator, in the one dimensional case, is $-i\hbar\dfrac{\partial}{\partial x}$, therefore

$$\left\langle\left[\hat{p},\hat{V}(\hat{x})\right]\right\rangle = \left\langle\left[-i\hbar\frac{\partial}{\partial x},\hat{V}(\hat{x})\right]\right\rangle = \left\langle\Psi\left|\left[-i\hbar\frac{\partial}{\partial x},\hat{V}(\hat{x})\right]\right|\Psi\right\rangle = -i\hbar\left\langle\Psi\left|\left[\frac{\partial}{\partial x},\hat{V}(\hat{x})\right]\right|\Psi\right\rangle =$$

$$= i\hbar\left\langle\Psi\left|\left[\hat{V}(\hat{x}),\frac{\partial}{\partial x}\right]\right|\Psi\right\rangle = i\hbar\left\langle\Psi\left|\hat{V}(\hat{x})\right|\frac{\partial}{\partial x}\Psi\right\rangle - i\hbar\left\langle\Psi\left|\frac{\partial}{\partial x}\left(\hat{V}(\hat{x})\Psi\right)\right\rangle =$$

$$= i\hbar\left\langle\Psi\left|\hat{V}(\hat{x})\right|\frac{\partial}{\partial x}\Psi\right\rangle - i\hbar\left\langle\Psi\left|\frac{\partial\hat{V}}{\partial x}(\hat{x})\right|\Psi\right\rangle - i\hbar\left\langle\Psi\left|\hat{V}(\hat{x})\right|\frac{\partial}{\partial x}\Psi\right\rangle = i\hbar\left\langle-\frac{\partial\hat{V}}{\partial x}(\hat{x})\right\rangle.$$

It results

$$\frac{d}{dt}\langle\hat{p}\rangle = \left\langle-\frac{\partial\hat{V}}{\partial x}(\hat{x})\right\rangle.$$

If we denote by $\hat{F}$ the "quantum force operator" induced by $-\dfrac{\partial\hat{V}}{\partial x}(\hat{x})$, we obtain the result of the statement. $\square$

In other words, we are considering an evolution, a changing rate in time of the expectation value of the momentum operator. This rate of change is determined by the rate of change in time of the expectation value of a force exercised on the quantum particle. The rate of change in time of the mean gives the trajectory for the dense part of the "cloud" where the particle is the most probably to be. value

---

Summary of Lecture 42. After proving the Ehrenfest theorem

$$\frac{d}{dt}\langle\hat{A}\rangle = \frac{1}{i\hbar}\left\langle[\hat{A},\hat{H}]\right\rangle + \left\langle\frac{\partial\hat{A}}{\partial t}\right\rangle$$

we can highlight some particular cases which give rise to the connection between the well known formulas of Classical Mechanics and their counterparts in Quantum Mechanics written in terms of quantum operators with respect to the expectation values. We have:

$$m\frac{dx}{dt} = mv = p \quad\rightarrow\quad m\frac{d}{dt}\langle\hat{x}\rangle = \langle\hat{p}\rangle$$

$$\frac{dp}{dt} = -\frac{dV}{dx} = F \quad\rightarrow\quad \frac{d}{dt}\langle\hat{p}\rangle = \left\langle-\frac{\partial\hat{V}}{\partial x}(\hat{x})\right\rangle = \langle\hat{F}\rangle$$

$$\frac{dA}{dt} = [A, H] \quad \rightarrow \quad \frac{d}{dt}\left\langle \hat{A} \right\rangle = \frac{1}{i\hbar}\left\langle [\hat{A}, \hat{H}] \right\rangle.$$

The deterministic formalism of Classical Mechanics is replaced by the probabilistic formalism of Quantum Mechanics.

10.2 Lecture 43: The Heisenberg General Uncertainty Principle

Starting from this lecture, we prefer to use the Dirac notation which is very useful in the applications. According to this notation, we adopt the ket vector $|\Psi\rangle$ instead of Ψ.

In the same way, it is easy to look at the Schrödinger equation in the form

$$\frac{\partial}{\partial t}|\Psi\rangle = \frac{1}{i\hbar}\hat{H}|\Psi\rangle$$

instead of the previous notation

$$\frac{\partial \Psi}{\partial t} = \frac{1}{i\hbar}\hat{H}\Psi.$$

Let us first prove the *Cauchy–Buniakowski–Schwarz inequality* corresponding to the context of the Hilbert spaces.

Theorem 10.2.1 *Let us consider two vector X and Y in the Hilbert space. Then*

$$|\langle X, Y \rangle| \le ||X|| \cdot ||Y||.$$

Proof Consider the axiom

$$\langle A, A \rangle \ge 0$$

applied to the ket vector

$$|A\rangle := |X\rangle - \frac{\langle Y, X \rangle}{||Y||^2} \cdot |Y\rangle = \left| X - \frac{\langle Y, X \rangle}{||Y||^2} \cdot Y \right\rangle.$$

Using the properties of the inner product, it results

$$\langle A, A \rangle = \left\langle X - \frac{\langle Y, X \rangle}{||Y||^2} \cdot Y \Big| X - \frac{\langle Y, X \rangle}{||Y||^2} \cdot Y \right\rangle =$$

$$= \langle X|X \rangle - \frac{\langle Y, X \rangle \cdot \langle X, Y \rangle}{||Y||^2} - \frac{\langle Y, X \rangle^* \cdot \langle Y, X \rangle}{||Y||^2} + \frac{\langle Y, X \rangle^* \cdot \langle Y, X \rangle \cdot \langle Y, Y \rangle}{||Y||^4} \geq 0$$

that is

$$||X||^2 - \frac{\langle Y, X \rangle \cdot \langle X, Y \rangle}{||Y||^2} - \frac{\langle Y, X \rangle^* \cdot \cancel{\langle Y, X \rangle}}{\cancel{||Y||^2}} + \frac{\langle Y, X \rangle^* \cdot \cancel{\langle Y, X \rangle}}{\cancel{||Y||^2}} \geq 0.$$

Using $\langle Y, X \rangle = \langle X, Y \rangle^*$, the statement follows. $\square$

Let us consider now the expectation value $\langle \hat{A} \rangle$ corresponding to the operator $\hat{A}$ and to the state Ψ and consider Ψ normalized. It is worth noticing that the expectation value can be defined for unnormalized states too, considering

$$\langle \hat{A} \rangle := \frac{\langle \Psi | \hat{A} | \Psi \rangle}{\langle \Psi | \Psi \rangle}$$

but we established to work with normalized states, therefore

$$\langle \hat{A} \rangle := \langle \Psi | \hat{A} | \Psi \rangle.$$

Let us take into account two Hermitian operators $\hat{A}$ and $\hat{B}$ such that their commutator is $i\hat{U}$, i.e.

$$[\hat{A}, \hat{B}] = i\hat{U}.$$

Exercise 10.2.2 Show that $\hat{U} = -i[\hat{A}, \hat{B}]$ is a Hermitian operator.

Hint: We start from $\langle \phi | \hat{U} \Psi \rangle$.

$$\langle \phi | -i[\hat{A}, \hat{B}]\Psi \rangle = -i\langle \phi | [\hat{A}, \hat{B}]\Psi \rangle = -i\langle \phi | (\hat{A}\hat{B} - \hat{B}\hat{A})\Psi \rangle = -i\langle \phi | \hat{A}\hat{B}\Psi \rangle + i\langle \phi | \hat{B}\hat{A}\Psi \rangle =$$

$$= -i\langle \hat{A}\phi | \hat{B}\Psi \rangle + i\langle \hat{B}\phi | \hat{A}\Psi \rangle = -i\langle \hat{B}\hat{A}\phi | \Psi \rangle + i\langle \hat{A}\hat{B}\phi | \Psi \rangle = \langle i\hat{B}\hat{A}\phi | \Psi \rangle + \langle -i\hat{A}\hat{B}\phi | \Psi \rangle =$$

$$= \langle i(\hat{B}\hat{A} - \hat{A}\hat{B})\phi | \Psi \rangle = \langle -i[\hat{A}, \hat{B}]\phi | \Psi \rangle = \langle \hat{U}\phi | \Psi \rangle. \square$$

It is important to observe that, since $\hat{U}$ is Hermitian, its expectation value $\langle \hat{U} \rangle$ is real. Therefore the expectation value of the commutator only is a pure imaginary number, i.e.

$$\langle [\hat{A}, \hat{B}] \rangle = \langle i\hat{U} \rangle = i\langle \hat{U} \rangle.$$

Consider also the operators

$$\hat{\bar{A}} := \hat{A} - \langle \hat{A} \rangle \quad \text{and} \quad \hat{\bar{B}} := \hat{B} - \langle \hat{B} \rangle.$$

It is easy to check that

Exercise 10.2.3

$$[\hat{\hat{A}}, \hat{\hat{B}}] = i\hat{U}.$$

Hint: Simple computations leads to

$$[\hat{\hat{A}}, \hat{\hat{B}}] = [\hat{A} - \langle\hat{A}\rangle, \hat{B} - \langle\hat{A}\rangle] = (\hat{A} - \langle\hat{A}\rangle)(\hat{B} - \langle\hat{A}\rangle) - (\hat{B} - \langle\hat{A}\rangle)(\hat{A} - \langle\hat{A}\rangle) =$$

$$= \hat{A}\hat{B} - \hat{B}\hat{A} = i\hat{U}. \quad \square$$

Exercise 10.2.4 Double hat operators are Hermitian if hat operators are Hermitian.

Hint: If $\hat{A}$ is Hermitian and $\langle\hat{A}\rangle$ is a real number, we have:

$$\langle\phi|\hat{\hat{A}}\Psi\rangle = \langle\phi|(\hat{A} - \langle\hat{A}\rangle)\Psi\rangle = \langle\phi|\hat{A}\Psi\rangle - \langle\phi|\langle\hat{A}\rangle\Psi\rangle = \langle\hat{A}\phi|\Psi\rangle - \langle\langle\hat{A}\rangle\phi|\Psi\rangle = \langle(\hat{A} - \langle\hat{A}\rangle)\phi|\Psi\rangle = \langle\hat{\hat{A}}\phi|\Psi\rangle.$$

$$\square$$

For a Hermitian operator $\hat{A}$, it is possible to define the *uncertainty of the expectation value* $\langle\hat{A}\rangle$ with respect the state Ψ as

$$(\Delta\hat{A}) := [\langle\Psi|(\hat{A} - \langle\hat{A}\rangle)^2|\Psi\rangle]^{1/2}.$$

This definition makes sense because the square root is extracted from a positive quantity. And this happens because $\hat{\hat{A}} = \hat{A} - \langle\hat{A}\rangle$ is a Hermitian operator and

$$\langle\Psi|(\hat{A} - \langle\hat{A}\rangle)^2|\Psi\rangle = \langle\Psi|\hat{\hat{A}}^2|\Psi\rangle = \langle\Psi|\hat{\hat{A}}^2\Psi\rangle = \langle\hat{\hat{A}}\Psi|\hat{\hat{A}}\Psi\rangle \geq 0.$$

Theorem 10.2.5 (The Heisenberg general uncertainty principle) *Consider two Hermitian operators* $\hat{A}, \hat{B}$ *such that* $[\hat{A}, \hat{B}] = i\hat{U}$ *and their double hat correspondents are* $\hat{\hat{A}} := \hat{A} - \langle\hat{A}\rangle$ *and* $\hat{\hat{B}} := \hat{B} - \langle\hat{B}\rangle$. *Denote by*

$$[\hat{\hat{A}}, \hat{\hat{B}}]_+ := \hat{\hat{A}}\hat{\hat{B}} + \hat{\hat{B}}\hat{\hat{A}}$$

their anticommutator. Then, the uncertainties of the expectation values of $\hat{A}$, *and* $\hat{B}$ *verify the inequality*

$$(\Delta\hat{A})^2 \cdot (\Delta\hat{B})^2 \geq \frac{1}{4}\left\langle[\hat{\hat{A}}, \hat{\hat{B}}]_+\right\rangle^2 + \frac{1}{4}\left\langle\hat{U}\right\rangle^2$$

Proof First, it is easy to check that the anticommutator of any two Hermitian operators is a Hermitian operator, that is

$$\langle\phi|[\hat{A}, \hat{B}]_+\Psi\rangle = \langle[\hat{A}, \hat{B}]_+\phi|\Psi\rangle.$$

This is an important property and we will use later, in the proof, a consequence of it, i.e. the expectation value of the anticommutator $\langle |[\hat{\hat{A}}, \hat{\hat{B}}]_+\rangle$, is a real number. Since

$$(\Delta\hat{A})^2 = \langle\Psi|(\hat{A} - \langle\hat{A}\rangle)^2|\Psi\rangle = \langle\Psi|\hat{\hat{A}}^2|\Psi\rangle,$$

we have

$$(\Delta\hat{A})^2 \cdot (\Delta\hat{B})^2 = \langle\Psi|\hat{\hat{A}}^2|\Psi\rangle \cdot \langle\Psi|\hat{\hat{B}}^2|\Psi\rangle = \langle\Psi|\hat{\hat{A}}|\hat{\hat{A}}\Psi\rangle \cdot \langle\Psi|\hat{\hat{B}}|\hat{\hat{B}}\Psi\rangle.$$

Since the double hat operators are Hermitian, it results

$$(\Delta\hat{A})^2 \cdot (\Delta\hat{B})^2 = \langle\hat{\hat{A}}\Psi|\hat{\hat{A}}\Psi\rangle \cdot \langle\hat{\hat{B}}\Psi|\hat{\hat{B}}\Psi\rangle \geq |\langle\hat{\hat{A}}\Psi|\hat{\hat{B}}\Psi\rangle|^2,$$

the last inequality is a consequence of the Cauchy–Buniakovski–Schwarz inequality. We use again the fact that we work with Hermitian operators, i.e.

$$(\Delta\hat{A})^2 \cdot (\Delta\hat{B})^2 \geq |\langle\Psi|\hat{\hat{A}}\hat{\hat{B}}|\Psi\rangle|^2 = \left|\left\langle\Psi\left|\frac{1}{2}(\hat{\hat{A}}\hat{\hat{B}} + \hat{\hat{B}}\hat{\hat{A}}) + \frac{1}{2}(\hat{\hat{A}}\hat{\hat{B}} - \hat{\hat{B}}\hat{\hat{A}})\right|\Psi\right\rangle\right|^2.$$

From both the commutator and anticommutator definitions, the previous inequality becomes

$$(\Delta\hat{A})^2 \cdot (\Delta\hat{B})^2 \geq |\langle\Psi|\hat{\hat{A}}\hat{\hat{B}}|\Psi\rangle|^2 = \frac{1}{4}\left|\left\langle\Psi\left|[\hat{\hat{A}}, \hat{\hat{B}}]_+ + [\hat{\hat{A}}, \hat{\hat{B}}]\right|\Psi\right\rangle\right|^2 = \frac{1}{4}\left|\left\langle\Psi\left|[\hat{\hat{A}}, \hat{\hat{B}}]_+ + i\hat{U}\right|\Psi\right\rangle\right|^2 =$$

$$= \frac{1}{4}\left|\left\langle\Psi\left|[\hat{\hat{A}}, \hat{\hat{B}}]_+\right|\Psi\right\rangle + i\left\langle\Psi\left|\hat{U}\right|\Psi\right\rangle\right|^2 = \frac{1}{4}\left|\left\langle[\hat{\hat{A}}, \hat{\hat{B}}]_+\right\rangle + i\left\langle\hat{U}\right\rangle\right|^2.$$

We have already discussed that, since $i\hat{U}$ is Hermitian, its expectation value is real that is the expectation value of the commutator is purely imaginary. The same, the expectation value of the anticommutator of two Hermitian operators is real. Taking into account that, for a complex number $z = a + ib$ is $|z|^2 = z \cdot z^* = (a + ib)(a - ib) = a^2 + b^2$, we can write

$$(\Delta\hat{A})^2 \cdot (\Delta\hat{B})^2 \geq \frac{1}{4}\left|\left\langle[\hat{\hat{A}}, \hat{\hat{B}}]_+\right\rangle + i\left\langle\hat{U}\right\rangle\right|^2 = \frac{1}{4}\left\langle[\hat{\hat{A}}, \hat{\hat{B}}]_+\right\rangle^2 + \frac{1}{4}\left\langle\hat{U}\right\rangle^2$$

$$\square$$

Let us remember the position operator $\hat{x} : L^2(\mathbb{R}) \to \mathbb{C}$ which associates to each state function $\Psi \in L^2(\mathbb{R})$, at each position x on the real line, the number $x\Psi(x) \in \mathbb{C}$. The position operator is Hermitian because

$$\langle\hat{x}\Psi|\Psi\rangle = \int_{-\infty}^{\infty} x\Psi^*(x)\Psi(x)dx = \int_{-\infty}^{\infty} \Psi^*(x)x\Psi(x)dx = \langle\Psi|\hat{x}\Psi\rangle.$$

The same for the momentum operator $\hat{p} : L^2(\mathbb{R}) \to \mathbb{C}$ which associates to each state function $\Psi \in L^2(\mathbb{R})$, at each position x on the real line, the number $-i\hbar \dfrac{\partial \Psi(x)}{\partial x} \in \mathbb{C}$. The momentum operator is a Hermitian operator too, because

$$\langle \hat{p}\Psi | \Psi \rangle = \left\langle -i\hbar \frac{\partial}{\partial x} \Psi | \Psi \right\rangle = i\hbar \left\langle \frac{\partial}{\partial x} \Psi | \Psi \right\rangle = i\hbar \int_{-\infty}^{\infty} \frac{\partial \Psi^*(x)}{\partial x} \Psi(x) dx =$$

$$= i\hbar \Psi^* \Psi |_{-\infty}^{\infty} - i\hbar \int_{-\infty}^{\infty} \Psi^*(x) \frac{\partial \Psi(x)}{\partial x} dx.$$

We have already used the argument that square integrable functions are limits of sets of $\mathcal{C}_0^\infty$. Here we obtain

$$\langle \hat{p}\Psi | \Psi \rangle = -i\hbar \int_{-\infty}^{\infty} \Psi^*(x) \frac{\partial \Psi(x)}{\partial x} dx = \int_{-\infty}^{\infty} \Psi^*(x) \left(-i\hbar \frac{\partial}{\partial x} \Psi(x) \right) dx = \langle \Psi | \hat{p}\Psi \rangle,$$

that is we succeeded to prove

$$\langle \hat{p}\Psi | \Psi \rangle = \langle \Psi | \hat{p}\Psi \rangle.$$

Now, having in mind the one-dimensional case, where between the Hermitian position operator $\hat{x}$ and the Hermitian momentum operator $\hat{p}$ there is the relation $[\hat{x}, \hat{p}] = i\hbar$, we can consider the following statement:

Corollary 10.2.6 (Heisenberg's uncertainty principle) *Consider two Hermitian operators $\hat{A}$, $\hat{B}$ such that $[\hat{A}, \hat{B}] = i\hbar$. Then*

$$(\Delta \hat{A}) \cdot (\Delta \hat{B}) \geq \frac{\hbar}{2}.$$

Proof In the statement of the theorem we replace $\hat{U}$ by $\hbar$. It follows

$$\langle \hat{U} \rangle = \langle \Psi | \hbar | \Psi \rangle = \langle \Psi | \hbar \Psi \rangle = \hbar \langle \Psi | \Psi \rangle = \hbar \quad \text{and}$$

$$(\Delta \hat{A})^2 \cdot (\Delta \hat{B})^2 \geq \frac{1}{4} \left\langle [\hat{A}, \hat{B}]_+ \right\rangle^2 + \frac{1}{4}\hbar^2 \geq \frac{1}{4}\hbar^2$$

which ends the proof. $\qquad \square$

Summary of Lecture 43. We presented some consequences of Quantum Mechanics postulates working in Dirac notation introduced in Lecture 28. We start by establishing the Cauchy–Buniakowski–Schwarz inequality corresponding to a Hilbert space:

$$|\langle X, Y \rangle| \leq ||X|| \cdot ||Y||.$$

Then, we prepare the statement and the proof of Heisenberg general uncertainty principle. The objects are two observables $\hat{A}$ and $\hat{B}$ such that their commutator is $i\hat{U}$, i.e.

$$[\hat{A}, \hat{B}] = i\hat{U}.$$

Under the previous conditions, $\hat{U}$ is a Hermitian operator. Some other operators can be defined,

$$\hat{\hat{A}} := \hat{A} - \langle \hat{A} \rangle \text{ and } \hat{\hat{B}} := \hat{B} - \langle \hat{B} \rangle,$$

where $\langle \hat{A} \rangle$ is the expectation value corresponding to the observable $\hat{A}$ and the normalized state Ψ:

$$\langle \hat{A} \rangle := \langle \Psi | \hat{A} | \Psi \rangle.$$

Under these circumstances, the expectation value is real and the double hat operators are Hermitian.

For a Hermitian operator $\hat{A}$ we define the uncertainty of the expectation value $\langle \hat{A} \rangle$ with respect to the state Ψ as

$$(\Delta \hat{A}) := [\langle \Psi | (\hat{A} - \langle \hat{A} \rangle)^2 | \Psi \rangle]^{1/2}.$$

Then we prove

Theorem (Heisenberg's general uncertainty principle) *Consider two Hermitian operators* $\hat{A}$, $\hat{B}$ *such that* $[\hat{A}, \hat{B}] = i\hat{U}$ *and their double hat correspondents* $\hat{\hat{A}} := \hat{A} - \langle \hat{A} \rangle$ *and* $\hat{\hat{B}} := \hat{B} - \langle \hat{B} \rangle$. *Denote by*

$$[\hat{\hat{A}}, \hat{\hat{B}}]_+ := \hat{\hat{A}}\hat{\hat{B}} + \hat{\hat{B}}\hat{\hat{A}}$$

their anticommutator. Then, the uncertainties of the expectation values of $\hat{A}$, $\hat{B}$ *verify the inequality*

$$(\Delta \hat{A})^2 \cdot (\Delta \hat{B})^2 \geq \frac{1}{4}\left\langle [\hat{\hat{A}}, \hat{\hat{B}}]_+ \right\rangle^2 + \frac{1}{4}\left\langle \hat{U} \right\rangle^2$$

Both position operator $\hat{x}$ *and momentum operator* $\hat{p}$ *are Hermitian and fulfill the relation* $[\hat{x}, \hat{p}] = i\hbar$. *As a consequence, the following statement holds:*
Corollary (Heisenberg's uncertainty principle) *Given two Hermitian operators* $\hat{A}$, $\hat{B}$ *such that* $[\hat{A}, \hat{B}] = i\hbar$, *then*

$$(\Delta \hat{A}) \cdot (\Delta \hat{B}) \geq \frac{\hbar}{2}.$$

10.3   Lecture 44: The Dirac Notation and the Meaning of a Quantum Mechanics Experiment

Let us start with a short review of *Dirac notation* in the case of finite dimensional complex vector spaces. Consider a *ket vector* $|u\rangle$. Suppose we work in a two dimensional complex vector space. Therefore we imagine it as

$$|u\rangle = \begin{pmatrix} u_1 \\ u_2 \end{pmatrix}$$

If we have another vector

$$|v\rangle = \begin{pmatrix} v_1 \\ v_2 \end{pmatrix}$$

the addition of the two ket vectors is given by the rule

$$|r\rangle = |u\rangle + |v\rangle = \begin{pmatrix} u_1 \\ u_2 \end{pmatrix} + \begin{pmatrix} v_1 \\ v_2 \end{pmatrix} = \begin{pmatrix} u_1 + v_1 \\ u_2 + v_2 \end{pmatrix}.$$

and the result is again a ket vector, here denoted by $|r\rangle \in \mathbb{V}$. The summation of ket vectors must determine a commutative group structure $(\mathbb{V}, +)$, that is the following properties hold:

$$(|u\rangle + |v\rangle) + |w\rangle = |u\rangle + (|v\rangle + |w\rangle) \text{ (associativity)}$$

$$|u\rangle + |0\rangle = |0\rangle + |u\rangle = |u\rangle \text{ (existence of zero vector for any } |u\rangle \in \mathbb{V})$$

$$\forall |u\rangle \in \mathbb{V} \; \exists | - u\rangle \in \mathbb{V} \text{ such that } |u\rangle + | - u\rangle = | - u\rangle + |u\rangle = |0\rangle \text{ (the additive inverse)}$$

$$|u\rangle + |v\rangle = |v\rangle + |u\rangle \text{ (commutativity)}$$

The multiplication by scalars from $\mathbb{C}$ follows the same type of definition:

$$\forall |u\rangle \in \mathbb{V}, \; \forall a \in \mathbb{C} \; \exists \, a|u\rangle \in \mathbb{V},$$

that is, in our case,

$$\alpha |u\rangle := \begin{pmatrix} \alpha u_1 \\ \alpha u_2 \end{pmatrix}$$

The multiplication by a scalar has the properties

$$a(b|u\rangle) = (ab)|u\rangle$$

$$1 \cdot |u\rangle = |u\rangle$$

$$a(|u\rangle + |v\rangle) = a|u\rangle + a|v\rangle$$

$$(a + b)|u\rangle = a|u\rangle + b|u\rangle$$

Some consequences of the complex vector space structure exist and can be easily verified:

$$0 \cdot |u\rangle = |0\rangle$$

$$(-1) \cdot |u\rangle = |-u\rangle$$

$$|u\rangle - |v\rangle = |u\rangle + |-v\rangle.$$

A basis is determined by a maximal set of linear independent elements. An important basis in V is determined by the vectors

$$|e_1\rangle = \begin{pmatrix} 1 \\ 0 \end{pmatrix} \text{ and } |e_2\rangle = \begin{pmatrix} 0 \\ 1 \end{pmatrix}.$$

Now, the *bra vector* corresponding to the ket vector $|u\rangle$ is, by definition,

$$\langle u| := (u_1^*, u_2^*)$$

where u_k^* is the complex conjugate of u_k. We remember that the complex conjugate of a complex number $z = a + ib$ is $z^* = a - ib$. The modulus of a complex number z is denoted by $|z|$. Its definition is $|z|^2 := a^2 + b^2 = z \cdot z^* = z^* \cdot z$ and we observe that it is a real positive number. The reason for Dirac to consider this bra and ket vectors is related to the inner product here denoted by the bracket $\langle u|v\rangle$. If we look at our arrangements, we have

$$\langle u|v\rangle = (u_1^*, u_2^*) \cdot \begin{pmatrix} v_1 \\ v_2 \end{pmatrix} = u_1^* v_1 + u_2^* v_2 = \sum_{k=1}^{2} u_k^* v_k.$$

Of course, one can increase the number of components of the vector space and one has the full picture presented in Lecture 28. Therefore, it is easy to check the properties of the inner product in this particular case:

$$\langle u|\beta_1 v_1 + \beta_2 v_2\rangle = \beta_1 \langle u|v_1\rangle + \beta_2 \langle u|v_2\rangle, \ \forall |v_1\rangle, |v_2\rangle, |u\rangle \in \mathbb{V}, \ \beta_1, \beta_2 \in \mathbb{C},$$

$$\langle v|u\rangle = \langle u|v\rangle^*, \ \forall |u\rangle, |v\rangle \in \mathbb{V},$$

$$\langle u|u\rangle \geq 0 \ \forall |u\rangle \in \mathbb{V}; \ \langle u, u\rangle = 0 \iff |u\rangle = |0\rangle.$$

The last property is the consequence of

$$\langle u|u \rangle = |u_1|^2 + |u_2|^2 = 0.$$

With a bra vector corresponding to the ket vector $|u\rangle$,

$$\langle u| := (u_1^*, u_2^*)$$

and a ket vector $|v\rangle$, another important operation is the outer product $|v\rangle\langle u|$. The result is a linear operator on $\mathbb{V}$. Indeed,

$$|v\rangle\langle u| := \begin{pmatrix} v_1 \\ v_2 \end{pmatrix} (u_1^*, u_2^*) = \begin{pmatrix} v_1 u_1^* & v_1 u_2^* \\ v_2 u_1^* & v_2 u_2^* \end{pmatrix}$$

So, in this particular case, the linear operators are 2×2 matrices with elements in $\mathbb{C}$. For a given linear operator

$$L = \begin{pmatrix} a_{11} & a_{12} \\ a_{21} & a_{22} \end{pmatrix}$$

its transpose is

$$L^T = \begin{pmatrix} a_{11} & a_{21} \\ a_{12} & a_{22} \end{pmatrix}.$$

The conjugate transpose (or the *Hermitian transpose*) of the matrix L is denoted by $L^\dagger$. Therefore, we have

$$L^\dagger = \begin{pmatrix} a_{11}^* & a_{21}^* \\ a_{12}^* & a_{22}^* \end{pmatrix}.$$

By definition, a linear operator is Hermitian if $L^\dagger = L$. If we look what happens when $L^\dagger = L$, we observe, from

$$\begin{pmatrix} a_{11}^* & a_{21}^* \\ a_{12}^* & a_{22}^* \end{pmatrix} = \begin{pmatrix} a_{11} & a_{12} \\ a_{21} & a_{22} \end{pmatrix}$$

that the diagonal of a Hermitian operator (here the matrix L) is given by real numbers and the second diagonal contains complex conjugate terms. You can check, it is a simple exercise, that

$$\langle Lu, v \rangle = \langle u, Lv \rangle,$$

therefore our previous definition of a Hermitian operator fits with the one in Lecture 31.

Consider now the matrix

$$A = \begin{pmatrix} 1 & 2+3i \\ 2-3i & -3 \end{pmatrix}.$$

It is easy to check that this matrix is Hermitian and has two real eigenvalues.

Can a Hermitian matrix in this case have to equal eigenvalues? First you can check that any 2×2 B matrix verifies the *Cayley relation*

$$B^2 - Tr B \cdot B + \det B \cdot I_2 = O_2,$$

where $Tr B$ is the sum of the elements of its main diagonal and the determinant of B is $\det B = b_{11}b_{22} - b_{12}b_{21}$, I_2 is the unit 2×2 matrix and O_2 is the 0 matrix. In fact, this is the matrix form of the characteristic equation used to determine the eigenvalues of B,

$$\det(B - \lambda I_2) = 0.$$

If B is a Hermitian matrix, the discriminant of

$$\lambda^2 - Tr B \cdot \lambda + \det B = 0,$$

is the positive number $(Tr B)^2 - 4 \det B = (b_{11} - b_{22})^2 + 4|b_{12}|^2$. So, the eigenvalues are real. Are they different or they can be equal? Equal eigenvalues happens when $(b_{11} - b_{22})^2 + 4|b_{12}|^2 = 0$, that is when $b_{11} = -b_{22}$ and $|b_{12}| = 0$. The last equality means that the complex number b_{12} has to be 0. The first equality means that the real numbers of the main diagonal are b_{11} and b_{11}. Therefore the only Hermitian 2×2 matrix which admits two equal eigenvalues is $b_{11}I_2$. So, the answer is yes but we observe that this matrix is a sort of "whatever happens nothing happens from the Physics point of view" because, for each possible vector $|u\rangle$, we have the same "output" b_{11}. We are interested in Hermitian operators having two different eigenvalues corresponding to some two different orthogonal eigenvectors.

Another class of important operators in Quantum Mechanics are the unitary operators. In this particular case, a *unitary operator* is a 2×2 matrix A such that $A^* \cdot A = A \cdot A^* = I_2$. It is easy to check

Exercise 10.3.1 Show that unitary operator U preserves the inner product, that is $\langle Ux|Uy\rangle = \langle x|y\rangle$ for any $|x\rangle, |y\rangle \in V$.

According to this formalism, the **Picture of a Quantum Mechanics Experiment** is the following.

For the **mathematical description of the experiment**, we need:

- a Hermitian operator $\hat{A}$;
- distinct eigenvalues $\lambda_n \in \mathbb{R}$ corresponding to the Hermitian operator $\hat{A}$;
- orthogonal eigenvectors $|\Psi_n\rangle$ corresponding to the previous eigenvalues $\lambda_n \in \mathbb{R}$ related by the well known equation

$$\hat{A}|\Psi_n\rangle = \lambda_n|\Psi_n\rangle.$$

On the other hand, the **physical description of the experiment** is given by:

- the Hermitian operator $\hat{A}$ is the measurable observable of the system;
- the eigenvectors $|\Psi_n\rangle$ are the possible states of the system;

- the distinct real eigenvalues λ_n are the results of the experiment.

In conclusion, we can measure only eigenvalues. Eigenvalues give information on the associated eigenvectors, that is on the state of the system.

> **Summary of Lecture 44.** After presenting the Dirac notation, we highlight the description of a Quantum Mechanics Experiment.
>   For the **mathematical description**, we need:
> - a Hermitian operator $\hat{A}$,
> - distinct eigenvalues $\lambda_n \in \mathbb{R}$ corresponding to the Hermitian operator $\hat{A}$;
> - orthogonal eigenvectors $|\Psi_n\rangle$, corresponding to the previous eigenvalues $\lambda_n \in \mathbb{R}$, related by the equation
>
> $$\hat{A}|\Psi_n\rangle = \lambda_n|\Psi_n\rangle.$$
>
> The **physical description** is related to the interpretation of the same equation
>
> $$\hat{A}|\Psi_n\rangle = \lambda_n|\Psi_n\rangle.$$
>
> It consists in the following statements:
> - the Hermitian operator $\hat{A}$ is the measurable observable of the system;
> - the eigenvectors $|\Psi_n\rangle$ are the possible states of the system;
> - distinct real eigenvalues λ_n are the results of the experiment and they correspond to the measurements.

10.4   Lecture 45: The Photon Polarization by Dirac's Notation

Let us consider a simple experiment. An unpolarized light beam is produced by a source S and it is sent to a horizontal polarizer device H_d^1. When the light comes out from the horizontal polarizer, it continues its trajectory as horizontally polarized light. If it meets another horizontal polarizer H_d^2, 100% of horizontally polarized light, which enters in the second horizontal polarizer, comes out from it, of course, horizontally polarized. This is an experiment which can be easily managed. By contrast, if the second is a vertical polarizer V_d^2, 0% of horizontally polarized light entering in V_d^2 comes out.

The experiment can be easily understood if one looks through two consecutive polaroid glasses lenses. Each surface of a glasses lens can be imagined as a rectangle having in its plane a length l on the x-axis and a height h on the y-axis. Two consecutive glasses lenses are arranged in the same way along the two parallel planes of their surfaces, if both the rectangles stay with l on the x-axis and h on the y-axis. In this case, one sees with only one lens exactly as one sees with the two consecutive lenses arranged in the same way. If the second lens is rotated by 90°, one cannot see through. There is no light coming out from the second lens.

We can continue the experiment considering a first horizontal polarizer H_d^1 and replacing the second polarizer with one which polarize at 45°. Only 50% of the vertical polarized light comes out from the second polarizer. The experiment is telling us the following result: If the second polarizer is polarizing light at α degrees with respect to the horizontal direction, only $\cos^2 \alpha\%$ photons, passing through this polarizer, can be measured. In this context, this is a general result and it fits the angles of 0°, 90° and 45° as seen before.

If we now turn again to the experiment, we can understand what happens with the color of light. The light beam consists of photons having the energy given by the formula $E = h\nu$. The color depends on frequency ν. When they reach the first polarizer, some of them go through, some not. All the photons going through become vertically polarized and again, according to the experiment, all or only a fraction of them pass through the second polarizer. There is no energy loss of photons which succeed in passing through both polarizers. So, they have the same frequency from the beginning to the end of their trajectory. It means that the color of light remains the same as it is.

Let us interpret the phenomenon according to Quantum Mechanics concepts presented in the previous lecture. We start considering the horizontal polarization described by the ket vector

$$|H\rangle = \begin{pmatrix} 1 \\ 0 \end{pmatrix}$$

and the vertical polarization described by the ket vector

$$|V\rangle = \begin{pmatrix} 0 \\ 1 \end{pmatrix}.$$

The choice is correct because, as we know from Lecture 41, the states has to be normalized. Let us look at the inner product between the ket vectors $|H\rangle$ and $|V\rangle$, that is $\langle H|V\rangle$. The physical meaning of this inner product is: If we prepare the light in the vertical polarized state, what are the chances to go through a horizontal polarizer? We know the answer: there are 0 chance to pass through.

Therefore if we prepare photons to be horizontally polarized and we send them to a vertical polarizer, the inner product $\langle V|H\rangle$ shows us the probability amplitude of this happening. To get the probability, we have to square the inner product. In fact, if we work in the complex space, we have to consider

$$|\langle V|H\rangle|^2 = \langle V|H\rangle^* \cdot \langle V|H\rangle.$$

The probability amplitude is a fundamental notion of Quantum Mechanics. Let us mathematically check the result. If we consider the ket vector

$$|V\rangle = \begin{pmatrix} 0 \\ 1 \end{pmatrix},$$

the corresponding bra vector is

$$\langle V| = (0\ 1),$$

that is

$$\langle V|H\rangle = (0\ 1) \cdot \begin{pmatrix} 1 \\ 0 \end{pmatrix} = 0.$$

In this case the probability amplitude is 0 and the probability is 0, which fits with the experiment. No photons horizontally polarized can passes through a vertical polarizer.

From these considerations, we have to frame the result in the conceptual scheme of Lecture 41: The state function $|H\rangle$ describes the set of all identically prepared systems, that is photons horizontally polarized by a horizontal polarizer. In all systems, identically prepared, the photon, horizontally polarized, do not pass through a vertical polarizer. The probability amplitude in this case is the number 0 and it is represented by $\langle V|H\rangle$.

The interpretation by Max Born is part of the principles of Quantum Mechanics: The probability amplitude provides the number corresponding to the relation between the state of the observed system into another possible state. The squared modulus of the probability amplitude is the probability to observe the initial state passed into the other possible state. In this case the probability is $|\langle V|H\rangle|^2 = 0$.

Let us now take into account another case, previously presented, under the same mathematical language.

Consider a photon polarized at 45°. Assuming a system of axes and supposing we measure the angle anticlockwise, i.e. from the x-axis to y-axis, the photon state vector can be represented by a superposition of the vertical polarized state and the horizontal polarized state,

$$|/\rangle = \frac{1}{\sqrt{2}}|V\rangle + \frac{1}{\sqrt{2}}|H\rangle.$$

First, let us ask if this state vector is mathematically well expressed. This means that the mathematical form has to express the following physical fact: 100% of the photons, which are first 45° polarized, comes out from a second 45° polarizer. In order to check if Mathematics gives us the right result, we have first to compute the probability amplitude

$$\langle /|/\rangle = \frac{1}{\sqrt{2}} \cdot \frac{1}{\sqrt{2}} \langle V + H|V + H\rangle = \frac{1}{2}\langle V|V\rangle + \frac{1}{2}\langle H|H\rangle = \frac{1}{2} + \frac{1}{2} = 1.$$

The probability is the square of this probability amplitude, that is 1. The result fits the experiment that is the mathematical form of $|/\rangle$ is well chosen.

A comment is necessary at this point: If one does not want to use the properties of the inner product, she can directly write

$$|/\rangle = \frac{1}{\sqrt{2}}|V\rangle + \frac{1}{\sqrt{2}}|H\rangle = \begin{pmatrix} \frac{1}{\sqrt{2}} \\ \frac{1}{\sqrt{2}} \end{pmatrix},$$

therefore

$$\langle/|/\rangle = \begin{pmatrix} \frac{1}{\sqrt{2}} & \frac{1}{\sqrt{2}} \end{pmatrix} \cdot \begin{pmatrix} \frac{1}{\sqrt{2}} \\ \frac{1}{\sqrt{2}} \end{pmatrix} = \frac{1}{2} + \frac{1}{2} = 1.$$

If we look at what happens if we prepare polarized photons at 45° and we allow them to pass through a vertical polarizer, we know, from the experiment that only 50% of photons will pass, that is the probability of this arrangement is $\frac{1}{2}$. Mathematics gives us the same result. We have first to compute the probability amplitude

$$\langle V|/\rangle = (0\ 1) \cdot \begin{pmatrix} \frac{1}{\sqrt{2}} \\ \frac{1}{\sqrt{2}} \end{pmatrix} = \frac{1}{\sqrt{2}}.$$

The probability is, in this case, $\langle V|/\rangle^2$, that is $\frac{1}{2}$.

If we consider the polarized photon at α degrees (assuming the angle measured with respect the x-axis) the corresponding state vector is

$$|\alpha\rangle = \cos\alpha \cdot |H\rangle + \sin\alpha \cdot |V\rangle = \begin{pmatrix} \cos\alpha \\ \sin\alpha \end{pmatrix}$$

If we look at what happens if we polarize photons at α degrees and we allow them to pass through a horizontal polarizer, we know, from the experiment, that only $\cos^2\alpha\%$ of photons will pass, that is the probability of this arrangement is $\cos^2\alpha$. Of course, Mathematics has to tell us the same thing. We first compute the probability amplitude described by

$$\langle H|\alpha\rangle = (1\ 0) \cdot \begin{pmatrix} \cos\alpha \\ \sin\alpha \end{pmatrix} = \cos\alpha,$$

therefore the probability is $\langle H|\alpha\rangle^2 = \cos^2\alpha$.

Now we are able to mathematically answer at the most general question: Which is the probability of photons, polarized at α degrees with respect the horizontal axis, to pass through a β polarizer? The state corresponding to α polarized photons is given by

$$|\alpha\rangle = \begin{pmatrix} \cos\alpha \\ \sin\alpha \end{pmatrix}$$

and the one corresponding to β polarized photons is described by

$$|\beta\rangle = \begin{pmatrix} \cos\beta \\ \sin\beta \end{pmatrix}.$$

The probability amplitude $\langle\beta|\alpha\rangle$ represents the number corresponding to the relation between the two states. In this case

$$\langle\beta|\alpha\rangle = (\cos\beta \ \sin\beta) \cdot \begin{pmatrix} \cos\alpha \\ \sin\alpha \end{pmatrix} = \cos\beta\cos\alpha + \sin\beta\sin\alpha = \cos(\beta - \alpha).$$

The probability for α degrees polarized photons to become β-degrees polarized photons is $\cos^2(\beta - \alpha)$.

We have to consider how this result is realized in an experiment implying vertical and horizontal polarized light. From the mathematical point of view, we have:

- a Hermitian operator $\hat{A}$;
- distinct eigenvalues $\lambda_n \in \mathbb{R}$ corresponding to the Hermitian operator $\hat{A}$;
- orthogonal eigenvectors $|\Psi_n\rangle$ corresponding to the previous eigenvalues $\lambda_n \in \mathbb{R}$ related by the equation

$$\hat{A}|\Psi_n\rangle = \lambda_n|\Psi_n\rangle.$$

Let us assume only two possible states for the polarized light described by the orthogonal ket vectors

$$|H\rangle = \begin{pmatrix} 1 \\ 0 \end{pmatrix} \quad \text{and} \quad |V\rangle = \begin{pmatrix} 0 \\ 1 \end{pmatrix}.$$

The Hermitian operator, such that $|V\rangle$ and $|V\rangle$ are the eigenvectors (eigenstates), is

$$\hat{A}_{HV} = \begin{pmatrix} 1 & 0 \\ 0 & -1 \end{pmatrix}$$

because

$$\hat{A}_{HV}|H\rangle = \begin{pmatrix} 1 & 0 \\ 0 & -1 \end{pmatrix}\begin{pmatrix} 1 \\ 0 \end{pmatrix} = \begin{pmatrix} 1 \\ 0 \end{pmatrix} = 1 \cdot |H\rangle,$$

and

$$\hat{A}_{HV}|V\rangle = \begin{pmatrix} 1 & 0 \\ 0 & -1 \end{pmatrix}\begin{pmatrix} 0 \\ 1 \end{pmatrix} = \begin{pmatrix} 0 \\ -1 \end{pmatrix} = -1 \cdot |V\rangle.$$

The distinct eigenvalues are $+1$ and -1. Now the physical meaning. The observable is given by a device which interacts only with horizontal and vertical polarized light and, from the mathematical point of view, it is the Hermitian operator $\hat{A}_{HV}$. Suppose the device is set on two light frequencies, a blue light and a red light. If the blue light is switched on and the red light is switched off, the horizontal polarized light passes through the device. This means that the blue light corresponds to the $+1$ eigenvalue. In this case, the blue light is giving us the image of the polarized light. It is horizontally polarized, therefore the corresponding state is $|H\rangle$. If the red light is switched on and the blue light is switched off, the corresponding eigenvalue is -1 and the corresponding eigenstate is $|V\rangle$.

Therefore in this Quantum Mechanics experiment, the device with two colored lights is mathematically described by the Hermitian operator $\hat{A}_{HV}$. The two possible states $|H\rangle$ and $|V\rangle$ of the system represent horizontal and vertical polarized photons which are recognized by the color of the light switched on. Each light switched on indicates the corresponding eigenvalue and this indicates the corresponding eigenvector. This is the only measurement we can make on the system.

Now we can ask about the Hermitian operator associated to the following ket vectors:

$$|/\rangle = \begin{pmatrix} \dfrac{1}{\sqrt{2}} \\ \dfrac{1}{\sqrt{2}} \end{pmatrix} \text{ and } |\backslash\rangle = \begin{pmatrix} -\dfrac{1}{\sqrt{2}} \\ \dfrac{1}{\sqrt{2}} \end{pmatrix}.$$

Let us remind the meaning of the forward slash ket vector: it is the state vector of $45°$ polarized photon. The backslash ket vector is the state vector of the $135°$ polarized photon. It was shown that $\langle /|/\rangle = 1$.

We left as an exercise to check the following relations:

$$\langle \backslash|\backslash\rangle = 1 \quad \text{and} \quad \langle \backslash|/\rangle = 0.$$

Which is the Hermitian matrix having these two eigenvectors? The following computations

$$\hat{A}_{\wedge}|/\rangle = \begin{pmatrix} 0 & 1 \\ 1 & 0 \end{pmatrix} \begin{pmatrix} \dfrac{1}{\sqrt{2}} \\ \dfrac{1}{\sqrt{2}} \end{pmatrix} = \begin{pmatrix} \dfrac{1}{\sqrt{2}} \\ \dfrac{1}{\sqrt{2}} \end{pmatrix} = 1 \cdot |/\rangle,$$

$$\hat{A}_{\wedge}|\backslash\rangle = \begin{pmatrix} 0 & 1 \\ 1 & 0 \end{pmatrix} \begin{pmatrix} -\dfrac{1}{\sqrt{2}} \\ \dfrac{1}{\sqrt{2}} \end{pmatrix} = -\begin{pmatrix} -\dfrac{1}{\sqrt{2}} \\ \dfrac{1}{\sqrt{2}} \end{pmatrix} = -1 \cdot |\backslash\rangle.$$

show that the Hermitian matrix is

$$\hat{A}_{\wedge} = \begin{pmatrix} 0 & 1 \\ 1 & 0 \end{pmatrix}$$

and the eigenvalues are $+1$ and -1. In this case, the physical meaning is similar to the one presented in the above case. The observable is a device which interacts only with $45°$ and $135°$ polarized light. From the mathematical point of view, the device is represented by the Hermitian operator $\hat{A}_{/\backslash}$. Suppose the device has light with two frequencies, green and yellow. If the green light is switched on and the yellow light is switched off, it means that the $45°$ polarized light passes through the device. The green light corresponds to the $+1$ eigenvalue. In this case, the green light indicates that the eigenstate has to be $|/\rangle$. If the yellow light is switched on and the green light is switched off, the corresponding eigenvalue is -1. Therefore the yellow light indicates that the corresponding eigenstate is $|\backslash\rangle$. The lights correspond to eigenvalues and the eigenvalues tell us the eigenstate.

The general case of polarization has two orthogonal eigenvectors corresponding to the states

$$|\alpha\rangle = \begin{pmatrix} \cos\alpha \\ \sin\alpha \end{pmatrix} \quad \text{and} \quad |\alpha_\perp\rangle = \begin{pmatrix} -\sin\alpha \\ \cos\alpha \end{pmatrix}.$$

The corresponding Hermitian matrix is

$$\hat{A}_{\alpha\alpha_\perp} = \begin{pmatrix} \cos 2\alpha & \sin 2\alpha \\ \sin 2\alpha & -\cos 2\alpha \end{pmatrix}$$

and the eigenvalues are $+1$ and -1.

We left as an exercise to show that

$$\hat{A}_{\alpha\alpha_\perp}|\alpha\rangle = 1 \cdot |\alpha\rangle \quad \text{and} \quad \hat{A}_{\alpha\alpha_\perp}|\alpha_\perp\rangle = -1 \cdot |\alpha_\perp\rangle$$

We can observe that $\alpha = 0$ corresponds to the first case studied when the Hermitian matrix is

$$\hat{A}_{HV} = \begin{pmatrix} 1 & 0 \\ 0 & -1 \end{pmatrix}$$

and the eigenstates are

$$|H\rangle = \begin{pmatrix} 1 \\ 0 \end{pmatrix} \quad \text{and} \quad |V\rangle = \begin{pmatrix} 0 \\ 1 \end{pmatrix}$$

while $\alpha = \dfrac{\pi}{4}$ corresponds to the second case studied when the Hermitian matrix is

$$\hat{A}_{/\backslash} = \begin{pmatrix} 0 & 1 \\ 1 & 0 \end{pmatrix}$$

and the eigenstate vectors are

$$|/\rangle = \begin{pmatrix} \dfrac{1}{\sqrt{2}} \\ \dfrac{1}{\sqrt{2}} \end{pmatrix} \quad \text{and} \quad |\backslash\rangle = \begin{pmatrix} -\dfrac{1}{\sqrt{2}} \\ \dfrac{1}{\sqrt{2}} \end{pmatrix}.$$

A planar wave is "circularly polarized" when it can be written as

$$|\Psi\rangle = \begin{pmatrix} \dfrac{1}{\sqrt{2}} \cdot e^{i\alpha_x} \\ \dfrac{1}{\sqrt{2}} \cdot e^{i\alpha_y} \end{pmatrix}.$$

with $|\alpha_x - \alpha_y| = \dfrac{\pi}{2}$. If we denote by

$$|\circlearrowleft\rangle = \begin{pmatrix} \dfrac{1}{\sqrt{2}} \\ \dfrac{i}{\sqrt{2}} \end{pmatrix} \quad \text{and} \quad |\circlearrowright\rangle = \begin{pmatrix} \dfrac{1}{\sqrt{2}} \\ \dfrac{-i}{\sqrt{2}} \end{pmatrix}$$

the right and left circularly polarized states, the wave can be described in the form

$$|\Psi\rangle = \Psi_R |\circlearrowright\rangle + \Psi_L |\circlearrowleft\rangle$$

where

$$\Psi_R = \langle\circlearrowright|\Psi\rangle \quad \text{and} \quad \Psi_L = \langle\circlearrowleft|\Psi\rangle.$$

Let us observe that

$$\langle\circlearrowright|\circlearrowleft\rangle = 0; \quad \langle\circlearrowright| = \begin{pmatrix} \dfrac{1}{\sqrt{2}} & \dfrac{-i}{\sqrt{2}} \end{pmatrix} \quad \text{and} \quad \langle\circlearrowleft| = \begin{pmatrix} \dfrac{1}{\sqrt{2}} & \dfrac{i}{\sqrt{2}} \end{pmatrix},$$

consequently

$$\langle\circlearrowright|\circlearrowright\rangle = 1; \quad \langle\circlearrowleft|\circlearrowleft\rangle = 1.$$

If we compute

$$\Psi_L = \langle\circlearrowleft|\Psi\rangle = \frac{1}{2}\left(e^{i\alpha_x} - i \cdot e^{i\alpha_y}\right)$$

and

$$\Psi_R = \langle\circlearrowright|\Psi\rangle = \frac{1}{2}\left(e^{i\alpha_x} + i \cdot e^{i\alpha_y}\right).$$

It results

$$|\Psi_L|^2 + |\Psi_R|^2 = 1.$$

The last relation shows the main constraint of the state function of Ψ which represents a polarized plane wave function. If we have in mind this representation, related to the electric field in the electromagnetic case, we may focus only at the right and left circularly polarized states

$$|\circlearrowright\rangle = \begin{pmatrix} \dfrac{1}{\sqrt{2}} \\ \dfrac{i}{\sqrt{2}} \end{pmatrix} \quad \text{and} \quad |\circlearrowleft\rangle = \begin{pmatrix} \dfrac{1}{\sqrt{2}} \\ \dfrac{-i}{\sqrt{2}} \end{pmatrix}.$$

The Hermitian matrix whose eigenvectors are the previous ones is

$$\hat{A}_{\circlearrowright\circlearrowleft} = \begin{pmatrix} 0 & -i \\ i & 0 \end{pmatrix}.$$

Exercise 10.4.1 Prove the following equalities:

1. $\hat{A}_{\circlearrowright\circlearrowleft}|\circlearrowright\rangle = |\circlearrowright\rangle$.
2. $\hat{A}_{\circlearrowright\circlearrowleft}|\circlearrowleft\rangle = -|\circlearrowleft\rangle$.
3. $\langle \alpha | \circlearrowright \rangle = \dfrac{1}{2}$.

Summary of Lecture 45. The present lecture is dedicated to the photon polarization as a straightforward application of the Dirac formalism. One of the most important concepts of Quantum Mechanics, the probability amplitude, is presented in detail. Specifically, we discuss how horizontal and vertical polarizations can be described from both physical and the mathematical points of view. Let us consider two possible states for the polarized light described by the orthogonal ket vectors

$$|H\rangle = \begin{pmatrix} 1 \\ 0 \end{pmatrix} \quad \text{and} \quad |V\rangle = \begin{pmatrix} 0 \\ 1 \end{pmatrix}.$$

The Hermitian operator, such that $|V\rangle$ and $|V\rangle$ are the eigenvectors (eigenstates), is

$$\hat{A}_{HV} = \begin{pmatrix} 1 & 0 \\ 0 & -1 \end{pmatrix}$$

because

$$\hat{A}_{HV}|H\rangle = \begin{pmatrix} 1 & 0 \\ 0 & -1 \end{pmatrix}\begin{pmatrix} 1 \\ 0 \end{pmatrix} = \begin{pmatrix} 1 \\ 0 \end{pmatrix} = 1 \cdot |H\rangle$$

and

$$\hat{A}_{HV}|V\rangle = \begin{pmatrix} 1 & 0 \\ 0 & -1 \end{pmatrix}\begin{pmatrix} 0 \\ 1 \end{pmatrix} = \begin{pmatrix} 0 \\ -1 \end{pmatrix} = -1 \cdot |V\rangle.$$

The distinct eigenvalues are $+1$ and -1. Now the physical meaning. The observable is a device which interacts only with horizontal and vertical polarized light and, from the mathematical point of view, it is the Hermitian operator $\hat{A}_{HV}$. Suppose the device works at two light frequencies, blue light and red light. If the blue light is switched on and the red light is switched off, the horizontal polarized light passes through the device. The blue light corresponds to the $+1$ eigenvalue. In this case, the blue light shows us the mathematical image of the polarized light. It is horizontally polarized, therefore the corresponding state is $|H\rangle$. If the red light is switched on and the blue light is switched off, the corresponding eigenvalue is -1 and the corresponding eigenstate is $|V\rangle$. We find similar descriptions both for $45°$; $135°$ polarized photons and circularly polarized photons. The corresponding Hermitian matrices are

$$\begin{pmatrix} 1 & 0 \\ 0 & -1 \end{pmatrix}; \begin{pmatrix} 0 & 1 \\ 1 & 0 \end{pmatrix}; \begin{pmatrix} 0 & -i \\ i & 0 \end{pmatrix};$$

they are the Pauli matrices. They represent an important theoretical support in the description not only of the photon polarization but also of the spin of electron description.

Chapter 11
Quantum Mechanics at the Next Level

Physicists have come to realize that Mathematics, when used with sufficient care, is a proven pathway to Truth.

Brian Greene

11.1 Lecture 46: The Electron Spin

Elementary particles, as electron, have an intrinsic characteristic named "spin". The analog of this property is a rotation. All rotations are made with respect one ore more than one axes of rotation. A rotation around an axis only is represented by a segment line (a vector) with an arrow at one end. The arrow gives information on the sense of rotation.

Before developing the Mathematics necessary to describe the electron spin, let us discuss how the spin of electron emerges from experiments and can be measured (see, for example [25]). Physicists established that measurements can be made by an electronic device with respect to a given direction, say z-direction. According to the electron spin structure which we will discuss below, there are only two possibilities: The device can indicate the possible "spin up" or "spin down" directions of the electron spin using two colored lights. If the blue light is switched on and the red light is switched off, the device indicates a "spin up" electron passing through it. If the red light is switched on and the blue light is switched off, the device indicates a "spin down" electron passing through it.

Using magnets, physicists can arrange electrons "spin up" with respect to z-axis. The measurements will indicate that 100% of this "spin up" electrons, passing through the device, are indicated as "spin up" and 0% are indicated as "spin down". A similar result will happen when the physicists will arrange the "spin down" electrons and the electrons pass through the device. The measurements will indicate that 100% of this "spin down" electrons passing through the device are indicated as "spin down" and 0% are indicated as "spin up".

© The Author(s), under exclusive license to Springer Nature Switzerland AG 2021
S. Capozziello and W.-G. Boskoff, *A Mathematical Journey to Quantum Mechanics*,
UNITEXT for Physics, https://doi.org/10.1007/978-3-030-86098-1_11

A first interesting "wired" result is the following. If we preserve the device in the position where it is and we send in it electrons prepared with the spin in the x-direction, the device will indicate 50% with "spin up" and 50% with "spin down". The intuition tells as that there are no "spin up" or "spin down" electrons in this situation, however half of them are "spin up" and half of them are "spin down". The same, if we prepare electrons with spin at a given angle with respect to the z-axis, we can have 32% with "spin up" and 68% with spin down. These results are "wired" because all the electrons are arranged with spin at "a given direction" and some of them will appear as "spin up" and all the others will appear as "spin down" electrons. Therefore the direction in which the electrons are "spin arranged" can be dealt under the standard of probability. According to this statement, there is an analog mathematics involved in the description of spin as the one adopted for the photon polarization.

In the Dirac notation, if we denote by $|\nearrow\rangle$ the state of identical electrons with spin prepared into a generic $\nearrow$ direction, we can decompose them as "spin up" electrons with $|\uparrow\rangle$ and "spin down" electron with $|\downarrow\rangle$. Then, it is

$$|\nearrow\rangle = a|\uparrow\rangle + b|\downarrow\rangle$$

where $a,\ b \in \mathbb{C}$ are probability amplitudes and

$$aa^* + bb^* = 1.$$

Let us insist on the language: a, that is $\langle\uparrow \mid \nearrow\rangle$, is the probability amplitude for electrons arranged with spin in a direction $\nearrow$ to be "spin up". The probability for electrons arranged with spin at direction $\nearrow$ to be spin up is aa^*. On the other hand, b, that is $\langle\downarrow \mid \nearrow\rangle$, is the probability amplitude for electrons with spin arranged in a direction $\nearrow$ to be "spin down". The probability for electrons with spin arranged in direction $\nearrow$ to be "spin down" is bb^*. Since there are only two possibilities, the relation $aa^* + bb^* = 1$ reflects the whole probability of the event to occur, the event being "to be spin up or spin down".

Let us consider now only a given electron among the identical prepared electrons at spin $\nearrow$. We understand that, after the electron interacts with the measurement device, it will be either "spin up" or "spin down''. Suppose the electron becomes a "spin up" one. Therefore, for this particular electron, the relation

$$|\nearrow\rangle = a|\uparrow\rangle + b|\downarrow\rangle$$

becomes

$$|\nearrow\rangle = 1|\uparrow\rangle + 0|\downarrow\rangle = |\uparrow\rangle.$$

An important point has to be remarked: Interfering with the equipment, the state $|\nearrow\rangle$, which is a superposition of "spin up" and "spin down", that is

$$|\nearrow\rangle = a|\uparrow\rangle + b|\downarrow\rangle,$$

collapses in one the two possible states, in our case the electron collapses in the "spin up" state $| \uparrow \rangle$.

Therefore the blue light of our device switches on when we send the electron, prepared at the state $| \nearrow \rangle$, into the equipment. This result means the collapse of the superposition state. It is worth noticing that, even if we observe the result for 1000, or for 1000000 electrons with spin arranged at $| \nearrow \rangle$, we can only ascertain the result with respect to the probability amplitudes. In other words, Nature imposes the probability amplitudes and we can only ascertain the realization of this probability. Neither the device nor the observer establish the probability amplitudes. The Nature Itself decides and it is important to underline this fundamental point.

Let us take into account again the relation

$$| \nearrow \rangle = a | \uparrow \rangle + b | \downarrow \rangle$$

where $a, \ b \in \mathbb{C}$ are probability amplitudes and $aa^* + bb^* = 1$.

In this relation, the ket vectors related to "spin up" and "spin down" are part of an orthonormal vector basis because no "spin up" electron is "spin down" and vice-versa. "Spin up" arranged electrons and all measured as "spin up". Therefore the relations

$$\langle \uparrow | \uparrow \rangle = 1; \quad \langle \downarrow | \downarrow \rangle = 1; \quad \langle \downarrow | \uparrow \rangle = 0;$$

hold. When identical electrons are arranged to spin in the direction of x-axis, the "spin right" direction is expressed as

$$| \rightarrow \rangle = \frac{1}{\sqrt{2}} | \uparrow \rangle + \frac{1}{\sqrt{2}} | \downarrow \rangle.$$

We have a similar relation for the electrons arranged in "spin left" direction, that is

$$| \leftarrow \rangle = \frac{1}{\sqrt{2}} | \uparrow \rangle - \frac{1}{\sqrt{2}} | \downarrow \rangle.$$

From this two relations, we can immediately deduce

$$| \uparrow \rangle = \frac{1}{\sqrt{2}} | \rightarrow \rangle + \frac{1}{\sqrt{2}} | \leftarrow \rangle$$

and

$$| \downarrow \rangle = \frac{1}{\sqrt{2}} | \rightarrow \rangle - \frac{1}{\sqrt{2}} | \leftarrow \rangle.$$

Therefore, we have the following situation:

- With respect to the z-axis, we have the "spin up" and "spin down" orthonormal basis $| \uparrow \rangle, \ | \downarrow \rangle$;

- With respect to the x-axis, we have the "spin right" and "spin left" orthonormal basis $|\rightarrow\rangle$, $|\leftarrow\rangle$;
- With respect to the y-axis we have the "spin in" and "spin out" orthonormal basis $|\times\rangle$, $|\circ\rangle$. The names for this two last possibilities to spin are related to the fact that the y-axis is perpendicular to the Oxz-plane.

If we make the same experiment with "spin in" direction arranged electrons and if we measure with respect to a "spin up-spin down" device, we obtain the same 50% "spin up" and 50% "spin down" electrons passing through the device. Therefore we have

$$|\times\rangle = \frac{1}{\sqrt{2}}|\uparrow\rangle + \frac{i}{\sqrt{2}}|\downarrow\rangle.$$

In the same way, it is

$$|\circ\rangle = \frac{1}{\sqrt{2}}|\uparrow\rangle - \frac{i}{\sqrt{2}}|\downarrow\rangle.$$

It is important to say that, in all situations presented above, we choose the $+$ sign to express the sense on the studied axis. Therefore here the $+$ is chosen to express the fact that "spin in" is oriented in the sense of y-axis orientation. As a consequence

$$|\uparrow\rangle = \frac{1}{\sqrt{2}}|\times\rangle + \frac{1}{\sqrt{2}}|\circ\rangle$$

and

$$|\downarrow\rangle = -\frac{i}{\sqrt{2}}|\times\rangle + \frac{i}{\sqrt{2}}|\circ\rangle.$$

Example 11.1.1 Let us check if we obtain $\frac{1}{2}$ probability when we measure "spin in" arranged electrons to be "spin left". First we write "spin in" arranged electrons in the form

$$|\times\rangle = \frac{1}{\sqrt{2}}|\uparrow\rangle + \frac{i}{\sqrt{2}}|\downarrow\rangle.$$

Then we consider the bra vector corresponding to the ket vector

$$|\leftarrow\rangle = \frac{1}{\sqrt{2}}|\uparrow\rangle - \frac{1}{\sqrt{2}}|\downarrow\rangle,$$

that is

$$\langle\leftarrow| = \frac{1}{\sqrt{2}}\langle\uparrow| - \frac{1}{\sqrt{2}}\langle\downarrow|.$$

We need to express the probability amplitude $\langle\leftarrow|\times\rangle$ and then to consider $|\langle\leftarrow|\times\rangle|^2$, that is $\langle\leftarrow|\times\rangle \cdot \langle\leftarrow|\times\rangle^*$. Using the orthonormality of the basis "spin up", "spin down" we have

$$\langle \leftarrow |\times\rangle = \left(\frac{1}{\sqrt{2}}\langle\uparrow| - \frac{1}{\sqrt{2}}\langle\downarrow|\right)\left(\frac{1}{\sqrt{2}}|\uparrow\rangle + \frac{i}{\sqrt{2}}|\downarrow\rangle\right) = \frac{1}{2} - \frac{i}{2}.$$

Therefore the probability is

$$|\langle\leftarrow|\times\rangle|^2 = \langle\leftarrow|\times\rangle \cdot \langle\leftarrow|\times\rangle^* = \left(\frac{1}{2} - \frac{i}{2}\right)\left(\frac{1}{2} + \frac{i}{2}\right) = \frac{1}{2}$$

which fits the experiment.

Exercise 11.1.2 Find the probability for "spin right" arranged electrons to be "spin out".

Hint. Take care at the bra vector attached to "spin out" ket vector. It is

$$\langle o| = \frac{1}{\sqrt{2}}\langle\uparrow| + \frac{i}{\sqrt{2}}\langle\downarrow|.$$

Then follow the line of the previous example and compute the probability amplitude $\langle o| \rightarrow\rangle$.

The "spin up" and "spin down" ket vectors can be seen as

$$|\uparrow\rangle = \begin{pmatrix} 1 \\ 0 \end{pmatrix} \quad \text{and} \quad |\downarrow\rangle = \begin{pmatrix} 0 \\ 1 \end{pmatrix}.$$

According to this formalism, the "spin right", "spin left", "spin in" and "spin out" ket vectors are

$$|\rightarrow\rangle = \begin{pmatrix} \frac{1}{\sqrt{2}} \\ \frac{1}{\sqrt{2}} \end{pmatrix}; \quad |\leftarrow\rangle = \begin{pmatrix} \frac{1}{\sqrt{2}} \\ \frac{-1}{\sqrt{2}} \end{pmatrix}; \quad |\times\rangle = \begin{pmatrix} \frac{1}{\sqrt{2}} \\ \frac{i}{\sqrt{2}} \end{pmatrix}; \quad |o\rangle = \begin{pmatrix} \frac{1}{\sqrt{2}} \\ \frac{-i}{\sqrt{2}} \end{pmatrix}.$$

Proposition 11.1.3 *The operator which allows the eigenvectors* $|\uparrow\rangle$ *and* $|\downarrow\rangle$ *corresponding to the eigenvalues* $+1$ *and* -1, *respectively, is*

$$\sigma_z = \begin{pmatrix} 1 & 0 \\ 0 & -1 \end{pmatrix}.$$

Proof We can check directly or we can identify the elements of the matrix from the two conditions:

$$\sigma_z|\uparrow\rangle = |\uparrow\rangle \quad \text{that is} \quad \begin{pmatrix} a & b \\ d & e \end{pmatrix}\begin{pmatrix} 1 \\ 0 \end{pmatrix} = \begin{pmatrix} 1 \\ 0 \end{pmatrix};$$

$$\sigma_z|\downarrow\rangle = -|\downarrow\rangle \quad \text{that is} \quad \begin{pmatrix} a & b \\ d & e \end{pmatrix}\begin{pmatrix} 0 \\ 1 \end{pmatrix} = \begin{pmatrix} 0 \\ -1 \end{pmatrix}.$$

These four equations have the solution $a = 1$; $b = 0$; $c = 0$; $d = -1$.   □

The observable is our device which establishes if an electron with the spin prepared in the j-direction "spins up" or "spins down". In this case, from the mathematical point of view, the observable is the Hermitian operator σ_z. The device has two lights, blue and red. If the blue light is switched on and the red light is switched off, a "spin up" electron passed through the device. The blue light corresponds to the $+1$ eigenvalue. In this case, the blue light is telling us that a "spin up" electron, corresponding to $| \uparrow \rangle$ state, passed. If the red light is switched on and the blue light is switched off, the corresponding eigenvalue is -1 and the corresponding eigenstate is $| \downarrow \rangle$.

At the end of all measurements, we find the probability P_u to be measured as "spin up" and the probability P_d to be measured as "spin down". If $|j\rangle = p| \uparrow \rangle + q| \downarrow \rangle$, then $P_u = pp^*$ and $P_d = qq^*$.

Following the proof of the above proposition, the reader can easily solve the following:

Exercise 11.1.4 Show that the operator which allows the eigenvectors $| \rightarrow \rangle$ and $| \leftarrow \rangle$ corresponding to the eigenvalues $+1$ and -1, respectively, is

$$\sigma_x = \begin{pmatrix} 0 & 1 \\ 1 & 0 \end{pmatrix}.$$

Exercise 11.1.5 Show that the operator which allows the eigenvectors $|\times\rangle$ and $|o\rangle$ corresponding to the eigenvalues $+1$ and -1, respectively, is

$$\sigma_y = \begin{pmatrix} 0 & -i \\ i & 0 \end{pmatrix}.$$

Therefore the same unitary Hermitian operators, the Pauli matrices, discussed in the case of photon polarization, can be adopted to study the electron spin.

Let us see the big picture of what we discovered until now. The eigenstates corresponding to Pauli matrices σ_z, σ_x, σ_y for the same $+1$ and -1 eigenvalues are

For the operator σ_z, the eigenvalue $+1$ corresponds to $| \uparrow \rangle = \begin{pmatrix} 1 \\ 0 \end{pmatrix}$ and -1 to $| \downarrow \rangle = \begin{pmatrix} 0 \\ 1 \end{pmatrix}$;

For the operator σ_x, the eigenvalue $+1$ corresponds to $| \rightarrow \rangle = \begin{pmatrix} \frac{1}{\sqrt{2}} \\ \frac{1}{\sqrt{2}} \end{pmatrix}$ and -1 to $| \leftarrow \rangle = \begin{pmatrix} \frac{1}{\sqrt{2}} \\ \frac{-1}{\sqrt{2}} \end{pmatrix}$;

For the operator σ_y, the eigenvalue $+1$ corresponds to $|\times\rangle = \begin{pmatrix} \frac{1}{\sqrt{2}} \\ \frac{i}{\sqrt{2}} \end{pmatrix}$ and -1 to $|o\rangle = \begin{pmatrix} \frac{1}{\sqrt{2}} \\ \frac{-i}{\sqrt{2}} \end{pmatrix}$.

If we look again at the formulas

$$|\rightarrow\rangle = \frac{1}{\sqrt{2}}|\uparrow\rangle + \frac{1}{\sqrt{2}}|\downarrow\rangle$$

or

$$|\times\rangle = \frac{1}{\sqrt{2}}|\uparrow\rangle + \frac{i}{\sqrt{2}}|\downarrow\rangle,$$

we find out that, for the orthogonal directions z and x, the probability to find an electron arranged to spin in the positive x direction to be measured as spinning in the positive z direction is $\frac{1}{2}$, that is $\cos^2\frac{\pi}{4}$. The same if we measure the probability to find an electron arranged to spin in the positive y direction to be measured as spinning in the positive z direction is $\frac{1}{2}$, that is $\cos^2\frac{\pi}{4}$. In fact if we denote by α the angle between the directions, the measured probability is $\cos^2\frac{\alpha}{2}$ and it is measured with respect to eigenstates corresponding to the same eigenvalue, $+1$ of the corresponding operators. In the first case, the probability is $|\langle\uparrow\,|\rightarrow\rangle|^2$.

This result can be generalized. Consider two vectors $u := (u_x, u_y, u_z)$ and $v := (v_x, v_y, v_z)$ such that

$$u_x^2 + u_y^2 + u_z^2 = v_x^2 + v_y^2 + v_z^2 = 1.$$

They represent directions in the three-dimensional Euclidean space. Denote by α the angle between the two unit vectors, that is

$$\cos\alpha = u_x v_x + u_y v_y + u_z v_z.$$

We can ask which is the probability amplitude, corresponding to electrons arranged with spin in the u direction, to be measured with spin in the v direction. Then, it is easy to find the probability, that is $\cos^2\frac{\alpha}{2}$.

Let us describe how to proceed to find out the probability amplitude and then the probability.

1. We construct the operators related to the given directions.
2. We find out the eigenvector corresponding to $+1$ eigenvalue for each of the two operators.
3. We compute the probability amplitude determined by the two eigenvectors.

Let us proceed according to the above steps. The operators related to the unit vectors u and v are

$$\sigma_u := u_x\sigma_x + u_y\sigma_y + u_z\sigma_z;$$

$$\sigma_v := v_x\sigma_x + v_y\sigma_y + v_z\sigma_z.$$

They are also called Pauli vectors. If we consider the first *Pauli vector*, we can observe that a particular choice as $u = (0, 0, 1)$ determines the σ_z Pauli matrix and we know the eigenvector corresponding to the eigenvalue $+1$. The same happens for the choices $(1, 0, 0)$ and $(0, 1, 0)$ which produce the other two Pauli matrices σ_x and σ_y. Therefore, except these three choices, we are interested to find the eigenvectors denoted by $|\Psi_u^+\rangle$ and $|\Psi_v^+\rangle$ corresponding to the eigenvalue $+1$ for each Pauli vector.

Theorem 11.1.6 *The eigenvector $|\Psi_u^+\rangle$, corresponding to the eigenvalue $+1$ of the Pauli vector σ_u, is*

$$|\Psi_u^+\rangle = \sqrt{\frac{1 + u_z}{2}} \begin{pmatrix} 1 \\ \dfrac{1 - u_z}{u_x - iu_y} \end{pmatrix}.$$

Proof Let us start from the definition of σ_u:

$$\sigma_u = u_x \begin{pmatrix} 0 & 1 \\ 1 & 0 \end{pmatrix} + u_y \begin{pmatrix} 0 & -i \\ i & 0 \end{pmatrix} + u_z \begin{pmatrix} 1 & 0 \\ 0 & -1 \end{pmatrix} = \begin{pmatrix} u_z & u_x - iu_y \\ u_x + iu_y & -u_z \end{pmatrix}.$$

We show that σ_u allows the eigenvector to have the form

$$|\Psi_u^+\rangle = a \begin{pmatrix} 1 \\ \beta \end{pmatrix}, \quad a \in \mathbb{R}, \ \beta \in \mathbb{C}.$$

The relation

$$\sigma_u \cdot |\Psi_u^+\rangle = |\Psi_u^+\rangle,$$

that is

$$\begin{pmatrix} u_z & u_x - iu_y \\ u_x + iu_y & -u_z \end{pmatrix} \begin{pmatrix} 1 \\ \beta \end{pmatrix} = \begin{pmatrix} 1 \\ \beta \end{pmatrix}$$

leads to the system

$$\begin{cases} u_z + \beta(u_x - iu_y) = 1 \\ (u_x + iu_y) - \beta u_z = \beta. \end{cases}$$

The two possible values of β

$$\beta = \frac{1 - u_z}{u_x - iu_y}$$

and

$$\beta = \frac{u_x + iu_y}{1 + u_z}$$

are the same because u is a unit vector. Therefore we can choose one of them, say the first one. It results

$$|\Psi_u^+\rangle = a \begin{pmatrix} 1 \\ \dfrac{1-u_z}{u_x - iu_y} \end{pmatrix}.$$

The corresponding bra vector is

$$\langle \Psi_u^+| = a \begin{pmatrix} 1 & \dfrac{1-u_z}{u_x + iu_y} \end{pmatrix}.$$

therefore we establish the constant a from $\langle \Psi_u^+|\Psi_u^+\rangle = 1$. It results

$$1 = \langle \Psi_u^+|\Psi_u^+\rangle = a^2 \begin{pmatrix} 1 & \dfrac{1-u_z}{u_x + iu_y} \end{pmatrix} \begin{pmatrix} 1 \\ \dfrac{1-u_z}{u_x - iu_y} \end{pmatrix} = a^2 \left(1 + \dfrac{(1-u_z)^2}{u_x^2 + u_y^2} \right) = \dfrac{2a^2}{1 + u_z}.$$

The sign is chosen to preserve the direction of the eigenvector, therefore

$$a = \sqrt{\dfrac{1+u_z}{2}}.$$

□

Corollary 11.1.7 *The eigenvector $|\Psi_u^-\rangle$ corresponding to the eigenvalue -1 of the Pauli vector σ_u is*

$$|\Psi_u^-\rangle = \sqrt{\dfrac{1-u_z}{2}} \begin{pmatrix} 1 \\ -\dfrac{1+u_z}{u_x - iu_y} \end{pmatrix}.$$

Let us continue to compute the probability amplitude $\langle \Psi_v^+|\Psi_u^+\rangle$ which expresses the number of electrons arranged with spin in the u direction to be measured with spin in the v direction.

Theorem 11.1.8 *If, for the unit vectors u and v, we have $\cos \alpha = u_x v_x + u_y v_y + u_z v_z$, then*

$$|\langle \Psi_u^+|\Psi_v^+\rangle|^2 = \cos^2 \dfrac{\alpha}{2}.$$

Proof If we compute $\langle \Psi_v^+|\Psi_u^+\rangle$ using

$$\langle \Psi_v^+| = \sqrt{\dfrac{1+v_z}{2}} \begin{pmatrix} 1 & \dfrac{1-v_z}{v_x + iv_y} \end{pmatrix}$$

and

$$|\Psi_u^+\rangle = \sqrt{\dfrac{1+u_z}{2}} \begin{pmatrix} 1 \\ \dfrac{1-u_z}{u_x - iu_y} \end{pmatrix}$$

and we want to express the result as depending on $\cos^2 \dfrac{\alpha}{2} = \dfrac{1 + \cos \alpha}{2} =$
$\dfrac{1 + u_x v_x + u_y v_y + u_z v_z}{2}$ the calculation is complicated. We can choose another way.
Let us perform a change of coordinates. In the plane determined by the two vectors
u and v, we choose one of the vectors as $(1, 0)$ and the other as $(\cos \alpha, \sin \alpha)$. The
third new coordinate is with respect to an axis perpendicular in the origin to the
plane determined by u and v. Without losing generality, we can say that, in the
new coordinate frame, the components of the two vectors u and v are $(1, 0, 0)$ and
$(\cos \alpha, \sin \alpha, 0)$. In the new system of coordinates, it is

$$\sigma_u = 1 \cdot \begin{pmatrix} 0 & 1 \\ 1 & 0 \end{pmatrix} + 0 \cdot \begin{pmatrix} 0 & -i \\ i & 0 \end{pmatrix} + 0 \cdot \begin{pmatrix} 1 & 0 \\ 0 & -1 \end{pmatrix} = \begin{pmatrix} 0 & 1 \\ 1 & 0 \end{pmatrix}$$

and

$$\sigma_v = \cos \alpha \begin{pmatrix} 0 & 1 \\ 1 & 0 \end{pmatrix} + \sin \alpha \cdot \begin{pmatrix} 0 & -i \\ i & 0 \end{pmatrix} + 0 \cdot \begin{pmatrix} 1 & 0 \\ 0 & -1 \end{pmatrix} = \begin{pmatrix} 0 & \cos \alpha - i \sin \alpha \\ \cos \alpha + i \sin \alpha & 0 \end{pmatrix}.$$

We know how computing the eigenvectors corresponding to the eigenvalue $+1$. They
are

$$|\Psi_u^+\rangle = \begin{pmatrix} \dfrac{1}{\sqrt{2}} \\ \dfrac{1}{\sqrt{2}} \end{pmatrix} \quad \text{and} \quad |\Psi_v^+\rangle = \begin{pmatrix} \dfrac{1}{\sqrt{2}} \\ \dfrac{1}{\sqrt{2}}(\cos \alpha + i \sin \alpha) \end{pmatrix}.$$

Therefore

$$\langle \Psi_v^+ | \Psi_u^+ \rangle = \begin{pmatrix} \dfrac{1}{\sqrt{2}} & \dfrac{1}{\sqrt{2}}(\cos \alpha - i \sin \alpha) \end{pmatrix} \begin{pmatrix} \dfrac{1}{\sqrt{2}} \\ \dfrac{1}{\sqrt{2}} \end{pmatrix} = \cos \dfrac{\alpha}{2} \left(\cos \dfrac{\alpha}{2} - i \sin \dfrac{\alpha}{2} \right)$$

and $\langle \Psi_u^+ | \Psi_v^+ \rangle = \cos \dfrac{\alpha}{2} \left(\cos \dfrac{\alpha}{2} + i \sin \dfrac{\alpha}{2} \right)$. The probability becomes

$$|\langle \Psi_u^+ | \Psi_v^+ \rangle|^2 = \cos^2 \dfrac{\alpha}{2} \left(\cos \dfrac{\alpha}{2} + i \sin \dfrac{\alpha}{2} \right) \left(\cos \dfrac{\alpha}{2} - i \sin \dfrac{\alpha}{2} \right) = \cos^2 \dfrac{\alpha}{2}.$$

$\square$

Let us, for the moment, consider only the mathematical point of view. We define
the following objects:

$$\sigma_z = \begin{pmatrix} 1 & 0 \\ 0 & -1 \end{pmatrix}; \quad \sigma_x = \begin{pmatrix} 0 & 1 \\ 1 & 0 \end{pmatrix}; \quad \sigma_y = \begin{pmatrix} 0 & -i \\ i & 0 \end{pmatrix};$$

without knowing, from the beginning, that they are used to describe both photon polarization and electron spin. We can find the eigenvectors and the eigenvalues of these *Pauli matrices* without knowing their physical meaning. Using these matrices in blocks of some other 3×3 or 4×4 matrices, we can create objects as

$$
\sigma_x = \begin{pmatrix} 0 & 1 & 0 \\ 1 & 0 & 1 \\ 0 & 1 & 0 \end{pmatrix}; \quad
\sigma_y = \begin{pmatrix} 0 & -i & 0 \\ i & 0 & -i \\ 0 & i & 0 \end{pmatrix}; \quad
\sigma_z = \begin{pmatrix} 1 & 0 & 0 \\ 0 & 0 & 0 \\ 0 & 0 & -1 \end{pmatrix};
$$

which admit, for the eigenvalues $1;\ 0,\ -1$, the orthogonal eigenvectors

$$
|x\rangle_1 = \frac{1}{2}\begin{pmatrix} 1 \\ \sqrt{2} \\ 1 \end{pmatrix} \quad
|x\rangle_0 = \frac{1}{\sqrt{2}}\begin{pmatrix} -1 \\ 0 \\ 1 \end{pmatrix} \quad
|x\rangle_{-1} = \frac{1}{2}\begin{pmatrix} 1 \\ -\sqrt{2} \\ 1 \end{pmatrix}
$$

$$
|y\rangle_1 = \frac{1}{2}\begin{pmatrix} -1 \\ -i\sqrt{2} \\ 1 \end{pmatrix} \quad
|y\rangle_0 = \frac{1}{\sqrt{2}}\begin{pmatrix} 1 \\ 0 \\ 1 \end{pmatrix} \quad
|y\rangle_{-1} = \frac{1}{2}\begin{pmatrix} -1 \\ i\sqrt{2} \\ 1 \end{pmatrix}
$$

$$
|z\rangle_1 = \begin{pmatrix} 0 \\ 0 \\ 1 \end{pmatrix} \quad
|z\rangle_0 = \begin{pmatrix} 0 \\ 1 \\ 0 \end{pmatrix} \quad
|z\rangle_{-1} = \begin{pmatrix} 1 \\ 0 \\ 0 \end{pmatrix};
$$

or the 4×4 matrices

$$
\sigma_x = \begin{pmatrix} 0 & \sqrt{3} & 0 & 0 \\ \sqrt{3} & -0 & 2 & 0 \\ 0 & 2 & 0 & \sqrt{3} \\ 0 & 0 & \sqrt{3} & 0 \end{pmatrix}; \quad
\sigma_y = \begin{pmatrix} 0 & -i\sqrt{3} & 0 & 0 \\ i\sqrt{3} & 0 & -2i & 0 \\ 0 & 2i & 0 & -i\sqrt{3} \\ 0 & 0 & i\sqrt{3} & 0 \end{pmatrix}; \quad
\sigma_z = \begin{pmatrix} 3 & 0 & 0 & 0 \\ 0 & 1 & 0 & 0 \\ 0 & 0 & -1 & 0 \\ 0 & 0 & 0 & -3 \end{pmatrix};
$$

which allow, for the eigenvalues $\dfrac{3}{2}, \dfrac{1}{2}, -\dfrac{1}{2}, -\dfrac{3}{2}$, the orthogonal eigenvectors

$$
|x\rangle_{3/2} = \frac{1}{2\sqrt{2}}\begin{pmatrix} 1 \\ \sqrt{3} \\ \sqrt{3} \\ 1 \end{pmatrix}; \quad
|x\rangle_{1/2} = \frac{1}{2\sqrt{2}}\begin{pmatrix} -\sqrt{3} \\ -1 \\ 1 \\ \sqrt{3} \end{pmatrix};
$$

$$
|x\rangle_{-1/2} = \frac{1}{2\sqrt{2}}\begin{pmatrix} \sqrt{3} \\ -1 \\ -1 \\ \sqrt{3} \end{pmatrix}; \quad
|x\rangle_{-3/2} = \frac{1}{2\sqrt{2}}\begin{pmatrix} -1 \\ \sqrt{3} \\ -\sqrt{3} \\ 1 \end{pmatrix};
$$

$$|y\rangle_{3/2} = \frac{1}{2\sqrt{2}} \begin{pmatrix} i \\ -\sqrt{3} \\ -i\sqrt{3} \\ 1 \end{pmatrix} ; \quad |y\rangle_{1/2} = \frac{1}{2\sqrt{2}} \begin{pmatrix} -i\sqrt{3} \\ 1 \\ -i \\ \sqrt{3} \end{pmatrix} ;$$

$$|y\rangle_{-1/2} = \frac{1}{2\sqrt{2}} \begin{pmatrix} i\sqrt{3} \\ 1 \\ i \\ \sqrt{3} \end{pmatrix} ; \quad |y\rangle_{-3/2} = \frac{1}{2\sqrt{2}} \begin{pmatrix} -i \\ -\sqrt{3} \\ i\sqrt{3} \\ 1 \end{pmatrix} ;$$

$$|z\rangle_{3/2} = \begin{pmatrix} 1 \\ 0 \\ 0 \\ 0 \end{pmatrix} ; \quad |z\rangle_{1/2} = \begin{pmatrix} 0 \\ 1 \\ 0 \\ 0 \end{pmatrix} ;$$

$$|z\rangle_{-1/2} = \begin{pmatrix} 0 \\ 0 \\ 1 \\ 0 \end{pmatrix} ; \quad |z\rangle_{-3/2} = \begin{pmatrix} 0 \\ 0 \\ 0 \\ 1 \end{pmatrix} .$$

It is possible to establish that the first group of 3×3 matrices are observables corresponding to particles with spin 1, while the second group of matrices are observables corresponding to particles with spin $3/2$. There are also matrices corresponding to particles with spin $5/2$. The electron is an example of particle having the spin $1/2$ with the two states $1/2$ and $-1/2$. These spin numbers will be understood under the standard of angular momentum in Quantum Mechanics.

Summary of Lecture 46. Spin is an intrinsic property of particles like electron. Something spinning is something which rotates with respect to one or more axes. The spin of electron cannot be imagined, it is wired with respect to our intuition. However, it can be mathematically described by Dirac notation using the Pauli matrices. In Dirac notation, if we denote by $| \nearrow \rangle$ the state of identical electrons with spin prepared into $\nearrow$ direction, we can write them with respect to "spin up", denoted by $| \uparrow \rangle$, and to "spin down", denoted by $| \downarrow \rangle$, as $| \nearrow \rangle = a| \uparrow \rangle + b| \downarrow \rangle$ where $a, \ b \in \mathbb{C}$ are the probability amplitudes with the constraint $aa^* + bb^* = 1$. Step by step, it is possible to advance in the construction of the spin big picture. We observe that "spin up" and "spin down" arrangements are in fact vectors along the z-axis and, using this idea, we can construct "spin right" and "spin left" along the x-axis and "spin in" and "spin out" along the y-axis. The last two pairs can be expressed with respect to "spin up" and "spin down" vectors. The eigenstates corresponding to the Pauli matrices

$$\sigma_z = \begin{pmatrix} 1 & 0 \\ 0 & -1 \end{pmatrix}; \sigma_x = \begin{pmatrix} 0 & 1 \\ 1 & 0 \end{pmatrix}; \sigma_y = \begin{pmatrix} 0 & -i \\ i & 0 \end{pmatrix};$$

for the same $+1$ and -1 eigenvalues, are respectively the vectors described earlier, that is

$$|\uparrow\rangle = \begin{pmatrix} 1 \\ 0 \end{pmatrix} \quad \text{and} \quad |\downarrow\rangle = \begin{pmatrix} 0 \\ 1 \end{pmatrix};$$

$$|\rightarrow\rangle = \begin{pmatrix} \frac{1}{\sqrt{2}} \\ \frac{1}{\sqrt{2}} \end{pmatrix} \quad \text{and} \quad |\leftarrow\rangle = \begin{pmatrix} \frac{1}{\sqrt{2}} \\ \frac{-1}{\sqrt{2}} \end{pmatrix};$$

$$|\times\rangle = \begin{pmatrix} \frac{1}{\sqrt{2}} \\ \frac{i}{\sqrt{2}} \end{pmatrix} \quad \text{and} \quad |\circ\rangle = \begin{pmatrix} \frac{1}{\sqrt{2}} \\ \frac{-i}{\sqrt{2}} \end{pmatrix}.$$

From a given unit vector $u = (u_x, u_y, u_z)$, we can construct the Pauli vector

$$\sigma_u = u_x \begin{pmatrix} 0 & 1 \\ 1 & 0 \end{pmatrix} + u_y \begin{pmatrix} 0 & -i \\ i & 0 \end{pmatrix} + u_z \begin{pmatrix} 1 & 0 \\ 0 & -1 \end{pmatrix} = \begin{pmatrix} u_z & u_x - iu_y \\ u_x + iu_y & -u_z \end{pmatrix}.$$

It can be proved that its eigenvector, corresponding to the eigenvalue $+1$, is

$$|\Psi_u^+\rangle = \sqrt{\frac{1 + u_z}{2}} \begin{pmatrix} 1 \\ \frac{u_x - iu_y}{1 - u_z} \end{pmatrix}$$

and, for two given unit vectors, the probability corresponding to electrons, arranged with spin in the u direction to be measured with spin in the v direction, depends on the angle α between the two vectors, that is

$$|\langle \Psi_u^+ | \Psi_v^+ \rangle|^2 = \cos^2 \frac{\alpha}{2}.$$

11.2 Lecture 47: Revisiting the Harmonic Oscillator. The Ladder Operators

In Lecture 25, we discussed how to solve the Schrödinger equation for the harmonic oscillator and, in the following ones, we described the whole mathematical structure related to the quantum harmonic oscillator. In particular, we discussed, besides the solutions, how to predict the possible energy levels.

We want to study how the mathematical apparatus of harmonic oscillator can be used to describe, in general, Quantum Mechanics. The result is a deep impact on elementary particles physics because this picture is related to ladder operators which allow us not only to have a comprehensive view on energy levels but also on creation and annihilation of matter. Specifically, in linear algebra, a ladder operator can increase or decrease the eigenvalues of a given operator. They are the creation and annihilation operators.

Let us start by understanding how states vary over time and then, using the Hermitian operator associated to harmonic oscillator, we construct two operators. One of them , successively applied, gives rise, starting from a first energy eigenvalue $\dfrac{\hbar\omega}{2}$, to all other possible energy levels $\left(n + \dfrac{1}{2}\right)\hbar\omega$ described in Lecture 25.

Consider a Hermitian operator $\hat{A}$ acting on the state of a system, denoted by $|\Psi\rangle$, giving a real eigenvalue a and representing the measurement with respect to the observable at the given state. We established already the mathematical and the physical meaning of an equation as

$$\hat{A}|\Psi\rangle = a|\Psi\rangle.$$

If we are interested to extract information on energy E of the system, the observable is the Hamiltonian $\hat{H}$. We want to see what happens if the state evolves over time.

Let us develop the 1-dimensional case. If there is no time dependence, the Hamiltonian operator is:

$$\hat{H}(x) = -\frac{\hbar^2}{2m}\frac{d^2}{dx^2} + \hat{V}(x).$$

The extracted information is the eigenvalue E coming from

$$\left[-\frac{\hbar^2}{2m}\frac{d^2}{dx^2} + \hat{V}(x)\right]|\Psi\rangle = E|\Psi\rangle.$$

Let us consider the state $|\Psi\rangle$ varying over time. This next step involves unitary operators which we have to take into account.

Denote by $\hat{I}$ the unit operator. The time-dependence of a state $|\Psi_1\rangle$ is the action of another operator $\hat{U}_t$ on $|\Psi_1\rangle$ transforming it into $\hat{U}_t|\Psi_1\rangle$.

If another state $|\Psi_2\rangle$ evolves in time simultaneously with $|\Psi_1\rangle$, we obtain, at the same time t, the state $\hat{U}_t|\Psi_2\rangle$. The bra vector, corresponding to $\hat{U}_t|\Psi_2\rangle$, is $\langle\Psi_2\hat{U}_t^\dagger|$, where $\hat{U}_t^\dagger$ is the transposed complex conjugate of $\hat{U}$.

Consider the probability amplitude of the two states, $\langle\Psi_2|\Psi_1\rangle$. It does not matter if we measure it at time $t = 0$ or at another chosen time. The probability amplitude does not evolve in time. Therefore

$$\langle \Psi_2 | \Psi_1 \rangle = \langle \Psi_2 \hat{U}_t^\dagger | \hat{U}_t \Psi_1 \rangle = \langle \Psi_2 | \hat{U}_t^\dagger \hat{U}_t | \Psi_1 \rangle,$$

that is at each considered t, it is

$$\hat{U}_t^\dagger \hat{U}_t = \hat{I}.$$

A similar relation

$$\hat{U}_t \hat{U}_t^\dagger = \hat{I}$$

appears when we compute

$$\langle \Psi_1 | \Psi_2 \rangle.$$

The evolution in time of a state happens only if $\hat{U}_t$ is a unitary operator. At time $t = 0$, it happens $\hat{U}_0 = \hat{I}$.

We can develop the new notion of ε-operator. The meaning of an ε-*operator* is related to a fact often used in physics: If we consider a sum of operators, each one having in front natural powers of ε, for small ε, all terms containing powers of ε strictly grater than one can be neglected under some circumstances.

For example, $\hat{B} = \hat{I} + k\varepsilon\hat{L}$ is an ε-operator because there are no terms with power strictly grater than one. We can say that the ε operator, attached to $\hat{B}$ is $\hat{B}_\varepsilon = \hat{B}$. If we have an operator as $\hat{C} = \hat{I} + f\varepsilon\hat{M} + j\varepsilon^2\hat{N}$, it becomes an ε-operator, if we cancel the term containing ε^2, that is if we consider only $\hat{C}_\varepsilon = \hat{I} + f\varepsilon\hat{M}$.

Consider now the ε-operator $\hat{M} = \hat{I} + ik\varepsilon\hat{T}$ and its corresponding dagger operator $\hat{M}^\dagger = \hat{I} - ik\varepsilon\hat{T}^\dagger$. The ε-operator, attached to the product $\hat{M}^\dagger\hat{M}$, is

$$(\hat{M}^\dagger\hat{M})_\varepsilon = \hat{I} + ik\varepsilon(\hat{T} - \hat{T}^\dagger).$$

Let us observe that the ε-operator, attached to the product $\hat{M}^\dagger\hat{M}$, is a unitary operator if and only if $\hat{T} = \hat{T}^\dagger$, that is when the operator $\hat{T}$ is Hermitian. Indeed

$$(\hat{M}^\dagger\hat{M})_\varepsilon = \hat{I} + ik\varepsilon(\hat{T} - \hat{T}^\dagger) = \hat{I} \text{ iff } \hat{T} - \hat{T}^\dagger = \hat{0}.$$

Now we replace unitary operators with ε-unitary operators to express the evolution in time of states, that is assume that evolution in time of states, for small values of ε, can be described by ε-unitary operators.

This assumption allows to describe in another way the time-dependent Schrödinger equation. Let $\hat{H}$ be the operator attached to the Hamiltonian of a physical system. We know that $\hat{H}$ is a Hermitian operator. According to the previous picture, we can construct the *unitary ε-operator*

$$\hat{U}_\varepsilon = \hat{I} - \frac{i\varepsilon}{\hbar}\hat{H}$$

and, for every small ε, it can be used to express an evolution in time of states.

Consider the state $|\Psi_t\rangle$ and its evolution in time $|\Psi_{t+\varepsilon}\rangle := \hat{U}_\varepsilon|\Psi_t\rangle$. It results

$$|\Psi_{t+\varepsilon}\rangle := \left(\hat{I} - \frac{i\varepsilon}{\hbar}\hat{H}\right)|\Psi_t\rangle = |\Psi_t\rangle - \frac{i\varepsilon}{\hbar}\hat{H}|\Psi_t\rangle.$$

We can write it in the form

$$\frac{|\Psi_{t+\varepsilon}\rangle - |\Psi_t\rangle}{\varepsilon} = -\frac{i}{\hbar}\hat{H}|\Psi_t\rangle.$$

In the limit $\varepsilon \to 0$ and for a convenient multiplication by $i\hbar$, it leads to

$$i\hbar\frac{d}{dt}|\Psi_t\rangle = \hat{H}|\Psi_t\rangle,$$

that is the time-dependent Schrödinger equation written in operator form. Here $i\hbar\dfrac{d}{dt}$ is the quantum operator related to the Hamiltonian $\hat{H}$, relation granted by the time evolution of quantum states.

When the quantum harmonic oscillator was studied in Lecture 25, the main step consisted in considering the wave function of the form

$$\Psi(t, x) = e^{-iEt/\hbar}\Psi(x),$$

that is with the spatial part separated by the time evolution. This form was replaced into the time-dependent Schrödinger equation

$$i\hbar\frac{\partial\Psi}{\partial t}(t, x) = -\frac{\hbar^2}{2m}\frac{\partial^2\Psi}{\partial x^2}(t, x) + V(t, x)\Psi(t, x)$$

corresponding to the potential which generates the restoring force of the classical harmonic oscillator,

$$V(t, x) = V(x) = \frac{1}{2}m\omega^2 x^2.$$

It resulted the equation we solved,

$$\left[-\frac{\hbar^2}{2m}\frac{d^2}{dx^2} + \frac{1}{2}m\omega^2 x^2\right]\Psi(x) = E\Psi(x),$$

with all possible energy levels $E = E_n = \left(1 + \dfrac{n}{2}\right)\hbar\omega$ extracted as eigenvalues.

The separation of time and space components suggests to consider a normalized state

$$|\Psi_t\rangle = e^{-iEt/\hbar}|\Psi_x\rangle.$$

Here t is a given parameter and the state variable is x. In this case, the complex number $e^{-iEt/\hbar}$ plays the role of unitary operator U_t, that is it expresses the evolution in time of a given state, here $|\Psi_x\rangle$. Then, we have

$$\hat{H}\, e^{-iEt/\hbar}|\Psi_x\rangle = \hat{H}|\Psi_t\rangle = i\hbar\frac{d}{dt}|\Psi_t\rangle = E \cdot e^{-iEt/\hbar}|\Psi_x\rangle.$$

It remains to solve $\hat{H}|\Psi_x\rangle = E|\Psi_x\rangle$, i.e.

$$\left[\frac{\hat{p}^2}{2m} + \frac{1}{2}m\omega^2\hat{x}^2\right]|\Psi_x\rangle = E|\Psi_x\rangle.$$

Taking into account the position and momentum operators, the previous operator equation becomes

$$\left[-\frac{\hbar^2}{2m}\frac{d^2}{dx^2} + \frac{1}{2}m\omega^2 x^2\right]|\Psi_x\rangle = E|\Psi_x\rangle.$$

Let us choose the normalized state

$$|\Psi_x\rangle := \sqrt[4]{\frac{m\omega}{\hbar\pi}}\, e^{-m\omega x^2/2\hbar}.$$

This choice comes from results of Lecture 25, that is

$$\Psi_n(x) = \frac{1}{\sqrt{2^n n!}}\left(\frac{m\omega}{\pi\hbar}\right)^{1/4} H_n\left(\sqrt{\frac{m\omega}{\hbar}}x\right)e^{-m\omega x^2/2\hbar}$$

for $n = 0$. We have

$$\left[-\frac{\hbar^2}{2m}\frac{d^2}{dx^2} + \frac{1}{2}m\omega^2 x^2\right]|\Psi_x\rangle = \left[-\frac{\hbar^2}{2m}\frac{d^2}{dx^2} + \frac{1}{2}m\omega^2 x^2\right]\sqrt[4]{\frac{m\omega}{\hbar\pi}}\, e^{-m\omega x^2/2\hbar} =$$

$$= \sqrt[4]{\frac{m\omega}{\hbar\pi}}\left[-\frac{\hbar^2}{2m}\frac{d^2}{dx^2}\left(e^{-mwx^2/2\hbar}\right) + \frac{1}{2}mw^2 x^2 e^{-mwx^2/2\hbar}\right] =$$

$$= \left[-\frac{1}{2}m\omega^2 x^2 + \frac{\hbar\omega}{2} + \frac{1}{2}m\omega^2 x^2\right]\sqrt[4]{\frac{m\omega}{\hbar\pi}}\, e^{-m\omega x^2/2\hbar} = \frac{\hbar\omega}{2}\sqrt[4]{\frac{m\omega}{\hbar\pi}}\, e^{-m\omega x^2/2\hbar}$$

which means

$$\hat{H}|\Psi_x\rangle = \frac{\hbar\omega}{2}|\Psi_x\rangle.$$

In this way, we obtained the energy ground state for the harmonic oscillator, $E_0 = \frac{\hbar\omega}{2}$, that is the minimum positive energy of the given Hamiltonian. Why this is the minimum energy? The ladder operators will help us to show this.

To continue, let us consider physical units such that $m = 1$ and $\hbar = 1$. It will be easier to understand the ladder operators, because on the contrary, we should carry a lot of constants in formulas. Let us write, in the new notations, some results obtained before. We have:

- The Hamiltonian of harmonic oscillator is $\hat{H} = \dfrac{1}{2}(\hat{p}^2 + \omega^2 \hat{x}^2)$;

- The normalized state is $|\Psi_x\rangle = \sqrt[4]{\dfrac{\omega}{\pi}} e^{-wx^2/2}$;

- The ground state energy for the Harmonic oscillator is $E_0 = \dfrac{\omega}{2}$;

- The commutator of the position and momentum operators gives $[\hat{x}, \hat{p}] = i$.

Let us start the computation considering

$$(\hat{p} + i\omega\hat{x})(\hat{p} - i\omega\hat{x}) = \hat{p}^2 + \omega^2 \hat{x}^2 + i\omega(\hat{x}\hat{p} - \hat{p}\hat{x}) = \hat{p}^2 + \omega^2 \hat{x}^2 - \omega$$

which leads to

$$\hat{H} = \frac{1}{2}(\hat{p} + i\omega\hat{x})(\hat{p} - i\omega\hat{x}) + \frac{\omega}{2}.$$

We define

$$\hat{a}^+ := \frac{1}{\sqrt{2\omega}}(\hat{p} + i\omega\hat{x}) \quad \text{and} \quad \hat{a}^- := \frac{1}{\sqrt{2\omega}}(\hat{p} - i\omega\hat{x}).$$

Here $\hat{a}^+$ is the *raising* or the *creation operator*, while $\hat{a}^-$ is the *lowering operator* or the *annihilation operator*. Another simple computation related to these operators is

$$[(\hat{p} + i\omega\hat{x}), (\hat{p} - i\omega\hat{x})] = i\omega(\hat{x}\hat{p} - \hat{p}\hat{x}) - i\omega(\hat{p}\hat{x} - \hat{x}\hat{p}) = -2\omega,$$

that is

$$[\hat{a}^+, \hat{a}^-] = -1.$$

The Hamiltonian of the harmonic oscillator, written in terms of *ladder operators*, is

$$\hat{H} = \omega(\hat{a}^+ \hat{a}^-) + \frac{\omega}{2}.$$

If we compute the action of the lowering operator on our chosen state $|\Psi_x\rangle$, we have

$$\hat{a}^- |\Psi_x\rangle = \left| \frac{1}{\sqrt{2\omega}} \left(-i\frac{d}{dx} - i\omega x \right) \sqrt[4]{\frac{\omega}{\pi}} e^{-\omega x^2/2} \right\rangle = |0\rangle,$$

i.e. this operator "destroys" the state function. Therefore

$$\hat{H}|\Psi_x\rangle = \omega(\hat{a}^+ \hat{a}^-)|\Psi_x\rangle + \frac{\omega}{2}|\Psi_x\rangle = \frac{\omega}{2}|\Psi_x\rangle = E_0|\Psi_x\rangle.$$

Another possible computation is related to $\hat{H}\hat{a}^+|\Psi_x\rangle$. It is

$$\hat{H}\hat{a}^+|\Psi_x\rangle = \left(\omega(\hat{a}^+\hat{a}^-) + \frac{\omega}{2}\right)\hat{a}^+|\Psi_x\rangle = \omega(\hat{a}^+\hat{a}^-)\hat{a}^+|\Psi_x\rangle + \frac{\omega}{2}\hat{a}^+|\Psi_x\rangle =$$

$$= \omega\hat{a}^+(\hat{a}^-\hat{a}^+)|\Psi_x\rangle + \frac{\omega}{2}\hat{a}^+|\Psi_x\rangle = \omega\hat{a}^+(\hat{a}^+\hat{a}^- + 1)|\Psi_x\rangle + \frac{\omega}{2}\hat{a}^+|\Psi_x\rangle =$$

$$= \omega\hat{a}^+|\Psi_x\rangle + \frac{\omega}{2}\hat{a}^+|\Psi_x\rangle = E_1\hat{a}^+|\Psi_x\rangle,$$

therefore $E_1 = \dfrac{3\omega}{2}$. Acting on $|\Psi_x\rangle$ with $\hat{a}^+$ we increase the energy, therefore we switched from level E_0 to level E_1. A simple exercise shows that the state $\hat{a}^+\hat{a}^+|\Psi_x\rangle$ leads to $E_2 = \dfrac{5\omega}{2}$, etc. Applying again and again $\hat{a}^+$ we build up the set of eigenvalues corresponding to the possible energy levels of the harmonic oscillator, $E_n = \left(n + \dfrac{1}{2}\right)\omega$. Restoring the standard units, we have $E_n = \left(n + \dfrac{1}{2}\right)\hbar\omega$.

In terms of states corresponding to energy levels, we can denote by $|0\rangle$ the basic state and by $|1\rangle$ the state corresponding to the energy E_0 when $\hat{a}^-|1\rangle = |0\rangle$. The operator $\hat{a}^+$ increases the energy and changes the state in the next superior state, and $\hat{a}^-$ decreases the energy to reach the previous inferior one. We have

$$\hat{a}^+|n\rangle = |n+1\rangle \quad \text{and} \quad \hat{a}^-|n+1\rangle = |n\rangle.$$

However, this statement does not work because

$$[\hat{a}^+, \hat{a}^-]|n\rangle = \hat{a}^+\hat{a}^-|n\rangle - \hat{a}^-\hat{a}^+|n\rangle = \hat{a}^+|n-1\rangle - \hat{a}^-|n+1\rangle = |n\rangle - |n\rangle = |0\rangle$$

is in disagreement with $[\hat{a}^+, \hat{a}^-] = -1$. But if we choose

$$\hat{a}^+|n\rangle = \sqrt{n+1}\,|n+1\rangle$$

$$\hat{a}^-|n\rangle = \sqrt{n}\,|n-1\rangle$$

it is simple to show that $[\hat{a}^+, \hat{a}^-] = -1$. Indeed,

$$[\hat{a}^+, \hat{a}^-]|n\rangle = \hat{a}^+\hat{a}^-|n\rangle - \hat{a}^-\hat{a}^+|n\rangle = \sqrt{n}\,\hat{a}^+|n-1\rangle - \sqrt{n+1}\,\hat{a}^-|n+1\rangle =$$

$$n|n\rangle - (n+1)|n\rangle = -|n\rangle.$$

Now it becomes obvious why we can call them the creation and the annihilation operators, respectively: The first one adds a quantum to a given energy level, while the second one extracts a quantum. In elementary particles physics, the first one creates and the second one destroys particles which corresponds exactly to the energy quanta $\hbar\omega$.

If we look at the formulas

$$E_n = \left(n + \frac{1}{2}\right)\hbar\omega$$

and

$$\hat{H} = \hbar\omega(\hat{a}^+\hat{a}^-) + \frac{\hbar\omega}{2}$$

restoring $\hbar$, we can define a "number operator" $\hat{N}$ considering

$$\hat{N} := \hat{a}^+\hat{a}^-.$$

The Hamiltonian of the harmonic oscillator becomes

$$\hat{H} = \hbar\omega\left(\hat{N} + \frac{1}{2}\right).$$

Since

$$\langle n|\hat{N}|n\rangle = \langle n|\hat{a}^+\hat{a}^-|n\rangle = \left(\hat{a}^-|n\rangle\right)^\dagger \hat{a}^-|n\rangle \geq 0,$$

it means that the smallest eigen-number is 0 and

$$\hat{a}^-|0\rangle = 0.$$

There are no other energy levels because the annihilation operator cannot extract other states "under" the basic $|0\rangle$ state which can be called the vacuum state.

This results constitute the modern view of harmonic oscillator. Dirac was credited with both the ladder operators and with this picture of harmonic oscillator. If we look back, the adventure started with the Hermite polynomials and the Mathematics behind them. Hilbert created the frame in which all quantum concepts defined before make sense and where the other physicists succeeded in creating models confirmed by experiments offering a self-consistent picture of Quantum World.

Summary of Lecture 47. Hermitian, unitary and ε-unitary operators are needed for a deeper viewpoint related to quantum harmonic oscillator. These formalism allows us to consider a new big picture of Quantum Mechanics. States representing the solutions of time-dependent Schrödinger equation are generated starting from a basic state using a special operator. The corresponding eigenvalues, related to the Hamilton operator, have a rule to be determined considering a state moving to another by the operator mentioned above. Consider physical units such that $m = 1$ and $\hbar = 1$ and let us define, in the new notations, expressions and results achieved before. We have:

- The Hamitonian of the harmonic oscillator is $\hat{H} = \frac{1}{2}(\hat{p}^2 + \omega^2 \hat{x}^2)$;

- The normalized state is $|\Psi_x\rangle = \sqrt[4]{\frac{\omega}{\pi}} e^{-\omega x^2/2}$;

- The ground state energy for the Harmonic oscillator is $E_0 = \frac{\omega}{2}$;

- The commutator of the position and momentum operators gives $[\hat{x}, \hat{p}] = i$. We start with a computation like:

$$(\hat{p} + i\omega\hat{x})(\hat{p} - i\omega\hat{x}) = \hat{p}^2 + \omega^2\hat{x}^2 + i\omega(\hat{x}\hat{p} - \hat{p}\hat{x}) = \hat{p}^2 + \omega^2\hat{x}^2 - \omega$$

which leads to

$$\hat{H} = \frac{1}{2}(\hat{p} + i\omega\hat{x})(\hat{p} - i\omega\hat{x}) + \frac{\omega}{2}.$$

We can denote

$$\hat{a}^+ := \frac{1}{\sqrt{2\omega}}(\hat{p} + i\omega\hat{x}) \quad \text{and} \quad \hat{a}^- := \frac{1}{\sqrt{2\omega}}(\hat{p} - i\omega\hat{x}).$$

Here $\hat{a}^+$ is called the raising or the creation operator, while $\hat{a}^-$, is called the lowering operator or the annihilation operator. Another simple computation, related to these operators, is

$$[(\hat{p} + i\omega\hat{x}), (\hat{p} - i\omega\hat{x})] = i\omega(\hat{x}\hat{p} - \hat{p}\hat{x}) - i\omega(\hat{p}\hat{x} - \hat{x}\hat{p}) = -2\omega,$$

that is

$$[\hat{a}^+, \hat{a}^-] = -1.$$

The Hamiltonian of the harmonic oscillator, written in terms of ladder operators, is

$$\hat{H} = \omega(\hat{a}^+\hat{a}^-) + \frac{\omega}{2}.$$

If we compute the action of the lowering operator on our chosen state $|\Psi_x\rangle$, we obtain

$$\hat{a}^-|\Psi_x\rangle = \left| \frac{1}{\sqrt{2\omega}}\left(-i\frac{d}{dx} - i\omega x\right)\sqrt[4]{\frac{\omega}{\pi}}e^{-\omega x^2/2}\right\rangle = |0\rangle,$$

i.e. this operator destroys the state function.

Therefore

$$\hat{a}^-|\Psi_x\rangle = |0\rangle.$$

Another computation leads to

$$\hat{H}\hat{a}^+|\Psi_x\rangle = \omega\hat{a}^+|\Psi_x\rangle + \frac{\omega}{2}\hat{a}^+|\Psi_x\rangle = E_1\hat{a}^+|\Psi_x\rangle,$$

with $E_1 = \frac{3\omega}{2}$. Acting on $|\Psi_x\rangle$ with $\hat{a}^+$, we increase the energy, therefore we switch from E_0 level to E_1 level.

A simple exercise shows that the state $\hat{a}^+\hat{a}^+|\Psi_x\rangle$ leads to $E_2 = \frac{5\omega}{2}$, etc. Applying again and again $\hat{a}^+$ we discover the set of eigenvalues corresponding to the possible energy levels of the harmonic oscillator, $E_n = \left(n + \frac{1}{2}\right)\omega$. If we study the states corresponding to the energy levels we find the formulas

$$\hat{a}^+|n\rangle = \sqrt{n+1}\,|n+1\rangle$$

$$\hat{a}^+|n\rangle = \sqrt{n}\,|n-1\rangle$$

which confirm the result $[\hat{a}^+, \hat{a}^-] = -1$.

If we look at the formulas

$$E_n = \left(n + \frac{1}{2}\right)\hbar\omega$$

and

$$\hat{H} = \hbar\omega(\hat{a}^+\hat{a}^-) + \frac{\hbar\omega}{2}$$

where we have restored $\hbar$, we can define the "number operator" $\hat{N}$ considering

$$\hat{N} := \hat{a}^+\hat{a}^-.$$

The Hamiltonian of the harmonic oscillator becomes

$$\hat{H} = \hbar\omega\left(\hat{N} + \frac{1}{2}\right).$$

Since

$$\langle n|\hat{N}|n\rangle = \langle n|\hat{a}^+\hat{a}^-|n\rangle = \left(\hat{a}^-|n\rangle\right)^\dagger \hat{a}^-|n\rangle \geq 0,$$

it means that the smallest eigen-number is 0 and

$$\hat{a}^-|0\rangle = 0.$$

There are no other energy levels because the annihilation operator cannot extract other states "under" the basic $|0\rangle$ state which is called the vacuum state.

11.3 Lecture 48: The Angular Momentum in Quantum Mechanics

In Classical Mechanics, the angular momentum **L** is defined with respect to the position and the momentum vectors, $\mathbf{r} = (x, y, z)$ and $\mathbf{p} = (p_x, p_y, p_z)$ respectively, by the formula $\mathbf{L} := \mathbf{r} \times \mathbf{p}$. Taking into account the formal development of the determinant

$$\begin{vmatrix} \vec{i} & \vec{j} & \vec{k} \\ x & y & z \\ p_x & p_y & p_z \end{vmatrix}$$

the components of the angular momentum are

$$L_x = yp_z - zp_y; \quad L_y = zp_x - xp_z; \quad L_z = xp_y - yp_x.$$

Therefore the operators corresponding to the quantum angular momentum are

$$\hat{L}_x = \frac{\hbar}{i}\left(y\frac{\partial}{\partial z} - z\frac{\partial}{\partial y}\right); \quad \hat{L}_y = \frac{\hbar}{i}\left(z\frac{\partial}{\partial x} - x\frac{\partial}{\partial z}\right); \quad \hat{L}_z = \frac{\hbar}{i}\left(x\frac{\partial}{\partial y} - y\frac{\partial}{\partial x}\right).$$

Let us recall that we proved, in Lecture 38, that in the 1-dimensional case it is

$$[\hat{x}, \hat{p}] = i\hbar.$$

This result can be easily extended to 3-dimension by the relations

$$[\hat{x}_k, \hat{p}_j] = i\hbar\delta_{kj}, \quad j, k = 1, 2, 3,$$

where δ_{jk} is the Kroneker delta symbol and $(x_1, x_2, x_3) = (x, y, z)$. The other possible relations are

$$[\hat{x}_k, \hat{x}_j] = [\hat{p}_k, \hat{p}_j] = 0.$$

It is a simple exercise to show how these relations among the position $\hat{x}_j$ and momentum $\hat{p}_k$ operators lead to the commutation relations for the quantum angular momentum

$$[\hat{L}_x, \hat{L}_y] = i\hbar\hat{L}_z; \quad [\hat{L}_y, \hat{L}_z] = i\hbar\hat{L}_x; \quad [\hat{L}_z, \hat{L}_x] = i\hbar\hat{L}_y.$$

We present only the computations necessary to prove the first equality, $[\hat{L}_x, \hat{L}_y] = i\hbar\hat{L}_z$.

Exercise 11.3.1 Prove that $[\hat{L}_x, \hat{L}_y] = i\hbar\hat{L}_z$.
  Hint: If we replace the operators, we obtain:

$$\hat{L}_x\hat{L}_y - \hat{L}_y\hat{L}_x = -\hbar^2\left[\left(y\frac{\partial}{\partial z} - z\frac{\partial}{\partial y}\right)\left(z\frac{\partial}{\partial x} - x\frac{\partial}{\partial z}\right) - \left(z\frac{\partial}{\partial x} - x\frac{\partial}{\partial z}\right)\left(y\frac{\partial}{\partial z} - z\frac{\partial}{\partial y}\right)\right] =$$

$$-\hbar^2\left[y\frac{\partial}{\partial z}\left(z\frac{\partial}{\partial x}\right) - y\frac{\partial}{\partial z}\left(x\frac{\partial}{\partial z}\right) - z\frac{\partial}{\partial y}\left(z\frac{\partial}{\partial x}\right) + z\frac{\partial}{\partial y}\left(x\frac{\partial}{\partial z}\right)\right] +$$

$$+\hbar^2\left[z\frac{\partial}{\partial x}\left(y\frac{\partial}{\partial z}\right) - z\frac{\partial}{\partial x}\left(z\frac{\partial}{\partial y}\right) - x\frac{\partial}{\partial z}\left(y\frac{\partial}{\partial z}\right) + x\frac{\partial}{\partial z}\left(z\frac{\partial}{\partial y}\right)\right] =$$

$$-\hbar^2\left[y\frac{\partial}{\partial x} + yz\frac{\partial^2}{\partial z\partial x} - yx\frac{\partial^2}{\partial z^2} - z^2\frac{\partial^2}{\partial y\partial x} + zx\frac{\partial^2}{\partial y\partial z}\right] +$$

$$+\hbar^2\left[zy\frac{\partial^2}{\partial x\partial z} - z^2\frac{\partial^2}{\partial x\partial y} - xy\frac{\partial^2}{\partial z^2} + x\frac{\partial}{\partial y} + xz\frac{\partial^2}{\partial z\partial y}\right] =$$

$$= -\hbar^2\left(-x\frac{\partial}{\partial y} + y\frac{\partial}{\partial x}\right) = i\hbar\frac{\hbar}{i}\left(x\frac{\partial}{\partial y} - y\frac{\partial}{\partial x}\right) = i\hbar\hat{L}_z. \quad \square$$

Exercise 11.3.2 Compute the operators $\hat{L}_x, \hat{L}_y, \hat{L}_z$ in spherical polar coordinates.
Solution. Consider

$$\begin{cases} x = r\sin\theta\cos\phi \\ y = r\sin\theta\sin\phi \\ z = r\cos\theta \end{cases}$$

Let us consider the Cartesian derivative operators $\dfrac{\partial}{\partial r}, \dfrac{\partial}{\partial\theta}, \dfrac{\partial}{\partial\phi}$ and develop them by the following chain rules:

$$\begin{cases} \dfrac{\partial}{\partial x} = \dfrac{\partial r}{\partial x}\dfrac{\partial}{\partial r} + \dfrac{\partial\theta}{\partial x}\dfrac{\partial}{\partial\theta} + \dfrac{\partial\phi}{\partial x}\dfrac{\partial}{\partial\phi} \\ \dfrac{\partial}{\partial y} = \dfrac{\partial r}{\partial y}\dfrac{\partial}{\partial r} + \dfrac{\partial\theta}{\partial y}\dfrac{\partial}{\partial\theta} + \dfrac{\partial\phi}{\partial y}\dfrac{\partial}{\partial\phi} \\ \dfrac{\partial}{\partial z} = \dfrac{\partial r}{\partial z}\dfrac{\partial}{\partial r} + \dfrac{\partial\theta}{\partial z}\dfrac{\partial}{\partial\theta} + \dfrac{\partial\phi}{\partial z}\dfrac{\partial}{\partial\phi} \end{cases}$$

Being $r = \sqrt{x^2 + y^2 + z^2}$, it results $\dfrac{\partial r}{\partial x} = \dfrac{x}{r} = \sin\theta\cos\phi$. In the same way we

obtain $\dfrac{\partial r}{\partial y} = \sin\theta\sin\phi$ and $\dfrac{\partial r}{\partial z} = \dfrac{x}{r} = \cos\theta$. Using

$$\theta = \arcsin\frac{\sqrt{x^2 + y^2}}{\sqrt{x^2 + y^2 + z^2}}$$

it results both $\dfrac{\partial\theta}{\partial x} = \dfrac{1}{r}\cos\theta\cos\phi$ and $\dfrac{\partial\theta}{\partial y} = \dfrac{1}{r}\cos\theta\sin\phi$.

From

$$\theta = \arccos\frac{z}{\sqrt{x^2 + y^2 + z^2}}$$

we have $\dfrac{\partial\theta}{\partial z} = -\dfrac{1}{r}\sin\theta$.

The last formulas we need are deduced from $\tan\phi = \dfrac{y}{x}$, that is from $\phi = \arctan\dfrac{y}{x}$.
We obtain

$$\frac{\partial\phi}{\partial x} = -\frac{1}{r}\frac{\sin\phi}{\sin\theta}; \quad \frac{\partial\phi}{\partial y} = \frac{1}{r}\frac{\cos\phi}{\sin\theta}; \quad \frac{\partial\phi}{\partial z} = 0.$$

Therefore we get

$$\begin{cases} \dfrac{\partial}{\partial x} = \sin\theta\cos\phi\dfrac{\partial}{\partial r} + \dfrac{1}{r}\cos\theta\cos\phi\dfrac{\partial}{\partial\theta} - \dfrac{1}{r}\dfrac{\sin\phi}{\sin\theta}\dfrac{\partial}{\partial\phi} \\[2mm] \dfrac{\partial}{\partial y} = \sin\theta\sin\phi\dfrac{\partial}{\partial r} + \dfrac{1}{r}\cos\theta\sin\phi\dfrac{\partial}{\partial\theta} + \dfrac{1}{r}\dfrac{\cos\phi}{\sin\theta}\dfrac{\partial}{\partial\phi} \\[2mm] \dfrac{\partial}{\partial z} = \cos\theta\dfrac{\partial}{\partial r} - \dfrac{1}{r}\sin\theta\dfrac{\partial}{\partial\theta} \end{cases}$$

Using the previous formulas and and the spherical representation of x, y and z, we can compute the angular momentum operators

$$\hat{L}_x = \frac{\hbar}{i}\left(y\frac{\partial}{\partial z} - z\frac{\partial}{\partial y}\right), \ \hat{L}_y = \frac{\hbar}{i}\left(z\frac{\partial}{\partial x} - x\frac{\partial}{\partial z}\right), \ \hat{L}_z = \frac{\hbar}{i}\left(x\frac{\partial}{\partial y} - y\frac{\partial}{\partial x}\right),$$

in spherical coordinates:

$$\begin{cases} \hat{L}_x = i\hbar\left(\sin\phi\dfrac{\partial}{\partial\theta} + \cot\theta\cos\phi\dfrac{\partial}{\partial\phi}\right) \\[2mm] \hat{L}_y = i\hbar\left(-\cos\phi\dfrac{\partial}{\partial\theta} + \cot\theta\sin\phi\dfrac{\partial}{\partial\phi}\right) \quad \square \\[2mm] \hat{L}_z = -i\hbar\dfrac{\partial}{\partial\phi} \end{cases}$$

We leave to the reader the computations, in spherical coordinates, of the operator:

$$\hat{L}^2 := \hat{L}_x^2 + \hat{L}_y^2 + \hat{L}_z^2,$$

where $\hat{L}_x^2 = \hat{L}_\nu(\hat{L}_\nu)$, $\nu \in \{x, y, z\}$. We obtain

$$\hat{L}^2 = -\hbar^2 \left(\frac{1}{\sin\theta} \frac{\partial}{\partial\theta} \left(\sin\theta \frac{\partial}{\partial\theta} \right) + \frac{1}{\sin^2\theta} \frac{\partial^2}{\partial\phi^2} \right).$$

A direct consequence is

$$[\hat{L}^2, \hat{L}_z] = 0.$$

Of course we can make the computations in Cartesian coordinates, but we did it in spherical coordinates because readers interested to continue the study of quantum mechanics can prove that the eigenvalues of $\hat{L}_z$ and $\hat{L}$ are determined with respect to the Laplace spherical harmonics Y_l^m which represent the solutions of Laplace equation in spherical domains. The spherical harmonics are written with respect to the Legendre polynomials.

$$Y_l^m(\theta, \phi) = \left[\frac{(2l+1)}{4\pi} \frac{(l-m)!}{(l+m)!} \right]^{1/2} (-1)^m e^{im\phi} P_l^m(\cos\theta), \ l \geq |m| \geq 0,$$

where, for $x = \cos\theta$, it is

$$P_l^m(x) = \sin^{|m|}\theta \frac{d^{|m|}}{dx^{|m|}} P_l(x)$$

with

$$P_l(x) = \left(2^l l!\right)^{-1} \frac{d^l}{dx^l} (x^2 - 1)^l.$$

The last formula is known as Rodrigues formula for the Legendre polynomial P_l of degree l. Some particular spherical harmonics are

$$Y_0^0(\theta, \phi) = \frac{1}{2\sqrt{\pi}}; \ Y_1^1(\theta, \phi) = -\frac{\sqrt{3}}{2\sqrt{2\pi}} \sin\theta e^{i\phi}; \ Y_2^1(\theta, \phi) = -\frac{1}{2}\sqrt{\frac{15}{2\pi}} \sin\theta \cos\theta e^{i\phi};$$

$$Y_1^{-1}(\theta, \phi) = \frac{\sqrt{3}}{2\sqrt{2\pi}} \sin\theta e^{-i\phi}; \ Y_2^{-1}(\theta, \phi) = -\frac{\sqrt{15}}{2\sqrt{2\pi}} \sin\theta \cos\theta e^{-i\phi}; \ \ldots\ldots$$

More about the Legendre polynomials and spherical harmonics will be discussed in Lecture 49.

Exercise 11.3.3 Show that

$$\hat{L}_z Y_l^m(\theta, \phi) = \hbar m Y_l^m(\theta, \phi) \text{ and } \hat{L}^2 Y_l^m(\theta, \phi) = \hbar^2 l(l+1) Y_l^m(\theta, \phi).$$

Hint. Use $\hat{L}_z Y_l^m(\theta, \phi)$ formula given above and the operators representation in spherical coordinates. In this way, we get the spectrum of angular momentum operators.

Now we can return to the Pauli matrices observing first that, in the case $n = 2$, the matrix related to an operator is given by the formula

$$\hat{A} = \begin{pmatrix} \langle 1|\hat{A}1 \rangle & \langle 1|\hat{A}2 \rangle \\ \langle 2|\hat{A}1 \rangle & \langle 2|\hat{A}2 \rangle \end{pmatrix},$$

where $|1\rangle$ and $|2\rangle$ are the eigenvectors of $\hat{A}$. If the operator $\hat{A}$ is the Pauli matrix $\hat{\sigma}_z$, the previous formula is verified for

$$\hat{\sigma}_z = \begin{pmatrix} \langle \uparrow |\hat{\sigma}_z| \uparrow \rangle & \langle \uparrow |\hat{\sigma}_z| \downarrow \rangle \\ \langle \downarrow |\hat{\sigma}_z| \uparrow \rangle & \langle \downarrow |\hat{\sigma}_z| \downarrow \rangle \end{pmatrix}.$$

The same for the other two Pauli matrices. With these considerations in mind, let us construct the *Pauli operators*:

$$\hat{M}_x = \frac{\hbar}{2}\hat{\sigma}_x; \quad \hat{M}_y = \frac{\hbar}{2}\hat{\sigma}_y; \quad \hat{M}_z = \frac{\hbar}{2}\hat{\sigma}_z.$$

We can compute how they "commute", that is

$$[\hat{M}_x, \hat{M}_y] = i\hbar\hat{M}_z; \quad [\hat{M}_y, \hat{M}_z] = i\hbar\hat{M}_x; \quad [\hat{M}_z, \hat{M}_x] = i\hbar\hat{M}_y.$$

exactly as the angular momentum operators, that is

$$[\hat{L}_x, \hat{L}_y] = i\hbar\hat{L}_z; \quad [\hat{L}_y, \hat{L}_z] = i\hbar\hat{L}_x; \quad [\hat{L}_z, \hat{L}_x] = i\hbar\hat{L}_y.$$

If

$$\hat{M}^2 := \hat{M}_x^2 + \hat{M}_y^2 + \hat{M}_z^2$$

a simple computation leads to

$$\hat{M}^2 = \frac{3\hbar^2}{4}\hat{I},$$

that is

$$[\hat{M}, \hat{M}_z] = 0$$

exactly as the formula proved before, i.e.

$$[\hat{L}, \hat{L}_z] = 0.$$

At the same time both the eigenvalues of $\hat{M}^2$ are $\frac{3}{4}\hbar^2$ and they corresponds to the eigenvectors $|\uparrow\rangle$; $|\uparrow\rangle$. The most important connection is related to the eigenvalues of $\hat{M}_z$. They are $+\frac{1}{2}\hbar$ and $-\frac{1}{2}\hbar$ corresponding to the realted eigenvectors. If we look at the eigenvalues of $\hat{L}_z$, we observe that they are an integer multiple of $\hbar$.

It is worth noticing that similitudes between the angular momentum operators and the momentum operators defined by the Pauli matrices cannot be taken into account when we are dealing with eigenvalues. In a case they are integers, in the other they are fractions. If one asks why we did not keep the original Pauli matrices and multiplied each one with a constant, the answer is because we want exactly the same commutation formulas as angular momentum operators. Therefore, this property of Pauli matrices operators is intrinsically related to the spin they represent. In other words, it is related to the *intrinsic angular momentum of electron* itself. We conclude that the *electron spin* corresponds to angular $\frac{1}{2}$ and $-\frac{1}{2}$ momenta.

Summary of Lecture 48. The operators corresponding to the quantum angular momentum are

$$\hat{L}_x = \frac{\hbar}{i}\left(y\frac{\partial}{\partial z} - z\frac{\partial}{\partial y}\right); \ \hat{L}_y = \frac{\hbar}{i}\left(z\frac{\partial}{\partial x} - x\frac{\partial}{\partial z}\right); \ \hat{L}_z = \frac{\hbar}{i}\left(x\frac{\partial}{\partial y} - y\frac{\partial}{\partial x}\right).$$

It can be proved that

$$[\hat{L}_x, \hat{L}_y] = i\hbar\hat{L}_z; \ [\hat{L}_y, \hat{L}_z] = i\hbar\hat{L}_x; \ [\hat{L}_z, \hat{L}_x] = i\hbar\hat{L}_y.$$

If we switch to spherical coordinates, they are

$$\begin{cases} \hat{L}_x = i\hbar\left(\sin\phi\frac{\partial}{\partial\theta} + \cot\theta\cos\phi\frac{\partial}{\partial\phi}\right) \\ \hat{L}_y = i\hbar\left(-\cos\phi\frac{\partial}{\partial\theta} + \cot\theta\sin\phi\frac{\partial}{\partial\phi}\right) \\ \hat{L}_z = -i\hbar\frac{\partial}{\partial\phi} \end{cases}$$

allowing the computation of the operator $\hat{L}$,

$$\hat{L}^2 := \hat{L}_x^2 + \hat{L}_y^2 + \hat{L}_z^2$$

in spheric coordinates as

$$\hat{L}^2 = -\hbar^2\left(\frac{1}{\sin\theta}\frac{\partial}{\partial\theta}\left(\sin\theta\frac{\partial}{\partial\theta}\right) + \frac{1}{\sin^2\theta}\frac{\partial^2}{\partial\phi^2}\right)$$

with the direct consequence

$$[\hat{L}^2, \hat{L}_z] = 0.$$

We can also prove that

$$\hat{L}_z Y_l^m(\theta, \phi) = \hbar m \, Y_l^m(\theta, \phi),$$

$$\hat{L}^2 Y_l^m(\theta, \phi) = \hbar^2 l(l+1) \, Y_l^m(\theta, \phi),$$

where the spherical harmonics $Y_l^m(\theta, \phi)$ are

$$Y_l^m(\theta, \phi) = \left[\frac{(2l+1)}{4\pi} \frac{(l-m)!}{(l+m)!} \right]^{1/2} (-1)^m e^{im\phi} P_l^m(\cos\theta), \; l \geq |m| \geq 0$$

and for $x = \cos\theta$, it is

$$P_l^m(x) = \sin^{|m|}\theta \, \frac{d^{|m|}}{dx^{|m|}} P_l(x),$$

written with respect to the Legendre polynomials

$$P_l(x) = \left(2^l l!\right)^{-1} \frac{d^l}{dx^l}(x^2 - 1)^l.$$

Let us construct the new operators

$$\hat{M}_x = \frac{\hbar}{2}\hat{\sigma}_x; \; \hat{M}_y = \frac{\hbar}{2}\hat{\sigma}_y; \; \hat{M}_z = \frac{\hbar}{2}\hat{\sigma}_z.$$

We can compute how they "commute", that is

$$[\hat{M}_x, \hat{M}_y] = i\hbar\hat{M}_z; \; [\hat{M}_y, \hat{M}_z] = i\hbar\hat{M}_x; \; [\hat{M}_z, \hat{M}_x] = i\hbar\hat{M}_y,$$

exactly as the angular momentum operators considered before. If

$$\hat{M}^2 := \hat{M}_x^2 + \hat{M}_y^2 + \hat{M}_z^2$$

a simple computation leads to

$$\hat{M}^2 = \frac{3\hbar^2}{4}\hat{I},$$

that is

$$[\hat{M}, \hat{M}_z] = 0$$

exactly as the formula proved before

$$[\hat{L}, \hat{L}_z] = 0.$$

The most important connection is related to the eigenvalues of $\hat{M}_z$. They are $+\frac{1}{2}\hbar$ and $-\frac{1}{2}\hbar$. If we look at the eigenvalues of $\hat{L}_z$ we observe that they are integer multiples of $\hbar$. Therefore the similitudes between the angular momentum operators and the momentum operators defined by the Pauli matrices stop when we are dealing with eigenvalues. In a case they are integers, in the other they are fractions. Therefore, this property of Pauli matrices operators is intrinsically related to the spin they represent: It must be the intrinsic angular momentum of electron itself. We conclude that the electron spin corresponds to angular $\frac{1}{2}$ and $-\frac{1}{2}$ momenta.

11.4   Lecture 49: From Differential Operators in Spherical Coordinates to the Hydrogen Atom

In Lecture 48 we presented the spherical harmonics $Y_l^m(\theta.\phi)$. Let us present now the way they appear. As we will see, they are fundamental for the correct formulation of the hydrogen atom problem. We first need to compute how the gradient operator looks like in spherical coordinates. Let us define the spherical coordinates in the form

$$\begin{cases} x = r\sin\theta\cos\phi \\ y = r\sin\theta\sin\phi \\ z = r\cos\theta \end{cases}$$

The position vector $\mathbf{r}$ is denoted, in Classical Mechanics, by $\vec{r}$ and

$$\vec{r} = r\sin\theta\cos\phi \ \vec{i} + r\sin\theta\sin\phi \ \vec{j} + r\cos\theta \ \vec{k}, \quad \theta \in [0, \pi), \quad \phi \in [0, 2\pi).$$

The entire surface of the sphere of radius r, centered at $O(0, 0, 0)$, is described. In Quantum Mechanics, $\vec{i}, \vec{j}, \vec{k}$ are replaced by the operator notations $\hat{x}, \hat{y}, \hat{z}$, therefore the previous form is replaced by

$$\mathbf{r} = r\sin\theta\cos\phi \ \hat{x} + r\sin\theta\sin\phi \ \hat{y} + r\cos\theta \ \hat{z}, \quad \theta \in [0, \pi), \quad \phi \in [0, 2\pi),$$

and the orthonormal Cartesian basis is $\{\hat{x}, \hat{y}, \hat{z}\}$. Denoting by $\{\hat{r}, \hat{\theta}, \hat{\phi}\}$ the spherical counterparts of $\hat{r}, \hat{\theta}, \hat{\phi}$, the first problem is how the Cartesian basis is related to its spherical counterparts.

Exercise 11.4.1 Show that

$$
\begin{cases}
\hat{x} = \hat{r} \sin\theta \cos\phi + \hat{\theta} \cos\theta \cos\phi - \hat{\phi} \sin\phi \\
\hat{y} = \hat{r} \sin\theta \sin\phi + \hat{\theta} \cos\theta \sin\phi + \hat{\phi} \cos\phi \\
\hat{z} = \hat{r} \cos\theta - \hat{\theta} \sin\theta
\end{cases}
$$

Solution. The most geometric way to prove the previous equalities is considering the sphere of radius **r**, centered at O, as a surface

$$f(\theta, \phi) = (r \sin\theta \cos\phi, r \sin\theta \sin\phi, r \cos\theta)$$

and considering the vectors $\dfrac{\partial f}{\partial \theta}, \dfrac{\partial f}{\partial \phi}, \dfrac{\partial f}{\partial \theta} \times \dfrac{\partial f}{\partial \phi}$. Using the Cartesian basis, we have

$$\hat{\theta} = \frac{\dfrac{\partial f}{\partial \theta}}{\left|\dfrac{\partial f}{\partial \theta}\right|} = \cos\theta \cos\phi \, \hat{x} + \cos\theta \sin\phi \, \hat{y} - \sin\theta \, \hat{z}.$$

In the same way

$$\hat{\phi} = \frac{\dfrac{\partial f}{\partial \phi}}{\left|\dfrac{\partial f}{\partial \phi}\right|} = -\sin\phi \, \hat{x} + \cos\phi \, \hat{y}$$

and

$$\hat{r} = \frac{\dfrac{\partial f}{\partial \theta} \times \dfrac{\partial f}{\partial \phi}}{\left|\dfrac{\partial f}{\partial \theta} \times \dfrac{\partial f}{\partial \phi}\right|} = \sin\theta \cos\phi \, \hat{x} + \sin\theta \sin\phi \, \hat{y} + \cos\theta \, \hat{z}.$$

In this way, we expressed, with respect to the Cartesian coordinates, the normalized Gauss frame, related to the sphere at each point determined by a given pair θ, ϕ. It remains only to solve the system using the Cramer rule:

$$\Delta = \begin{vmatrix} \sin\theta \cos\phi & \sin\theta \sin\phi & \cos\theta \\ \cos\theta \cos\phi & \cos\theta \sin\phi & -\sin\theta \\ -\sin\phi & \cos\phi & 0 \end{vmatrix} = 1; \quad \Delta_x = \begin{vmatrix} \hat{r} & \sin\theta \sin\phi & \cos\theta \\ \hat{\theta} & \cos\theta \sin\phi & -\sin\theta \\ \hat{\phi} & \cos\phi & 0 \end{vmatrix}; \hat{x} = \frac{\Delta_x}{\Delta}, \ldots$$

We obtain

$$\begin{cases} \hat{x} = \hat{r}\sin\theta\cos\phi + \hat{\theta}\cos\theta\cos\phi - \hat{\phi}\sin\phi \\ \hat{y} = \hat{r}\sin\theta\sin\phi + \hat{\theta}\cos\theta\sin\phi + \hat{\phi}\cos\phi \\ \hat{z} = \hat{r}\cos\theta - \hat{\theta}\sin\theta, \end{cases}$$

that is the desired result. Before finishing the exercise, let us remember that $\hat{r}$ plays the role of Gauss map. On the other hand, $\vec{N}$, $\hat{\theta}$ and $\hat{\phi}$ are orthonormal vectors in the Euclidean 3-dimensional space. □

It is worth noticing that, in the previous lecture, we obtained the following formulas in spherical coordinates

$$\begin{cases} \dfrac{\partial}{\partial x} = \sin\theta\cos\phi\dfrac{\partial}{\partial r} + \dfrac{1}{r}\cos\theta\cos\phi\dfrac{\partial}{\partial\theta} - \dfrac{1}{r}\dfrac{\sin\phi}{\sin\theta}\dfrac{\partial}{\partial\phi} \\ \dfrac{\partial}{\partial y} = \sin\theta\sin\phi\dfrac{\partial}{\partial r} + \dfrac{1}{r}\cos\theta\sin\phi\dfrac{\partial}{\partial\theta} + \dfrac{1}{r}\dfrac{\cos\phi}{\sin\theta}\dfrac{\partial}{\partial\phi} \\ \dfrac{\partial}{\partial z} = \cos\theta\dfrac{\partial}{\partial r} - \dfrac{1}{r}\sin\theta\dfrac{\partial}{\partial\theta}, \end{cases}$$

while, the formula we used to describe the gradient, in Cartesian coordinates, is:

$$\nabla = \hat{x}\frac{\partial}{\partial x} + \hat{y}\frac{\partial}{\partial y} + \hat{z}\frac{\partial}{\partial z}.$$

Exercise 11.4.2 Show that the gradient in spherical coordinates has the form

$$\nabla = \hat{r}\frac{\partial}{\partial r} + \hat{\theta}\frac{1}{r}\frac{\partial}{\partial\theta} + \hat{\phi}\frac{1}{r\sin\theta}\frac{\partial}{\partial\phi}.$$

Solution. We replace $\hat{x}$, $\hat{y}$, $\hat{z}$, $\dfrac{\partial}{\partial x}$, $\dfrac{\partial}{\partial y}$, $\dfrac{\partial}{\partial z}$ in the Cartesian formula of the gradient.

$$\nabla = \hat{x}\frac{\partial}{\partial x} + \hat{y}\frac{\partial}{\partial y} + \hat{z}\frac{\partial}{\partial z} =$$

$$= \left(\hat{r}\sin\theta\cos\phi + \hat{\theta}\cos\theta\cos\phi - \hat{\phi}\sin\phi\right)\left(\sin\theta\cos\phi\frac{\partial}{\partial r} + \frac{1}{r}\cos\theta\cos\phi\frac{\partial}{\partial\theta} - \frac{1}{r}\frac{\sin\phi}{\sin\theta}\frac{\partial}{\partial\phi}\right) +$$

$$+ \left(\hat{r}\sin\theta\sin\phi + \hat{\theta}\cos\theta\sin\phi + \hat{\phi}\cos\phi\right)\left(\sin\theta\sin\phi\frac{\partial}{\partial r} + \frac{1}{r}\cos\theta\sin\phi\frac{\partial}{\partial\theta} + \frac{1}{r}\frac{\cos\phi}{\sin\theta}\frac{\partial}{\partial\phi}\right) +$$

$$+ \left(\hat{r}\cos\theta - \hat{\theta}\sin\theta\right)\left(\cos\theta\frac{\partial}{\partial r} - \frac{1}{r}\sin\theta\frac{\partial}{\partial\theta}\right).$$

The coefficient of $\hat{r}$ is

$$\sin^2\theta\cos^2\theta\frac{\partial}{\partial r} + \frac{1}{r}\sin\theta\cos\theta\cos^2\phi\frac{\partial}{\partial\theta} - \frac{1}{r}\sin\phi\cos\phi\frac{\partial}{\partial\phi} + \sin^2\phi\cos^2\phi\frac{\partial}{\partial r} + \frac{1}{r}\sin\theta\cos\theta\frac{\partial}{\partial\theta} +$$

$$+ \frac{1}{r}\sin\phi\cos\phi\frac{\partial}{\partial\phi} + \cos^2\theta\frac{\partial}{\partial r} - \frac{1}{r}\sin\theta\cos\theta\frac{\partial}{\partial\theta},$$

which gives $\dfrac{\partial}{\partial r}$. The coefficient of $\hat{\theta}$ is

$$\sin\theta\cos\theta\cos^2\phi\frac{\partial}{\partial r} + \frac{1}{r}\cos^2\theta\cos^2\phi\frac{\partial}{\partial\theta} - \frac{1}{r}\frac{\cos\theta\cos\phi\sin\phi}{\sin\theta}\frac{\partial}{\partial\phi} + \sin\theta\cos\theta\sin^2\phi\frac{\partial}{\partial r} +$$

$$+ \frac{1}{r}\cos^2\theta\sin^2\phi\frac{\partial}{\partial\theta} + \frac{1}{r}\frac{\cos\theta\cos\phi\sin\phi}{\sin\theta}\frac{\partial}{\partial\phi} - \sin\theta\cos\theta\frac{\partial}{\partial r} + \frac{1}{r}\sin^2\theta\frac{\partial}{\partial\theta},$$

which gives $\dfrac{1}{r}\dfrac{\partial}{\partial\theta}$. In the same way the coefficient of $\hat{\phi}$ is

$$-\sin\phi\sin\theta\cos\phi\frac{\partial}{\partial r} - \frac{1}{r}\sin\phi\cos\theta\cos\phi\frac{\partial}{\partial\theta} + \frac{1}{r}\frac{\sin^2\phi}{\sin\theta}\frac{\partial}{\partial\phi} + \sin\phi\sin\theta\cos\phi\frac{\partial}{\partial r} +$$

$$+ \frac{1}{r}\sin\phi\cos\theta\cos\phi\frac{\partial}{\partial\theta} + \frac{1}{r}\frac{\cos^2\phi}{\sin\theta}\frac{\partial}{\partial\phi},$$

which gives $\dfrac{1}{r}\dfrac{1}{\sin\theta}\dfrac{\partial}{\partial\phi}$. It results

$$\nabla = \hat{r}\frac{\partial}{\partial r} + \hat{\theta}\frac{1}{r}\frac{\partial}{\partial\theta} + \hat{\phi}\frac{1}{r\sin\theta}\frac{\partial}{\partial\phi}.$$

Let us observe that, from

$$\begin{cases} \hat{r} = \hat{x}\sin\theta\cos\phi + \hat{y}\sin\theta\sin\phi + \hat{z}\cos\theta \\ \hat{\theta} = \hat{x}\cos\theta\cos\phi + \hat{y}\cos\theta\sin\phi - \hat{z}\sin\theta \\ \hat{\phi} = -\hat{x}\sin\phi + \hat{y}\cos\phi, \end{cases}$$

we can obtain

$$\begin{cases} \hat{x} = \hat{r}\sin\theta\cos\phi + \hat{\theta}\cos\theta\cos\phi - \hat{\phi}\sin\phi \\ \hat{y} = \hat{r}\sin\theta\sin\phi + \hat{\theta}\cos\theta\sin\phi + \hat{\phi}\cos\phi \\ \hat{z} = \hat{r}\cos\theta - \hat{\theta}\sin\theta, \end{cases}$$

but also

$$\frac{\partial\hat{r}}{\partial r} = 0, \quad \frac{\partial\hat{r}}{\partial\theta} = \hat{x}\cos\theta\cos\phi + \hat{y}\cos\theta\sin\phi - \hat{z}\sin\theta = \hat{\theta}, \quad \frac{\partial\hat{r}}{\partial\phi} = -\hat{x}\sin\theta\sin\phi + \hat{y}\sin\theta\cos\phi = \sin\theta\hat{\phi}$$

$$\frac{\partial\hat{\theta}}{\partial r} = 0, \quad -\frac{\partial\hat{\theta}}{\partial\theta} = \hat{x}\sin\theta\cos\phi + \hat{y}\sin\theta\sin\phi + \hat{z}\cos\theta = \hat{r}, \quad \frac{\partial\hat{\theta}}{\partial\phi} = \cos\theta(-\hat{x}\sin\phi + \hat{y}\cos\phi) = \cos\theta\hat{\phi}$$

$$\frac{\partial \hat{\phi}}{\partial r} = 0, \quad -\frac{\partial \hat{\phi}}{\partial \theta} = 0, \quad \frac{\partial \hat{\phi}}{\partial \phi} = -\hat{r}\sin\theta - \hat{\theta}\cos\theta.$$

The last formula is obtained after replacing $\hat{x}$ and $\hat{y}$ in the equality $\dfrac{\partial \hat{\phi}}{\partial \phi} = -\hat{x}\cos\phi -$ $\hat{y}\sin\phi$.

In order to obtain the Laplace operator ∇^2, we have to take into account that $\nabla^2 = \nabla \cdot \nabla$, that is

$$\nabla^2 = \left(\hat{r}\frac{\partial}{\partial r} + \hat{\theta}\frac{1}{r}\frac{\partial}{\partial \theta} + \hat{\phi}\frac{1}{r\sin\theta}\frac{\partial}{\partial \phi} \right) \cdot \left(\hat{r}\frac{\partial}{\partial r} + \hat{\theta}\frac{1}{r}\frac{\partial}{\partial \theta} + \hat{\phi}\frac{1}{r\sin\theta}\frac{\partial}{\partial \phi} \right).$$

Theorem 11.4.3 *The Laplace operator in spherical coordinates is*

$$\nabla^2 = \frac{1}{r^2}\left[\frac{\partial}{\partial r}\left(r^2 \frac{\partial}{\partial r} \right) + \frac{1}{\sin\theta}\frac{\partial}{\partial \theta}\left(\sin\theta\frac{\partial}{\partial \theta} \right) + \frac{1}{\sin^2\theta}\frac{\partial^2}{\partial \phi^2} \right].$$

Proof The formula

$$\left(\hat{r}\frac{\partial}{\partial r} + \hat{\theta}\frac{1}{r}\frac{\partial}{\partial \theta} + \hat{\phi}\frac{1}{r\sin\theta}\frac{\partial}{\partial \phi} \right) \cdot \left(\hat{r}\frac{\partial}{\partial r} + \hat{\theta}\frac{1}{r}\frac{\partial}{\partial \theta} + \hat{\phi}\frac{1}{r\sin\theta}\frac{\partial}{\partial \phi} \right)$$

gives

$$\hat{r}\frac{\partial}{\partial r}\left(\hat{r}\frac{\partial}{\partial r} + \hat{\theta}\frac{1}{r}\frac{\partial}{\partial \theta} + \hat{\phi}\frac{1}{r\sin\theta}\frac{\partial}{\partial \phi} \right) + \hat{\theta}\frac{1}{r}\frac{\partial}{\partial \theta}\left(\hat{r}\frac{\partial}{\partial r} + \hat{\theta}\frac{1}{r}\frac{\partial}{\partial \theta} + \hat{\phi}\frac{1}{r\sin\theta}\frac{\partial}{\partial \phi} \right) +$$

$$+\hat{\phi}\frac{1}{r\sin\phi}\frac{\partial}{\partial \phi}\left(\hat{r}\frac{\partial}{\partial r} + \hat{\theta}\frac{1}{r}\frac{\partial}{\partial \theta} + \hat{\phi}\frac{1}{r\sin\theta}\frac{\partial}{\partial \phi} \right)$$

To continue, the partial derivatives are applied to each term, but we have to consider

$$\hat{r}\cdot\hat{r} = \hat{\theta}\cdot\hat{\theta} = \hat{\phi}\cdot\hat{\phi} = 1; \quad \hat{r}\cdot\hat{\theta} = \hat{\theta}\cdot\hat{r} = \hat{\theta}\cdot\hat{\phi} = \hat{\phi}\cdot\hat{\theta} = \hat{\phi}\cdot\hat{r} = \hat{r}\cdot\hat{\phi} = 0$$

and

$$\frac{\partial \hat{r}}{\partial r} = 0, \quad \frac{\partial \hat{r}}{\partial \theta} = \hat{\theta}, \quad \frac{\partial \hat{r}}{\partial \phi} = \hat{\phi}\sin\theta$$

$$\frac{\partial \hat{\theta}}{\partial r} = 0, \quad \frac{\partial \hat{\theta}}{\partial \theta} = -\hat{r}, \quad \frac{\partial \hat{\theta}}{\partial \phi} = \hat{\phi}\cos\theta$$

$$\frac{\partial \hat{\phi}}{\partial r} = 0, \quad -\frac{\partial \hat{\phi}}{\partial \theta} = 0, \quad \frac{\partial \hat{\phi}}{\partial \phi} = -\hat{r}\sin\theta - \hat{\theta}\cos\theta.$$

Leaving between brackets only non-zero terms, we have, for the Laplace operator ∇^2, the form

$$\left[\hat{r}\frac{\partial \hat{r}}{\partial r}\frac{\partial}{\partial r}+\frac{\partial^2}{\partial r^2}+\hat{r}\frac{1}{r}\frac{\partial \hat{\theta}}{\partial r}\frac{\partial}{\partial \theta}+\hat{r}\frac{1}{r\sin\theta}\frac{\partial \hat{\phi}}{\partial r}\frac{\partial}{\partial \phi}\right]+\left[\hat{\theta}\frac{1}{r}\frac{\partial \hat{r}}{\partial \theta}\frac{\partial}{\partial r}+\hat{\theta}\frac{1}{r^2}\frac{\partial \hat{\theta}}{\partial \theta}\frac{\partial}{\partial \theta}+\frac{1}{r^2}\frac{\partial^2}{\partial \theta^2}+\hat{\theta}\frac{1}{r\sin\theta}\frac{\partial \hat{\phi}}{\partial \theta}\frac{\partial}{\partial \phi}\right]$$

$$+\left[\frac{\hat{\phi}}{r\sin\theta}\frac{\partial \hat{r}}{\partial \phi}\frac{\partial}{\partial r}+\frac{\hat{\phi}}{r^2\sin\theta}\frac{\partial \hat{\theta}}{\partial \phi}\frac{\partial}{\partial \theta}+\frac{\hat{\phi}}{r^2\sin^2\theta}\frac{\partial \hat{\phi}}{\partial \phi}\frac{\partial}{\partial \phi}+\frac{1}{r^2\sin^2\theta}\frac{\partial^2}{\partial \phi^2}\right].$$

Let us now cancel all the terms which are 0 from the partial derivatives of $\hat{r}, \hat{\theta}, \hat{\phi}$ and from dot products:

$$\left[\cancel{\hat{r}\frac{\partial \hat{r}}{\partial r}\frac{\partial}{\partial r}}+\frac{\partial^2}{\partial r^2}+\cancel{\hat{r}\frac{1}{r}\frac{\partial \hat{\theta}}{\partial r}\frac{\partial}{\partial \theta}}+\cancel{\hat{r}\frac{1}{r\sin\theta}\frac{\partial \hat{\phi}}{\partial r}\frac{\partial}{\partial \phi}}\right]+\left[\hat{\theta}\frac{1}{r}\hat{\theta}\frac{\partial}{\partial r}-\cancel{\hat{\theta}\frac{1}{r^2}\hat{r}\frac{\partial}{\partial \theta}}+\frac{1}{r^2}\frac{\partial^2}{\partial \theta^2}+\cancel{\hat{\theta}\frac{1}{r\sin\theta}\frac{\partial \hat{\phi}}{\partial \theta}\frac{\partial}{\partial \phi}}\right]$$

$$+\left[\frac{\hat{\phi}}{r\sin\theta}\hat{\phi}\sin\theta\frac{\partial}{\partial r}+\frac{\hat{\phi}}{r^2\sin\theta}\hat{\phi}\cos\theta\frac{\partial}{\partial \theta}-\cancel{\frac{\hat{\phi}}{r^2\sin^2\theta}(\hat{r}\sin\theta+\hat{\theta}\cos\theta)\frac{\partial}{\partial \phi}}+\frac{1}{r^2\sin^2\theta}\frac{\partial^2}{\partial \phi^2}\right]=$$

$$=\frac{\partial^2}{\partial r^2}+\frac{2}{r}\frac{\partial}{\partial r}+\frac{1}{r^2}\frac{\cos\theta}{\sin\theta}\frac{\partial}{\partial \theta}+\frac{1}{r^2}\frac{\partial^2}{\partial \theta^2}+\frac{1}{r^2\sin^2\theta}\frac{\partial^2}{\partial \phi^2}=$$

$$=\frac{1}{r^2}\left[\left(r^2\frac{\partial^2}{\partial r^2}+2r\frac{\partial}{\partial r}\right)+\left(\frac{\cos\theta}{\sin\theta}\frac{\partial}{\partial \theta}+\frac{\partial^2}{\partial \theta^2}\right)+\frac{1}{\sin^2\theta}\frac{\partial^2}{\partial \phi^2}\right]=$$

$$=\frac{1}{r^2}\left[\frac{\partial}{\partial r}\left(r^2\frac{\partial}{\partial r}\right)+\frac{1}{\sin\theta}\frac{\partial}{\partial \theta}\left(\sin\theta\frac{\partial}{\partial \theta}\right)+\frac{1}{\sin^2\theta}\frac{\partial^2}{\partial \phi^2}\right].$$

$\square$

Considering now the *spherical harmonics*

$$Y_l^m(\theta,\phi)=\left[\frac{(2l+1)}{4\pi}\frac{(l-m)!}{(l+m)!}\right]^{1/2}(-1)^m e^{im\phi}P_l^m(\cos\theta),\ l\geq |m|\geq 0$$

with $x=\cos\theta$, we have

$$P_l^m(x)=\sin^{|m|}\theta\frac{d^{|m|}}{dx^{|m|}}P_l(x)$$

which are written with respect to the *Legendre polynomials*

$$P_l(x)=\left(2^l l!\right)^{-1}\frac{d^l}{dx^l}(x^2-1)^l.$$

Let us discuss how they appear. The constant depending on m and l appears from a normalization condition exactly as in the case of Hermite polynomials. The fact that for $|m| \leq l$ the formula still has sense is the consequence of the mathematical consistency of polynomials. This constraint has physical implication too. Let us concentrate on the mathematical consistency showing that these polynomials are solutions of an equation related to the Schrödinger equation.

Let us start from the time-independent Schrödinger equation

$$\left(-\frac{\hbar^2}{2m}\nabla^2 + V(\mathbf{r})\right)\Psi(\mathbf{r}) = E\Psi(\mathbf{r})$$

which can be written in the form

$$-\frac{\hbar^2}{2m}\nabla^2\Psi(\mathbf{r}) + V(\mathbf{r})\Psi(\mathbf{r}) = E\Psi(\mathbf{r}).$$

We denote $\Psi(\mathbf{r})$ by Ψ and keep in mind that it depends on (x, y, z). Explicitly, the equation to solve is

$$-\frac{\hbar^2}{2m}\left(\frac{\partial^2}{\partial x^2} + \frac{\partial^2}{\partial y^2} + \frac{\partial^2}{\partial z^2}\right)\Psi + V(x, y, z)\Psi = E\Psi,$$

written with respect to Cartesian coordinates. The solution of this equation depends on V. If we separate the variables assuming

$$\Psi(x, y, z) = X(x)Y(y)Z(z),$$

which can be simply written as $\Psi = XYZ$, we can transform the Schrödinger equation in

$$-\frac{\hbar^2}{2m}\left(YZ\frac{\partial^2 X}{\partial x^2} + XZ\frac{\partial^2 Y}{\partial y^2} + YZ\frac{\partial^2 Z}{\partial z^2}\right) + V(x, y, z)XYZ = EXYZ.$$

After dividing by XYZ, it results

$$-\frac{\hbar^2}{2m}\left(\frac{1}{X}\frac{\partial^2 X}{\partial x^2} + \frac{1}{Y}\frac{\partial^2 Y}{\partial y^2} + \frac{1}{Z}\frac{\partial^2 Z}{\partial z^2}\right) + V(x, y, z) = E.$$

In Lecture 23 and Lecture 24, we solved the problem in 1-dimensional case for $V = 0$ and $V = $ constant. Even here it is simple to solve for these cases because we have a sum of functions depending only on x, y and z, respectively, and this sum is a constant. This result is possible only when each function is a constant, therefore there exist the constants c_1, c_2, c_3 such that we have to solve

$$-\frac{\hbar^2}{2m}\frac{1}{X}\frac{\partial^2 X}{\partial x^2} = c_1; \quad -\frac{\hbar^2}{2m}\frac{1}{Y}\frac{\partial^2 Y}{\partial y^2} = c_2; \quad -\frac{\hbar^2}{2m}\frac{1}{Z}\frac{\partial^2 Z}{\partial z^2} = c_3.$$

Each equation has the form

$$f'' + kf = 0$$

and the solution depends on the constant k, that is we have one kind of solution when $k < 0$ and another kind when $k > 0$. This situation realized also in the mentioned 23 and 24 lectures. We can solve also for $V(x, y, z) = V_x(x) + V_y(y) + V_z(z)$. In this case, the form of the equation is $f'' + kf = g$ where k is a constant and g is a function.

Now, let us return to our main question: How do the spherical harmonics appear? The idea is related to the Schrödinger equation written in spherical coordinates. Therefore it is related to the Laplace operator written also in spherical coordinates.

We start from

$$-\frac{\hbar^2}{2m}\frac{1}{r^2}\left[\frac{\partial}{\partial r}\left(r^2\frac{\partial\Psi}{\partial r}\right) + \frac{1}{\sin\theta}\frac{\partial}{\partial\theta}\left(\sin\theta\frac{\partial\Psi}{\partial\theta}\right) + \frac{1}{\sin^2\theta}\frac{\partial^2\Psi}{\partial\phi^2}\right] + V(r)\Psi = E\Psi.$$

Assuming $\Psi(r, \theta, \phi) = R(r)Y(\theta, \phi)$, we can write the previous equation as

$$-\frac{\hbar^2}{2m}\frac{1}{r^2}\left[Y\frac{\partial}{\partial r}\left(r^2\frac{\partial R}{\partial r}\right) + \frac{R}{\sin\theta}\frac{\partial}{\partial\theta}\left(\sin\theta\frac{\partial Y}{\partial\theta}\right) + \frac{R}{\sin^2\theta}\frac{\partial^2 Y}{\partial\phi^2}\right] + V(r)RY = ERY.$$

After dividing by $-\dfrac{\hbar^2}{mr^2}RY$, we obtain

$$\left[\frac{1}{R}\frac{\partial}{\partial r}\left(r^2\frac{\partial R}{\partial r}\right) + \frac{1}{Y\sin\theta}\frac{\partial}{\partial\theta}\left(\sin\theta\frac{\partial Y}{\partial\theta}\right) + \frac{1}{Y\sin^2\theta}\frac{\partial^2 Y}{\partial\phi^2}\right] - \frac{2mr^2}{\hbar^2}(V(r) - E) = 0.$$

If we separate the equation in the part depending on r and the part depending on θ, ϕ, we have

$$\left[\frac{1}{R}\frac{\partial}{\partial r}\left(r^2\frac{\partial R}{\partial r}\right) - \frac{2mr^2}{\hbar^2}(V(r) - E)\right] + \frac{1}{Y}\left[\frac{1}{\sin\theta}\frac{\partial}{\partial\theta}\left(\sin\theta\frac{\partial Y}{\partial\theta}\right) + \frac{1}{\sin^2\theta}\frac{\partial^2 Y}{\partial\phi^2}\right] = 0$$

and consequently, the part depending on r has to be a constant K and the part depending on θ, ϕ has to be $-K$. We can choose the constant to be $l(l + 1)$. The reason is related to the second equation

$$\frac{1}{Y}\left[\frac{1}{\sin\theta}\frac{\partial}{\partial\theta}\left(\sin\theta\frac{\partial Y}{\partial\theta}\right) + \frac{1}{\sin^2\theta}\frac{\partial^2 Y}{\partial\phi^2}\right] = -l(l + 1)$$

which allows a solution depending on the Legendre polynomials. The other equation to solve remains

$$\frac{1}{R}\frac{\partial}{\partial r}\left(r^2\frac{\partial R}{\partial r}\right) - \frac{2mr^2}{\hbar^2}(V(r) - E) = l(l+1).$$

Let us consider the equation in Y after multiplying by $Y\sin^2\theta$:

$$\sin\theta\frac{\partial}{\partial\theta}\left(\sin\theta\frac{\partial Y}{\partial\theta}\right) + \frac{\partial^2 Y}{\partial\phi^2} = -l(l+1)Y\sin^2\theta.$$

To solve it, we make another split of variables,

$$Y(\theta,\phi) = \Theta(\theta)\Lambda(\phi)$$

and then we divide by $\Theta\Lambda$:

$$\frac{1}{\Theta}\sin\theta\frac{\partial}{\partial\theta}\left(\sin\theta\frac{\partial\Theta}{\partial\theta}\right) + \frac{1}{\Lambda}\frac{\partial^2\Lambda}{\partial\phi^2} = -l(l+1)\sin^2\theta.$$

If we split in a part depending on θ and a part depending on ϕ, we have

$$\left[\frac{1}{\Theta}\sin\theta\frac{\partial}{\partial\theta}\left(\sin\theta\frac{\partial\Theta}{\partial\theta}\right) + l(l+1)\sin^2\theta\right] + \frac{1}{\Lambda}\frac{\partial^2\Lambda}{\partial\phi^2} = 0.$$

Based on the same reasoning, we can consider two opposite constants such that

$$\frac{1}{\Theta}\sin\theta\frac{\partial}{\partial\theta}\left(\sin\theta\frac{\partial\Theta}{\partial\theta}\right) + l(l+1)\sin^2\theta = m^2$$

and

$$\frac{1}{\Lambda}\frac{\partial^2\Lambda}{\partial\phi^2} = -m^2.$$

The last equation has the solution $\Lambda(\phi) = e^{im\phi}$. Since $\Lambda(0) = \Lambda(2\pi)$, it results $e^{i2m\pi} = 1$, that is $m \in \mathbb{Z}$.

The other equation can be written in the form

$$\sin\theta\frac{\partial}{\partial\theta}\left(\sin\theta\frac{\partial\Theta}{\partial\theta}\right) + [l(l+1)\sin^2\theta - m^2]\Theta = 0.$$

After we consider $\Theta(\theta) = P_l^m(\cos\theta)$ and $x = \cos\theta$, this equation is equivalent to

$$\frac{d}{dx}\left[(1-x^2)\frac{d}{dx}\right]P_l^m(x) + \left[l(l+1) - \frac{m^2}{1-x^2}\right]P_l^m(x) = 0.$$

This happens because

$$\sin\theta \frac{d}{d\theta} P_l^m(\cos\theta) = -\sin^2\theta \frac{d}{dx} P_l^m(x) = -(1-x^2)\frac{d}{dx} P_l^m(x)$$

and

$$\sin\theta \frac{d}{d\theta} = \sin\theta \frac{dx}{d\theta}\frac{d}{dx} = -(1-x^2)\frac{d}{dx}.$$

Dividing by $1-x^2$, we get the above equation above. For $m=0$ this is the Legendre equation in the original form, that is

$$\frac{d}{dx}\left[(1-x^2)\frac{d}{dx}\right] P_l(x) + l(l+1)P_l(x) = 0,$$

with the solution given by the Rodrigues formula

$$P_l(x) = \left(2^l l!\right)^{-1} \frac{d^l}{dx^l}(x^2-1)^l.$$

Before checking it, let us understand how the Legendre polynomials were discovered.

We can start from the Legendre idea to see a gravitational potential as a series depending (via polynomials) on the angle determined by the two position vectors $\mathbf{r}, \mathbf{r_1}$ associated to the bodies between which the gravitational force acts.

If $\dfrac{1}{|\mathbf{r}-\mathbf{r_1}|} = \dfrac{1}{\sqrt{r^2 + r_1^2 - 2rr_1\cos\theta}} = \dfrac{1}{r\sqrt{1+h^2-2h\cos\theta}}$ where $h := \dfrac{r_1}{r}$, Legendre defined

the polynomials P_n which satisfy the relation

$$\frac{1}{\sqrt{1+h^2-2h\cos\theta}} = \sum_{n\in\mathbb{N}} h^n P_n(\cos\theta).$$

Denote $x := \cos\theta$, the previous relation becomes

$$(1-2xh+h^2)^{-1/2} = \sum_{n\in\mathbb{N}} h^n P_n(x).$$

Theorem 11.4.4 *Legendre's polynomials satisfy the following recurrence relation*

$$xP_n(x) = \frac{n+1}{2n+1}P_{n+1}(x) + \frac{n}{2n+1}P_{n-1}(x), \quad n \geq 1.$$

Proof We consider the derivative with respect to h of the formula

$$(1-2xh+h^2)^{-1/2} = \sum_{n\in\mathbb{N}} h^n P_n(x).$$

It results

$$(x - h)(1 - 2xh + h^2)^{-3/2} = \sum_{n \in \mathbb{N}} nh^{n-1} P_n(x).$$

Multiplying by $(1 - 2xh + h^2)$ both sides, we have

$$(x - h)(1 - 2xh + h^2)^{-1/2} = (1 - 2xh + h^2) \sum_{n \in \mathbb{N}} nh^{n-1} P_n(x), \text{ i.e}$$

$$(x - h) \sum_{n \in \mathbb{N}} h^n P_n(x) = (1 - 2xh + h^2) \sum_{n \in \mathbb{N}} nh^{n-1} P_n(x).$$

From the relation

$$x \sum_{n \in \mathbb{N}} h^n P_n(x) - \sum_{n \in \mathbb{N}} h^{n+1} P_n(x) = \sum_{n \in \mathbb{N}} nh^{n-1} P_n(x) - 2x \sum_{n \in \mathbb{N}} nh^n P_n(x) + \sum_{n \in \mathbb{N}} nh^{n+1} P_n(x)$$

we identify the coefficients of h^n from both members and we obtain

$$x P_n(x) - P_{n-1}(x) = (n + 1) P_{n+1}(x) - 2xn P_n(x) + (n - 1) P_{n-1}(x),$$

that is

$$(n + 1) P_{n+1}(x) = (2n + 1)x P_n(x) - n P_{n-1}(x)$$

which represents the recurrence from the above statement.   □

Theorem 11.4.5 *Legendre's polynomials satisfy the following recurrence relation*

$$n P_n(x) = x P'_n(x) - P'_{n-1}(x), \ \ n \geq 1$$

where $P'_n(x) := \dfrac{d}{dx} P_n(x).$

Proof The derivative with respect to h of the formula

$$(1 - 2xh + h^2)^{-1/2} = \sum_{n \in \mathbb{N}} h^n P_n(x)$$

lead us to

$$(x - h)(1 - 2xh + h^2)^{-3/2} = \sum_{n \in \mathbb{N}} nh^{n-1} P_n(x).$$

The derivative with respect to x of the same formula is

$$h(1 - 2xh + h^2)^{-3/2} = \sum_{n \in \mathbb{N}} h^n P'_n(x),$$

therefore

$$\frac{h}{x - h} \sum_{n \in \mathbb{N}} n h^{n-1} P_n(x) = \sum_{n \in \mathbb{N}} h^n P_n'(x).$$

The coefficients of h^n of both members of the equality

$$\sum_{n \in \mathbb{N}} n h^n P_n(x) = x \sum_{n \in \mathbb{N}} h^n P_n'(x) - \sum_{n \in \mathbb{N}} h^{n+1} P_n'(x)$$

satisfy the relation

$$n P_n(x) = x P_n'(x) - P_{n-1}'(x)$$

of our statement. □

Using the two statements before we can cancel out x in a new relation for the Legendre polynomials.

Corollary 11.4.6 *The Legendre polynomials satisfy the equality*

$$P_{n+1}'(x) = (2n + 1) P_n(x) + P_{n-1}'(x).$$

Proof The derivative with respect to x of the relation

$$x P_n(x) = \frac{n + 1}{2n + 1} P_{n+1}(x) + \frac{n}{2n + 1} P_{n-1}(x)$$

is

$$P_n(x) + x P_n'(x) = \frac{n + 1}{2n + 1} P_{n+1}'(x) + \frac{n}{2n + 1} P_{n-1}'(x), \quad n \geq 1.$$

By replacing

$$x P_n'(x) = n P_n(x) + P_{n-1}'(x)$$

in the previous relation, we obtain

$$P_{n+1}'(x) = (2n + 1) P_n(x) + P_{n-1}'(x)$$

□

This relation with no x in front of terms is verified by the polynomials

$$P_l(x) = \left(2^l l!\right)^{-1} \frac{d^l}{dx^l} (x^2 - 1)^l.$$

This is a simple computation and we left to the reader as exercise.

Let us return to the equation these polynomials verify by proving the following theorem. This step is necessary when we check that the functions (implied in spherical harmonics)

$$P_l^m(x) = (1 - x^2)^{m/2} \frac{d^m}{dx^m} P_l(x)$$

are solutions of the equation

$$\frac{d}{dx}\left[(1 - x^2)\frac{d}{dx}\right] P_l^m(x) + \left[l(l+1) - \frac{m^2}{1 - x^2}\right] P_l^m(x) = 0.$$

Theorem 11.4.7 *The Legendre equation*

$$(1 - x^2)\frac{d^2Y}{dx^2} - 2x\frac{dY}{dx} + l(l+1)Y = 0$$

has the solution

$$P_l(x) = (2^l l!)^{-1} \frac{d^l}{dx^l}[(x^2 - 1)^l].$$

Proof Let us denote

$$y := (x^2 - 1)^l.$$

It results

$$\frac{dy}{dx} = 2lx(x^2 - 1)^{l-1}, \quad \text{that is} \quad (x^2 - 1)\frac{dy}{dx} = 2lxy.$$

The Leibniz formula for the nth-order derivative of a product of two functions states that

$$D^{(n)}(uv) = uD^{(n)}v + nD(u)D^{(n-1)}(v) + \frac{n(n+1)}{2}D^{(2)}(u)D^{(n-2)}(v) + \cdots,$$

the binomial coefficients being in front of each term. Applying it in the equality

$$D^{(l+1)}\left[(x^2 - 1)\frac{dy}{dx}\right] = 2l D^{(l+1)}[xy],$$

it results

$$(x^2 - 1)\frac{d^{l+2}y}{dx^{l+2}} + 2x(l+1)\frac{d^{l+1}y}{dx^{l+1}} + l(l+1)\frac{d^l y}{dx^l} = 2l\left[x\frac{d^{l+1}y}{dx^{l+1}} + (l+1)\frac{d^l y}{dx^l}\right],$$

that is

$$(1 - x^2)\frac{d^{l+2}y}{dx^{l+2}} - 2x\frac{d^{l+1}y}{dx^{l+1}} + l(l+1)\frac{d^l y}{dx^l} = 0.$$

If we denote

$$Y := \frac{d^l y}{dx^l},$$

the previous equation becomes

$$(1 - x^2)\frac{d^2Y}{dx^2} - 2x\frac{dY}{dx} + l(l+1)Y = 0,$$

i.e. the equation coming from the statement of the theorem. Therefore the general solution is

$$Y = K\frac{d^l}{dx^l}[(x^2 - 1)^l],$$

where K is a constant. If we impose the condition $P_l(1) = 1$, the constant K can be identified by applying the Leibniz formula for $(x + 1)^l(x - 1)^l$. Therefore

$$K = \frac{d^l}{dx^l}\left[(x + 1)^l(x - 1)^l\right]|_{x=1} = l!(x + 1)^l|_{x=1} = l!2^l.$$

For this constant only the previous recurrence relations are verified.   □

Let us highlight that the Legendre equation

$$(1 - x^2)\frac{d^2Y}{dx^2} - 2x\frac{dY}{dx} + l(l+1)Y = 0$$

allows the particular solution

$$Y_l(x) = \frac{d^l}{dx^l}[(x^2 - 1)^l],$$

that is we can cancel the multiplicative constant. Now, denote by

$$Y_{l,m} := \frac{d^m Y_l}{dx^m},$$

we have

Theorem 11.4.8 *The functions $Y_{l,m}$ satisfy the relation*

$$(1 - x^2)\frac{d^2Y_{l,m}}{dx^2} = 2m(m+1)\frac{dY_{l,m}}{dx} + [m(m+1) - l(l+1)]Y_{l,m}.$$

Proof Consider the m derivative of both members of Legendre's equation

$$(1 - x^2)\frac{d^2Y}{dx^2} - 2x\frac{dY}{dx} + l(l+1)Y = 0$$

whose solution is now Y_l. It results

$$\frac{d^m}{dx^m}\left[(1-x^2)\frac{d^2Y_l}{dx^2} - 2x\frac{dY_l}{dx} + l(l+1)Y_l\right] = 0.$$

The Leibniz formula leads to

$$\frac{d^m}{dx^m}\left[(1-x^2)\frac{d^2Y_l}{dx^2}\right] = (1-x^2)\frac{d^{m+2}Y_l}{dx^{m+2}} - 2mx\frac{d^{m+1}Y_l}{dx^{m+1}} - m(m+1)\frac{d^mY_l}{dx^m}$$

and

$$\frac{d^m}{dx^m}\left[x\frac{dY_l}{dx}\right] = x\frac{d^{m+1}Y_l}{dx^{m+1}} + m\frac{d^mY_l}{dx^m}.$$

Replacing in the initial formula, we obtain the result from the statement of the theorem. $\qquad\square$

Theorem 11.4.9 *The functions*

$$P_l^m(x) = (1-x^2)^{m/2}\frac{d^m}{dx^m}P_l(x)$$

verify the equality

$$\frac{d}{dx}\left[(1-x^2)\frac{d}{dx}\right]P_l^m(x) + \left[l(l+1) - \frac{m^2}{1-x^2}\right]P_l^m(x) = 0.$$

Proof Let us first observe that multiplicative constant can be canceled out and the theorem statement can be reduced to:

$$Q_l^m(x) := (1-x^2)^{m/2}\frac{d^mY_l}{dx^m}(x).$$

It verifies the equality

$$\frac{d}{dx}\left[(1-x^2)\frac{d}{dx}\right]Q_l^m(x) + \left[l(l+1) - \frac{m^2}{1-x^2}\right]Q_l^m(x) = 0.$$

We prefer to write Q_l^m as

$$Q_l^m(x) := (1-x^2)^{m/2}Y_{l,m}.$$

We have first

$$\frac{dQ_l^m}{dx} = \frac{d}{dx}\left[(1-x^2)^{m/2}Y_{l,m}\right] = -mx(1-x^2)^{m/2-1}Y_{l,m} + (1-x^2)^{m/2}\frac{dY_{l,m}}{dx},$$

i.e.

$$(1-x^2)\frac{dQ_l^m}{dx} = -mx(1-x^2)^{m/2}Y_{l,m} + (1-x^2)^{m/2+1}\frac{dY_{l,m}}{dx}.$$

Then

$$\frac{d}{dx}\left[(1-x^2)\frac{dQ_l^m}{dx}\right] = -m(1-x^2)^{m/2}Y_{l,m} + \frac{m^2x^2}{1-x^2}(1-x^2)^{m/2}Y_{l,m} - mx(1-x^2)^{m/2}\frac{dY_{l,m}}{dx} +$$

$$+ \left(\frac{m}{2}+1\right)(1-x^2)^{m/2}(-2x)\frac{dY_{l,m}}{dx} + (1-x^2)^{m/2}\left[(1-x^2)\frac{d^2Y_{l,m}}{dx^2}\right].$$

If we replace $\left[(1-x^2)\dfrac{d^2Y_{l,m}}{dx^2}\right]$ with $2m(m+1)\dfrac{dY_{l,m}}{dx} + [m(m+1)-l(l+1)]Y_{l,m}$

and we cancel the two contrary sign terms $2(m+1)x(1-x^2)\dfrac{dY_{l,m}}{dx}$, we obtain the equality

$$\frac{d}{dx}\left[(1-x^2)\frac{dQ_l^m}{dx}\right] = -\cancel{mQ_l^m} + \frac{m^2x^2}{1-x^2}(1-x^2)^{m/2}Y_{l,m} + m^2Q_l^m + \cancel{mQ_l^m} - l(l+1)Q_l^m$$

which ends the proof. □

These considerations have important physical applications as we are going to discuss. In Lecture 19, we presented the Bohr model of hydrogen atom. The first part of this lecture offers us the possibility to revisit this subject using the time-independent Schrödinger equation solution in spherical coordinates. As we know, the hydrogen atom consists in a nucleus with a single proton inside and an electron "orbiting" the nucleus. If we denote by $\mathbf{m}$ the mass of the electron and by M the mass of nucleus, we have to consider the *reduced mass* $\mu := \dfrac{M\mathbf{m}}{M+\mathbf{m}}$ which is equivalent to the mass located at the centre of gravity of the system electron-nucleus. The Schrödinger equation can be written considering a potential $V(r)$ described by the Coulomb term

$$V(r) := -\frac{e^2}{4\pi\varepsilon_0 r},$$

where e is the charge of the electron, ε_0 is the permittivity of vacuum, r is the distance between the nucleus and electron. The Schrödinger equation is

$$-\frac{\hbar^2}{2\mu}\frac{1}{r^2}\left[\frac{\partial}{\partial r}\left(r^2\frac{\partial\Psi}{\partial r}\right) + \frac{1}{\sin\theta}\frac{\partial}{\partial\theta}\left(\sin\theta\frac{\partial\Psi}{\partial\theta}\right) + \frac{1}{\sin^2\theta}\frac{\partial^2\Psi}{\partial\phi^2}\right] - \frac{e^2}{4\pi\varepsilon_0 r}\Psi = E\Psi.$$

After the separation of variables, we have the following equations:

$$\frac{1}{R}\frac{\partial}{\partial r}\left(r^2\frac{\partial R}{\partial r}\right) + \frac{2\mu r^2}{\hbar^2}\left(\frac{e^2}{4\pi\varepsilon_0 r} + E\right) = l(l+1),$$

called the *radial equation* with solution $R(r)$;

$$\sin\theta \frac{\partial}{\partial\theta}\left(\sin\theta\frac{\partial\Theta}{\partial\theta}\right) + [l(l+1)\sin^2\theta - m^2]\Theta = 0,$$

called the *polar equation*, with solution $\Theta(\theta)$;

$$\frac{1}{\Lambda}\frac{\partial^2\Lambda}{\partial\phi^2} = -m^2,$$

called the *azimuthal equation*, with solution $\Lambda(\phi)$. The total solution is $\Psi(r,\theta,\phi) = R(r)\Theta(\theta)\Lambda(\phi)$.

We have solved both the polar and the azimuthal equations and we have obtained the solutions: $\Theta(\theta) = P_l^m(\cos\theta)$ and $\Lambda(\phi) = e^{im\phi}$. The numbers m and l, $|m| \leq l$, which are used to separate the variables in a convenient form are called *quantum numbers* related to the Schrödinger equation solutions.

The *radial equation* can be written in the form

$$\frac{\partial}{\partial r}\left(r^2\frac{\partial R}{\partial r}\right) + \frac{2\mu r^2}{\hbar^2}\left(\frac{e^2}{4\pi\varepsilon_0 r} + E\right)R = l(l+1)R,$$

or equivalently

$$\frac{d^2 R}{dr^2} + \frac{2}{r}\frac{dR}{dr} + \left[\frac{2\mu}{\hbar^2}\left(\frac{e^2}{4\pi\varepsilon_0 r} + E\right) - \frac{l(l+1)}{r^2}\right]R = 0.$$

To solve it, let us consider what happens for large r. The equation to solve remains

$$\frac{d^2 R_\infty}{dr^2} + \left[\frac{2\mu E}{\hbar^2}\right]R_\infty = 0.$$

For an electron far from the nucleus, the energy E approaches to 0. The atom is stable when the positive charge is close (equal) to the negative charge, therefore we have to look for a solution where the energy E is strictly negative. It is easy to check that a solution is

$$R_\infty = ce^{-r\sqrt{-2\mu E/\hbar^2}}.$$

If we choose

$$R_n(r) = R_\infty(r)e^{r\mu e^2/2n\pi\varepsilon_0\hbar^2}$$

in the case when the energy levels are $E_n = -\dfrac{\mu e^4}{8n^2\varepsilon_0^2 h^2}$, we have the radial solution depending on another *quantum number n*. According to this model, the structure of the hydrogen atom is given by the quantum numbers n, l, m. This is a more detailed and self-consistent picture of the hydrogen atom. For a detailed discussion from an experimental point of view, the reader can consult the books in Refs. [12, 13, 16, 25].

Summary of Lecture 49. We first obtain the formula of the gradient written in spherical coordinates

$$\nabla = \hat{r}\frac{\partial}{\partial r} + \hat{\theta}\frac{1}{r}\frac{\partial}{\partial \theta} + \hat{\phi}\frac{1}{r\sin\theta}\frac{\partial}{\partial \phi}.$$

For this purpose, we need the following formulas

$$\begin{cases} \dfrac{\partial}{\partial x} = \sin\theta\cos\phi\dfrac{\partial}{\partial r} + \dfrac{1}{r}\cos\theta\cos\phi\dfrac{\partial}{\partial \theta} - \dfrac{1}{r}\dfrac{\sin\phi}{\sin\theta}\dfrac{\partial}{\partial \phi} \\[2mm] \dfrac{\partial}{\partial y} = \sin\theta\sin\phi\dfrac{\partial}{\partial r} + \dfrac{1}{r}\cos\theta\sin\phi\dfrac{\partial}{\partial \theta} + \dfrac{1}{r}\dfrac{\cos\phi}{\sin\theta}\dfrac{\partial}{\partial \phi} \\[2mm] \dfrac{\partial}{\partial z} = \cos\theta\dfrac{\partial}{\partial r} - \dfrac{1}{r}\sin\theta\dfrac{\partial}{\partial \theta} \end{cases}$$

and

$$\begin{cases} \hat{x} = \hat{r}\sin\theta\cos\phi + \hat{\theta}\cos\theta\cos\phi - \hat{\phi}\sin\phi \\ \hat{y} = \hat{r}\sin\theta\sin\phi + \hat{\theta}\cos\theta\sin\phi + \hat{\phi}\cos\phi \\ \hat{z} = \hat{r}\cos\theta - \hat{\theta}\sin\theta. \end{cases}$$

From the relations

$$\hat{r}\cdot\hat{r} = \hat{\theta}\cdot\hat{\theta} = \hat{\phi}\cdot\hat{\phi} = 1; \ \hat{r}\cdot\hat{\theta} = \hat{\theta}\cdot\hat{r} = \hat{\theta}\cdot\hat{\phi} = \hat{\phi}\cdot\hat{\theta} = \hat{\phi}\cdot\hat{r} = \hat{r}\cdot\hat{\phi} = 0$$

$$\frac{\partial\hat{r}}{\partial r} = 0, \ \frac{\partial\hat{r}}{\partial\theta} = \hat{\theta}, \ \frac{\partial\hat{r}}{\partial\phi} = \hat{\phi}\sin\theta$$

$$\frac{\partial\hat{\theta}}{\partial r} = 0, \ \frac{\partial\hat{\theta}}{\partial\theta} = -\hat{r}, \ \frac{\partial\hat{\theta}}{\partial\phi} = \hat{\phi}\cos\theta$$

$$\frac{\partial\hat{\phi}}{\partial r} = 0, \ -\frac{\partial\hat{\phi}}{\partial\theta} = 0, \ \frac{\partial\hat{\phi}}{\partial\phi} = -\hat{r}\sin\theta - \hat{\theta}\cos\theta$$

the Laplace operator ∇^2 becomes

$$\nabla^2 = \frac{1}{r^2}\left[\frac{\partial}{\partial r}\left(r^2\frac{\partial}{\partial r}\right) + \frac{1}{\sin\theta}\frac{\partial}{\partial\theta}\left(\sin\theta\frac{\partial}{\partial\theta}\right) + \frac{1}{\sin^2\theta}\frac{\partial^2}{\partial\phi^2}\right].$$

Laplace operator is involved in the time-independent Schrödinger equation

$$\left(-\frac{\hbar^2}{2m}\nabla^2 + V(\mathbf{r})\right)\Psi(\mathbf{r}) = E\Psi(\mathbf{r}).$$

We discuss how to solve it, first in Cartesian coordinates and then in spherical coordinates. In spherical coordinates, we have

$$-\frac{\hbar^2}{2m}\frac{1}{r^2}\left[\frac{\partial}{\partial r}\left(r^2\frac{\partial\Psi}{\partial r}\right) + \frac{1}{\sin\theta}\frac{\partial}{\partial\theta}\left(\sin\theta\frac{\partial\Psi}{\partial\theta}\right) + \frac{1}{\sin^2\theta}\frac{\partial^2\Psi}{\partial\phi^2}\right] + V(r)\Psi = E\Psi.$$

Separating the variables, we have $\Psi(r,\theta,\phi) = R(r)Y(\theta,\phi)$. We can write the previous equation as

$$-\frac{\hbar^2}{2m}\frac{1}{r^2}\left[Y\frac{\partial}{\partial r}\left(r^2\frac{\partial R}{\partial r}\right) + \frac{R}{\sin\theta}\frac{\partial}{\partial\theta}\left(\sin\theta\frac{\partial Y}{\partial\theta}\right) + \frac{R}{\sin^2\theta}\frac{\partial^2 Y}{\partial\phi^2}\right] + V(r)RY = ERY.$$

After dividing by $-\dfrac{\hbar^2}{mr^2}RY$, we obtain

$$\left[\frac{1}{R}\frac{\partial}{\partial r}\left(r^2\frac{\partial R}{\partial r}\right) + \frac{1}{Y\sin\theta}\frac{\partial}{\partial\theta}\left(\sin\theta\frac{\partial Y}{\partial\theta}\right) + \frac{1}{Y\sin^2\theta}\frac{\partial^2 Y}{\partial\phi^2}\right] - \frac{2mr^2}{\hbar^2}(V(r) - E) = 0.$$

If we separate the equation in the part depending on r and the part depending on θ, ϕ, we have

$$\left[\frac{1}{R}\frac{\partial}{\partial r}\left(r^2\frac{\partial R}{\partial r}\right) - \frac{2mr^2}{\hbar^2}(V(r) - E)\right] + \frac{1}{Y}\left[\frac{1}{\sin\theta}\frac{\partial}{\partial\theta}\left(\sin\theta\frac{\partial Y}{\partial\theta}\right) + \frac{1}{\sin^2\theta}\frac{\partial^2 Y}{\partial\phi^2}\right] = 0$$

and consequently, the part depending on r has to be a constant K and the part depending on θ, ϕ has to be $-K$. We can choose the constant to be $l(l+1)$. The reason is related to the second equation

$$\frac{1}{Y}\left[\frac{1}{\sin\theta}\frac{\partial}{\partial\theta}\left(\sin\theta\frac{\partial Y}{\partial\theta}\right) + \frac{1}{\sin^2\theta}\frac{\partial^2 Y}{\partial\phi^2}\right] = -l(l+1)$$

which allows a solution depending on the Legendre polynomials. The other equation to be solved is

$$\frac{1}{R}\frac{\partial}{\partial r}\left(r^2\frac{\partial R}{\partial r}\right) - \frac{2mr^2}{\hbar^2}(V(r) - E) = l(l+1),$$

whose solution depends on $V(r)$. Let us consider the equation in Y after we multiply by $Y\sin^2\theta$. It is

$$\sin\theta\frac{\partial}{\partial\theta}\left(\sin\theta\frac{\partial Y}{\partial\theta}\right)+\frac{\partial^2 Y}{\partial\phi^2}=-l(l+1)Y\sin^2\theta.$$

To solve it, we perform another split of variables,

$$Y(\theta,\phi)=\Theta(\theta)\Lambda(\phi)$$

and then we divide both terms by $\Theta\Lambda$. It is

$$\frac{1}{\Theta}\sin\theta\frac{\partial}{\partial\theta}\left(\sin\theta\frac{\partial\Theta}{\partial\theta}\right)+\frac{1}{\Lambda}\frac{\partial^2\Lambda}{\partial\phi^2}=-l(l+1)\sin^2\theta.$$

If we split in a part depending on θ and a part depending on ϕ, i.e.

$$\left[\frac{1}{\Theta}\sin\theta\frac{\partial}{\partial\theta}\left(\sin\theta\frac{\partial\Theta}{\partial\theta}\right)+l(l+1)\sin^2\theta\right]+\frac{1}{\Lambda}\frac{\partial^2\Lambda}{\partial\phi^2}=0,$$

based on the same reasoning, we can consider two opposite constants such that

$$\frac{1}{\Theta}\sin\theta\frac{\partial}{\partial\theta}\left(\sin\theta\frac{\partial\Theta}{\partial\theta}\right)+l(l+1)\sin^2\theta=m^2$$

and

$$\frac{1}{\Lambda}\frac{\partial^2\Lambda}{\partial\phi^2}=-m^2.$$

The last equation has the solution $\Lambda(\phi)=e^{im\phi}$. Since $\Lambda(0)=\Lambda(2\pi)$, it results $e^{i2m\pi}=1$, that is $m\in\mathbb{Z}$.

The other equation can be written in the form

$$\sin\theta\frac{\partial}{\partial\theta}\left(\sin\theta\frac{\partial\Theta}{\partial\theta}\right)+[l(l+1)\sin^2\theta-m^2]\Theta=0.$$

After considering $\Theta(\theta)=P_l^m(\cos\theta)$ and $x=\cos\theta$, this equation is equivalent to

$$\frac{d}{dx}\left[(1-x^2)\frac{d}{dx}\right]P_l^m(x)+\left[l(l+1)-\frac{m^2}{1-x^2}\right]P_l^m(x)=0.$$

To define $P_l^m(x)$, we have first to solve the equation

$$\frac{d}{dx}\left[(1-x^2)\frac{d}{dx}\right]Y+[l(l+1)]Y=0$$

which is known as the Legendre equation. We can prove that the solutions are the Legendre polynomials

$$P_l(x) = \left(2^l l!\right)^{-1} \frac{d^l}{dx^l}[(x^2 - 1)^l].$$

Then we show that the functions

$$P_l^m(x) = (1 - x^2)^{m/2} \frac{d^m}{dx^m} P_l(x)$$

verify the equality

$$\frac{d}{dx}\left[(1 - x^2)\frac{d}{dx}\right] P_l^m(x) + \left[l(l+1) - \frac{m^2}{1 - x^2}\right] P_l^m(x) = 0.$$

In the case of the hydrogen atom, we denote by $\mathbf{m}$ the mass of the electron and by M the mass of nucleus we have to consider the reduced mass $\mu := \dfrac{M\mathbf{m}}{M + \mathbf{m}}$ which is equivalent to the mass located at the centre of gravity of the system electron-nucleus.

The Schrödinger equation can be written considering a potential $V(r)$ described by the Coulomb term

$$V(r) := -\frac{e^2}{4\pi \varepsilon_0 r},$$

where e is the charge of the electron, ε_0 is the permittivity of vacuum, r is the distance between the nucleus and electron. The Schrödinger equation is

$$-\frac{\hbar^2}{2\mu} \frac{1}{r^2} \left[\frac{\partial}{\partial r}\left(r^2 \frac{\partial \Psi}{\partial r}\right) + \frac{1}{\sin \theta} \frac{\partial}{\partial \theta}\left(\sin \theta \frac{\partial \Psi}{\partial \theta}\right) + \frac{1}{\sin^2 \theta} \frac{\partial^2 \Psi}{\partial \phi^2}\right] - \frac{e^2}{4\pi \varepsilon_0 r}\Psi = E\Psi.$$

Therefore, after the separation of variables we have the following equations:

$$\frac{1}{R} \frac{\partial}{\partial r}\left(r^2 \frac{\partial R}{\partial r}\right) + \frac{2\mu r^2}{\hbar^2}\left(\frac{e^2}{4\pi \varepsilon_0 r} + E\right) = l(l+1),$$

called the radial equation having as solution $R(r)$;

$$\sin \theta \frac{\partial}{\partial \theta}\left(\sin \theta \frac{\partial \Theta}{\partial \theta}\right) + [l(l+1)\sin^2 \theta - m^2]\Theta = 0,$$

called the polar equation, having as solution $\Theta(\theta)$;

$$\frac{1}{\Lambda}\frac{\partial^2 \Lambda}{\partial \phi^2} = -m^2,$$

called the azimuthal equation, having as solution $\Lambda(\phi)$.

The total solution is

$$\Psi(r, \theta, \phi) = R(r)\Theta(\theta)\Lambda(\phi).$$

$R(r)$ is in fact

$$R_n(r) = R_\infty(r)e^{r\mu e^2/2n\pi\varepsilon_0\hbar^2}$$

which depends on the energy levels $E_n = -\dfrac{\mu e^4}{8n^2\varepsilon_0^2h^2}$ and the quantum number n,

$$\Theta(\theta) = P_l^m(\cos\theta)$$

and

$$\Lambda(\phi) = e^{im\phi}.$$

The numbers m and l, with $|m| \leq l$, are used to separate the variables in a convenient form. They are also quantum numbers related to the Schrödinger equation solutions.

11.5 Lecture 50: The Pauli Matrices and the Dirac Equation. Towards the Relativistic Quantum Mechanics

In Lecture 41, it is stated the third postulate of Quantum Mechanics:
3. The state function Ψ evolves in time according to the time-dependent Schrödinger equation

$$i\hbar\frac{\partial}{\partial t}\Psi(t, \mathbf{r}) = \left(-\frac{\hbar^2}{2m}\nabla^2 + V(t, \mathbf{r})\right)\Psi(t, \mathbf{r}).$$

It is the basic rule under which the quantum world works. All the examples we gave in Lecture 41 showed how important is the Schrödinger equation and why it has to be part of the "axiomatic frame" of Quantum Mechanics.

However the Schrödinger equation can replaced by evolution equations having the same structure but a different Hamiltonian. In general, the above equation can be written in the symbolic form

$$i\hbar \frac{\partial}{\partial t} \Psi = \hat{H} \Psi$$

where

$$\hat{H} = -\frac{\hbar^2}{2m} \nabla^2 + \hat{V} = \frac{\hat{\mathbf{p}}^2}{2m} + \hat{V}.$$

Dirac considered the same form of the Schrödinger equation but changed the Hamilton operator assuming the total energy derived from Special Relativity (see Lecture 11), that is

$$E^2 = m^2 c^4 + p^2 c^2.$$

Therefore he considered

$$H^2 = m^2 c^4 + p^2 c^2$$

and proposed a Hamiltonian operator of the form

$$\hat{H} = c\, \mathbf{a} \cdot \hat{\mathbf{p}} + bmc^2$$

where $\mathbf{a} := (a_1, a_2, a_3)$ is a constant vector and b is a scalar constant. This *Dirac Hamiltonian* must fulfill the condition

$$m^2 c^4 + p^2 c^2 = \left(c\, \mathbf{a} \cdot \hat{\mathbf{p}} + bmc^2\right)\left(c\, \mathbf{a} \cdot \hat{\mathbf{p}} + bmc^2\right).$$

Identifying the terms from both sides with equal sign, it results

$$a_1^2 = a_2^2 = a_3^2 = b^2 = 1$$

$$a_1 a_2 = a_1 a_3 = a_2 a_3 = ba_1 = ba_2 = ba_3 = 0.$$

Dirac considered the first line $a_1^2 = a_2^2 = a_3^2 = b^2 = 1$ and wrote the second lines having in mind that the solutions could be matrices:

$$a_1 a_2 + a_2 a_1 = a_1 a_3 + a_3 a_1 = a_3 a_2 + a_2 a_3 = a_1 a_2 + a_2 a_1 = a_1 a_2 + a_2 a_1 = a_1 a_2 + a_2 a_1 = 0.$$

Dirac realized that the conventional choice of solutions is related to the Pauli matrices. Written in terms of Pauli matrices, we have

$$a_1 = \begin{pmatrix} 0 & \sigma_x \\ \sigma_x & 0 \end{pmatrix}; \quad a_2 = \begin{pmatrix} 0 & \sigma_y \\ \sigma_y & 0 \end{pmatrix}; \quad a_3 = \begin{pmatrix} 0 & \sigma_z \\ \sigma_z & 0 \end{pmatrix}; \quad b = \begin{pmatrix} I & 0 \\ 0 & -I \end{pmatrix}$$

and they are called the *Dirac matrices*. Written as 4×4 matrices, we have

$$a_1 = \begin{pmatrix} 0 & 0 & 0 & 1 \\ 0 & 0 & -1 & 0 \\ 0 & 1 & 0 & 0 \\ -1 & 0 & 0 & 0 \end{pmatrix}; \quad a_2 = \begin{pmatrix} 0 & 0 & 0 & -i \\ 0 & 0 & i & 0 \\ 0 & -i & 0 & 0 \\ i & 0 & 0 & 0 \end{pmatrix}; \quad a_3 = \begin{pmatrix} 0 & 0 & 1 & 0 \\ 0 & 0 & 0 & -1 \\ 1 & 0 & 0 & 0 \\ 0 & -1 & 0 & 0 \end{pmatrix};$$

$$\text{and} \quad b = \begin{pmatrix} 1 & 0 & 0 & 0 \\ 0 & 1 & 0 & 0 \\ 0 & 0 & -1 & 0 \\ 0 & 0 & 0 & -1 \end{pmatrix}; \quad \text{the Dirac wave state is a } 4 \times 1 \text{vector } \Psi = \begin{pmatrix} \Psi_1 \\ \Psi_2 \\ \Psi_3 \\ \Psi_4 \end{pmatrix}.$$

Therefore the Dirac equation is a system of four partial differential equations dependending on the total energy of Special Relativity. It can be written in the form:

$$\frac{i}{c}\frac{\partial}{\partial t}\Psi = \left(a_1 \frac{1}{i}\frac{\partial}{\partial x} + a_2 \frac{1}{i}\frac{\partial}{\partial y} + a_3 \frac{1}{i}\frac{\partial}{\partial z} + b\frac{mc}{\hbar} \right)\Psi,$$

and rearranged as

$$\frac{i}{c}\frac{\partial}{\partial t}\Psi = \left(-ia_1 \frac{\partial}{\partial x} - ia_2 \frac{\partial}{\partial y} - ia_3 \frac{\partial}{\partial z} + b\frac{mc}{\hbar} \right)\Psi,$$

or, equivalently,

$$i\hbar\frac{\partial}{\partial t}\Psi = \left(-i\hbar c\, \mathbf{a} \cdot \nabla + bmc^2 \right)\Psi,$$

where $\mathbf{a}$ is a vector with components the Dirac matrices a_k, b is the matrix written above and ∇ is the standard gradient. If we multiply by the matrix b, it results

$$i\hbar b \frac{\partial}{\partial t}\Psi = \left(-i\hbar c\, b\mathbf{a} \cdot \nabla + b^2 mc^2 \right)\Psi.$$

Let us denote $\gamma^0 := b$ and $\gamma^j := ba_j$. The Dirac equation becomes

$$i\hbar\gamma^0 \frac{\partial}{\partial t}\Psi = \left(-i\hbar c\, \gamma^j \cdot \nabla + mc^2 \right)\Psi.$$

Another elegant form in which we can arrange it is

$$\frac{1}{c}\gamma^0 \frac{\partial}{\partial t}\Psi + (\gamma^j \cdot \nabla)\Psi + \frac{imc}{\hbar}\Psi = 0.$$

Finally, the Dirac γ matrices are

$$\gamma^1 = \begin{pmatrix} 0 & \sigma_x \\ -\sigma_x & 0 \end{pmatrix}; \quad \gamma^2 = \begin{pmatrix} 0 & \sigma_y \\ -\sigma_y & 0 \end{pmatrix}; \quad \gamma^3 = \begin{pmatrix} 0 & \sigma_z \\ -\sigma_z & 0 \end{pmatrix}; \quad \gamma^0 = \begin{pmatrix} I & 0 \\ 0 & -I \end{pmatrix}.$$

Simple computations show that $(\gamma^0)^2 = I$ while $(\gamma^j)^2 = -I$ for $j \in \{1, 2, 3\}$; $(\gamma^j)^\dagger = \gamma^j$ for $j \in \{0, 1, 2, 3\}$. And more, $\gamma^j \gamma^l - \gamma^l \gamma^j = 0$ for $j \neq l$, $j, l \in \{0, 1, 2, 3\}$. The solution

$$\Psi = \begin{pmatrix} \Psi_1 \\ \Psi_2 \\ \Psi_3 \\ \Psi_4 \end{pmatrix}$$

of the Dirac equation consists of four components. Each component is a plane wave

$$\Psi_j(t, \mathbf{r}) = A_j e^{-i\hbar(\mathbf{p} \cdot \mathbf{r} - Et)}, \quad j \in \{1, 2, 3, 4\},$$

with A_j the amplitudes expressed by positive constants. This solution is called the *spinor* wave function, or simply, the spinor.

As a final consideration, let us consider a squared Dirac operator. We have

$$\left(\frac{1}{c} \gamma^0 \frac{\partial}{\partial t} + (\gamma^j \cdot \nabla) + \frac{imc}{\hbar} \right)^2 \Phi =$$

$$= \left(\frac{1}{c} (\gamma^0)^\dagger \frac{\partial}{\partial t} + ((\gamma^j)^\dagger \cdot \nabla) - \frac{imc}{\hbar} \right) \left(\frac{1}{c} \gamma^0 \frac{\partial}{\partial t} + (\gamma^j \cdot \nabla) + \frac{imc}{\hbar} \right) \Phi =$$

$$= \left(\frac{1}{c^2} \frac{\partial^2}{\partial t^2} - \nabla^2 + \frac{m^2 c^2}{\hbar^2} \right) \Phi$$

which gives the *Klein-Gordon equation*

$$\left(\frac{1}{c^2} \frac{\partial^2}{\partial t^2} - \nabla^2 \right) \Phi = -\frac{m^2 c^2}{\hbar^2} \Phi.$$

We can observe that all wave solutions of the Dirac equation are solutions of the Klein-Gordon equation. However the Klein-Gordon one is a relativistic wave equation for scalar particles with spin zero. They can be composite particles, like $\Phi = \Psi^* \Psi$, derived from spinors. Example of such particles can be the pion or the Higgs boson. Relativistic Quantum Mechanics is the argument of more advanced courses over the basic approach of this book. We refer the reader to more advanced text like Refs. [28, 29].

Summary of Lecture 50. Starting from the formal form of Schrödinger equation

$$i\hbar \frac{\partial}{\partial t} \Psi = \hat{H} \Psi,$$

Dirac thought to change the classical Hamilton operator using the formula of total energy derived in Special Relativity, that is $H^2 = m^2c^4 + p^2c^2$. Therefore he considered a Hamiltonian operator in the form of "square root" of the previous expression, i.e.

$$\hat{H} = c\,\mathbf{a} \cdot \hat{\mathbf{p}} + bmc^2.$$

Computations lead to the final form of Dirac equation

$$i\hbar\frac{\partial}{\partial t}\Psi = \left(-i\hbar c\,\mathbf{a} \cdot \nabla + bmc^2\right)\Psi$$

written with the 4×4 matrices

$$a_1 = \begin{pmatrix} 0 & \sigma_x \\ \sigma_x & 0 \end{pmatrix}; \quad a_2 = \begin{pmatrix} 0 & \sigma_y \\ \sigma_y & 0 \end{pmatrix}; \quad a_3 = \begin{pmatrix} 0 & \sigma_z \\ \sigma_z & 0 \end{pmatrix}; \quad b = \begin{pmatrix} I & 0 \\ 0 & -I \end{pmatrix},$$

which involves the Pauli matrices as blocks inside them. If we multiply by the matrix b, it results

$$i\hbar b\frac{\partial}{\partial t}\Psi = \left(-i\hbar c\,\mathbf{ba} \cdot \nabla + b^2mc^2\right)\Psi.$$

A more appealing form is obtained after we denote $\gamma^0 := b$ and $\gamma^j := ba_j$,

$$\frac{1}{c}\gamma^0\frac{\partial}{\partial t}\Psi + (\gamma^j \cdot \nabla)\Psi + \frac{imc}{\hbar}\Psi = 0.$$

The solution of this equation is a 4-dimensional spinor. It is possible to derive also the Klein-Gordon equation from the squared Dirac operator. We have

$$\left(\frac{1}{c^2}\frac{\partial^2}{\partial t^2} - \nabla^2\right)\Phi = -\frac{m^2c^2}{\hbar^2}\Phi$$

whose solutions are relativistic scalar particles.

Chapter 12
Conclusions

....I have not failed. I've just found 10,000 ways that won't work.

Thomas Alva Edison

In simple words, someone can say that Quantum Mechanics can be divided in two parts: *Before Principles* and *After Principles*.

Before Principles means that several experimental facts point out the inadequacy of Classical Mechanics to describe, for example, light and particles at a fundamental level. This statement needs also a new conceptualization that, eventually, brings to formulate an adequate Mathematics to describe them.

As said, light cannot be described in the context of Classical Mechanics. Its fundamental nature is represented by electromagnetic waves which need the formulation of Special Relativity. Besides, light can be made of particles so we have to accept the fact that light is wave and particles at same time. The appearance of this *duality* means that only Maxwell's equations are not enough to catch its second nature. The "particles of light", called photons, have energy. The energy formula related to these "particles' depends on the Planck constant which revealed one of the most important (probably the most fundamental) constant of Physics.

This line of thinking immediately brings to the question if any material element, also particles with mass, have a wave nature: de Broglie answered this question showing that any particle have a wave counterpart. This means that the wave-particle behavior is an intrinsic feature of Nature. Any massive body has a related wavelength.

The conceptual development of this idea led to the Schrödinger equation which has not only a wave interpretation. Its solutions have the meaning of probability to find a particle in a region of space at a given time. For example, a wave itself cannot represent an electron if its speed is greater than the speed of light. So, we need to

© The Author(s), under exclusive license to Springer Nature Switzerland AG 2021
S. Capozziello and W.-G. Boskoff, *A Mathematical Journey to Quantum Mechanics*,
UNITEXT for Physics, https://doi.org/10.1007/978-3-030-86098-1_12

replace waves by wave packets when we study electrons and other particles different from photons.

From a conceptual point of view, this approach shift the standard deterministic interpretation of Classical Mechanics to a probabilistic one. Concepts like non-locality and indeterminism find a place in Physics and need a new mathematical formulation. Step by step, a new theory can be formulated using the Mathematics necessary to correctly understand the results of experiments. For example, the Heisenberg uncertainty principle appears in its elementary formulation because we use the Gauss wave packets. In fact, using them, through the Fourier transform and the Pinsky theorem, the Heisenberg principle naturally emerges.

This fundamental step depends on the Plancharel–Parceval formula established for functions belonging to the intersection between $L^1(R)$ and $L^2(R)$. Here $L^2(R)$ is a Hilbert space involved in the theory of one-dimensional quantum harmonic oscillator. In almost all models, Hilbert spaces can be used. If we look at the time-dependent Schrödinger equation, we see an eigenvector-eigenvalue problem in which a Hermitian operator is involved. It is possible to prove that the Hamiltonian itself is a Hermitian operator.

At least locally, unitary operators can be associated to the time-dependent Schrödinger equation. In a Hilbert space, we can find a particular case of Cauchy–Buniakowski–Schwarz inequality leading to the Heisenberg uncertainty principle. More, operators associated to the Schrödinger equation allow us to derive their expectation values. The commutator of two operators indicates if they can be simultaneously measured and their evolution oblige us to rearrange the way in which quantum objects have to be considered. With these considerations in mind and with these new mathematical tools, Principles of Quantum Mechanics appear.

After their formulation, the big picture of Quantum Mechanics can be achieved and this allows to have another perspective for any subject already considered. Ladder operators offer another view for the harmonic oscillator problem. Quantized energy levels, obtained from the classical Hermite polynomials, are now obtained using the creation and the annihilation operators. The obsolete Bohr hydrogen model is replaced by considering the gradient and Laplace operators in spherical coordinates. The solution depends on spherical harmonics and so quantum numbers appear. Pauli matrices are involved in photon polarization description. Angular operators and Pauli operators are related to the electron spin. And again, the Pauli matrices are involved in obtaining the Dirac equation starting from the Scrödinger equation. We step to Klein–Gordon equation necessary for the relativistic description of other particles besides the electron. With this trend, we can go beyond the non-relativistic Quantum Mechanics up to the full formulation of the Quantum Field Theory.

In summary, we presented the basic conceptual tools of the so called "Copenhagen Interpretation of Quantum Mechanics" [26] stressing, in particular, the mathematical aspects of the construction. It is worth noticing that, in literature, such mathematical aspects are sometimes overlooked with respect to the experimental and axiomatic aspects. On the contrary, we believe that a full understanding of Mathematics behind the theory constitutes a straightforward way to realize the big picture of Quantum Mechanics.

Clearly, the material presented in this book is not exhaustive at all. We intended to present only the basic aspects of Quantum Mechanics without claiming for completeness. We have to stress also that we did not take into account other important topics like entanglement or alternative interpretation of Quantum Mechanics.

Quantum entanglement occurs when quantum objects like particles are generated, interact, or share spatial proximity in a way such that the quantum state of each component of the system cannot be described independently of the state of the others. This happens also if particles are separated by large distances. In other words, an entangled state cannot be described by the standard superposition principle of states. Entangled states are solutions of the Schrödinger equation but, unlike pure states, has no classical counterpart. Entanglement is a genuine quantum phenomenon at the core of Quantum Mechanics [30].

Furthermore, other interpretations of Quantum Mechanics are possible. They originate in the attempt to explain how Quantum Mechanics and its mathematical apparatus corresponds to the "Reality" despite of the fact that the theory has been experimentally tested by a large number of extremely accurate experiments. In other words, no final consensus has been reached on how measurements, quantum states, wave functions and operators can represent consistently the physical world.

For example, the so called *Many Worlds Interpretation* considers a universal wave function (the so-called Wave Function of the Universe) where no wave function collapse is associated with measurements. Measurements are explained by the decoherence, which occurs when interacting states produce entanglement and "split" the universe into mutually unobservable histories. The final effect is that distinct universes (distinct realities and causally distinct histories) emerge within the framework of a comprehensive Multiverse [31]. In this interpretation, the main role in evolution of systems is played by the Schrödinger equation (or other related evolutionary equations like the Wheeler–de Witt equation adopted in Quantum Cosmology)to describe these distinct universes (realities) on the ground of quantum systems [32]. The main result is the probability to realize a given universe without the necessity of measurements implying the collapse of the wave function.

Several approach are possible: we have the de Broglie–Bohm theory, the von Neumann–Wigner interpretation, the Quantum Information Theory and many others. See [26] for a detailed discussion. As concluding remark, we can say that Quantum Mechanics, together with General Relativity, is one of the most active research areas not only for Physics but for the whole Knowledge.

Finally, we humbly hope to have stimulated the curiosity of the Reader offering the first necessary steps towards further insights into the fascinating world of Quantum Theory.

Bibliography

1. Boskoff, W.G., Capozziello, S.: A Mathematical Journey to Relativity. Springer, Dordrecht (2020)
2. Arnol'd, V.I.: Mathematical Methods of Classical Mcchanics. Springer, New York (1989)
3. Goldstein, H.: Classical Mechanics. Addison-Wesley, Boston (1959)
4. Debnath, L., Mikusiński, P.: Hilbert Spaces with Applications, 3rd edn. Academic, Cambridge (2005)
5. Conway, J.: A Course in Functional Analysis, 2nd edn. Springer, New York (1990)
6. Griffel, D.H.: Applied Functional Analysis. Dover Publications, New York (2012)
7. Lang, S.: Real and Functional Analysis. Springer, New York (1993)
8. Stone, J.V.: The Fourier Transform - A Tutorial Introduction. Sebtel Press (2021)
9. Sneddon, I.N.: Fourier Transforms. Dover Publications, New York (2010)
10. Bunge, M.: Quantum Theory and Reality. Springer, Berlin (1967)
11. Bunge, M.: Mach's Critique of Newtonian Mechanics. Amer. J. Phys. **34**, 585 (1966)
12. Susskind, L., Friedmann, A.: Quantum Mechanics: The Theoretical Minimum. Penguin Books Ltd, London (2015)
13. Griffiths, D.J., Schroeter, D.F.: Introduction to Quantum Mechanics. Cambridge University Press, Cambridge (2018)
14. Dirac, P.A.M.: Principles of Quantum Mechanics. Oxford University Press, Oxford (1930)
15. Shankar, R.: Principles of Quantum Mechanics. Springer, Berlin (1994)
16. Sakurai, J.J., Napolitano, J.: Modern Quantum Mechanics. Pearson, London (2010)
17. Zettili, N.: Quantum Mechanics: Concepts and Applications. Wiley, New York (2009)
18. Bransden, B.H., Joachain, C.J.: Quantum Mechanics. Longman Group UK Ltd, Harlow (1990)
19. Townsend, J.: A Modern Approach to Quantum Mechanics. University Science Books, Sausalito (2000)
20. Breman, P.R.: Introductory Quantum Mechanics. Springer International Publishing AG, New York (2019)
21. Norsen, T.: Foundations of Quantum Mechanics. Springer International Publishing AG, New York (2017)
22. von Neumann, J.: Mathematical Foundations of Quantum Mechanics. Princeton University Press, Princeton (2018)
23. Rae, A.I.M., Napolitano, J.: Quantum Mechanics. CRC Press Taylor & Francis Group, Milton Park (2016)

S. Capozziello and W.-G. Boskoff, *A Mathematical Journey to Quantum Mechanics*, UNITEXT for Physics, https://doi.org/10.1007/978-3-030-86098-1

24. La Rana, A., Rossi, P.: The blossoming of quantum mechanics in Italy: the roots, the context and the first spreading in Italian universities (1900–1947). Eur. Phys. J. H **45**(4–5) (2020)

25. Messiah, A.: Quantum Mechanics, vol. 1. North-Holland Publishing Company, Amsterdam (1961)

26. D'Espagnat, B.: Conceptual Foundations of Quantum Mechanics. Addison Wesley, Boston (1976)

27. Dahmen, H.D., Brandt, S.: The Picture Book of Quantum Mechanics. Springer, Heidelberg (2012)

28. Kaku, M.: Quantum Field Theory: A Modern Introduction. Oxford University Press, Oxford (1994)

29. Sakurai, J.J.: Advanced Quantum Mechanics. Addison Wesley, Boston (1967)

30. Horodecki, R., Horodecki, P., Horodecki, M., Horodecki, K.: Quantum entanglement. Rev. Mod. Phys. **81**, 865 (2009)

31. Everett, H.: Relative state formulation of quantum mechanics. Rev. Mod. Phys. **29**, 454 (1957)

32. Halliwell, J.J.: Introductory Lectures on Quantum Cosmology. In: Coleman, S., Hartle, J.B., Piran, T., Weinberg, S.: Quantum Cosmology and Baby Universes, Proceedings of the 7th Jerusalem Winter School for Theoretical Physics, Jerusalem, Israel. World Scientific, Singapore (1991)

Index